# AN ENGLISH CAR DESIGNER ABROAD

## Designing for GM, Audi, Porsche and Mazda

PETER BIRTWHISTLE

*More from Veloce*

*Biographies*

A Chequered Life – Graham Warner and the Chequered Flag (Hesletine)
A Life Awheel – The 'auto' biography of W de Forte (Skelton)
Amédée Gordini ... a true racing legend (Smith)
André Lefebvre, and the cars he created at Voisin and Citroën (Beck)
Bunty – Remembering a gentleman of noble Scottish-Irish descent (Schrader)
Chris Carter at Large – Stories from a lifetime in motorcycle racing (Carter & Skelton)
Cliff Allison, The Official Biography of – From the Fells to Ferrari (Gauld)
Edward Turner – The Man Behind the Motorcycles (Clew)
Driven by Desire – The Desiré Wilson Story
First Principles – The Official Biography of Keith Duckworth (Burr)
Inspired to Design – F1 cars, Indycars & racing tyres: the autobiography of Nigel Bennett (Bennett)
Jack Sears, The Official Biography of – Gentleman Jack (Gauld)
Jim Redman – 6 Times World Motorcycle Champion: The Autobiography (Redman)
John Chatham – 'Mr Big Healey' – The Official Biography (Burr)
The Lee Noble Story (Wilkins)
Mason's Motoring Mayhem – Tony Mason's hectic life in motorsport and television (Mason)
Raymond Mays' Magnificent Obsession (Apps)
Pat Moss Carlsson Story, The – Harnessing Horsepower (Turner)
'Sox' – Gary Hocking – the forgotten World Motorcycle Champion (Hughes)
Tony Robinson – The biography of a race mechanic (Wagstaff)
Virgil Exner – Visioneer: The Official Biography of Virgil M Exner Designer Extraordinaire (Grist)

*General*

Austin Cars 1948 to 1990 – A Pictorial History (Rowe)
Chevrolet Corvette (Starkey)
Chrysler 300 – America's Most Powerful Car 2nd Edition (Ackerson)
Chrysler PT Cruiser (Ackerson)
Dodge Challenger & Plymouth Barracuda (Grist)
Dodge Charger – Enduring Thunder (Ackerson)
Dodge Dynamite! (Grist)
Ferrari 288 GTO, The Book of the (Sackey)
Ferrari 333 SP (O'Neil)
Fiat & Abarth 124 Spider & Coupé (Tipler)
Fiat & Abarth 500 & 600 – 2nd Edition (Bobbitt)
Fiats, Great Small (Ward)
Inside the Rolls-Royce & Bentley Styling Department – 1971 to 2001 (Hull)
Mazda MX-5/Miata 1.6 Enthusiast's Workshop Manual (Grainger & Shoemark)
Mazda MX-5/Miata 1.8 Enthusiast's Workshop Manual (Grainger & Shoemark)
Mazda MX-5 Miata, The book of the – The 'Mk1' NA-series 1988 to 1997 (Long)
Mazda MX-5 Miata, The book of the – The 'Mk2' NB-series 1997 to 2004 (Long)
Mazda MX-5 Miata Roadster (Long)
Mazda Rotary-engined Cars (Cranswick)
Mike the Bike – Again (Macauley)
Porsche 356 (2nd Edition) (Long)
Porsche 908 (Födisch, Neßhöver, Roßbach, Schwarz & Roßbach)
Porsche 911 Carrera – The Last of the Evolution (Corlett)
Porsche 911R, RS & RSR, 4th Edition (Starkey)
Porsche 911, The Book of the (Long)
Porsche 911 – The Definitive History 2004-2012 (Long)
Porsche – The Racing 914s (Smith)
Porsche 911SC 'Super Carrera' – The Essential Companion (Streather)
Porsche 914 & 914-6: The Definitive History of the Road & Competition Cars (Long)
Porsche 924 (Long)
The Porsche 924 Carreras – evolution to excellence (Smith)
Porsche 928 (Long)
Porsche 930 to 935: The Turbo Porsches (Starkey)
Porsche 944 (Long)
Porsche 964, 993 & 996 Data Plate Code Breaker (Streather)
Porsche 993 'King Of Porsche' – The Essential Companion (Streather)
Porsche 996 'Supreme Porsche' – The Essential Companion (Streather)
Porsche 997 2004-2012 – Porsche Excellence (Streather)
Porsche Boxster – The 986 series 1996-2004 (Long)
Porsche Boxster & Cayman – The 987 series (2004-2013) (Long)
Porsche Racing Cars – 1953 to 1975 (Long)
Porsche Racing Cars – 1976 to 2005 (Long)
Porsche – The Rally Story (Meredith)
Porsche: Three Generations of Genius (Meredith)
Powered by Porsche (Smith)

www.veloce.co.uk

First published in February 2020 by Veloce Publishing Limited, Veloce House, Parkway Farm Business Park, Middle Farm Way, Poundbury, Dorchester DT1 3AR, England. Tel +44 (0)1305 260068 / Fax 01305 250479 / e-mail info@veloce.co.uk / web www.veloce.co.uk or www.velocebooks.com.
ISBN: 978-1-787114-70-8; UPC: 6-36847-01470-4.

Readers with ideas for automotive books, or books on other transport or related hobby subjects, are invited to write to the editorial director of Veloce Publishing at the above address. British Library Cataloguing in Publication Data – A catalogue record for this book is available from the British Library. Typesetting, design and page make-up all by Veloce Publishing Ltd on Apple Mac. Printed in India by Parksons Graphics.

# AN ENGLISH CAR DESIGNER ABROAD

## Designing for GM, Audi, Porsche and Mazda

VELOCE PUBLISHING
THE PUBLISHER OF FINE AUTOMOTIVE BOOKS

For my wonderful kids, Denys and Charlotte – it's important you know what your dad got up to, and if anyone else is interested then they're welcome to look over your shoulders.

And in memory of Geoff Lawson, a truly great designer, who left this world far too soon. You got me on my way as a car designer, Geoff – I learnt everything from you.

# Contents

# Preface

My late mother-in-law from my first marriage, Antje, once said to me:

"Peter, you're such a sensible person, it is beyond my belief why you are so interested in cars."

Well, I suppose she had a point – it is a bit of a sickness. But it's something I was just born with, and if you talk to most car designers, they will say the same thing.

My interest, however, went beyond a simple fascination. I knew right from the start that I wanted to design cars – or at least draw them – and from a very early age, that was my personal goal.

As I start to write this book in late 2014, I have come to the end of a 40-year career in Automobile Design. It has spanned an era where the art of designing a car has gone from being a truly artistic and romantic occupation, to a cut-throat business dictated by product planners, marketeers and that invisible curse: deadlines.

During my 40 years in the industry, the art of designing cars has changed so much that I really feel the need to document my experiences. If I don't, then, like many crafts, the knowledge and information will just drift into history and be forgotten.

Peter Birtwhistle may never be a name remembered as a significant or great car designer, and I have no problems living with that. However, I have had some amazing experiences, gathered many stories and anecdotes, in this business, and it would be a pity if they went untold.

This book, however, is not just dedicated to my life in the car industry; it is inseperable from my personal story, so they are woven together here, too.

Designers have big egos, but I have tried to tell it as it happened. I apologise if events are recorded in a way that doesn't accurately reflect reality, or differs from another's point of view, but I can only tell the story the way I saw it.

So, let's begin, in Reading, England, March 7, 1951.

***Peter Birtwhistle***

# Chapter 1

## *The early years*

**I came into the world not alone. My twin brother, Roger, preceded me by 11 hours. Having spent nine months with him in my mother's womb, I needed time to reflect on how I could get the better of him right from the start. We were not identical, and the bond so often associated with twins was not to be. Though we would grow up together and were encouraged by our parents to share, this was not part of my personal plan; right from the start I was competitive, and a twin brother would not get in my way.**

Our parents, Denys and Valerie, were very caring, but unprepared for the challenges of raising twins.

My father, Denys, or 'Birty' as he was known to all his friends (a nickname I would also take on later in life), was born in Barton-upon-Humber, Lincolnshire, the son of a well-known town doctor called Percy Birtwhistle. His mother, Ella, was very gifted artistically, but as a doctor's wife in the late 19th century, her artistic skills were mainly limited to hand crafts. I've seen some of the beautiful wildlife watercolours she did, and it is clear to me from whom I inherited my creative talents.

*Proud parents, Denys and Valerie.*

Continuing the medical tradition in the Birtwhistle family, my father attended the London University hospital Guy's during the war years, where he studied to be a dentist. He qualified just before the end of the war and was then drafted into the Royal Navy, where he served as a Dental Officer aboard an aircraft carrier named HMS Patroller. After a tour to Australia via the Suez Canal, he returned to the UK. When he realised that life in 'The South' had got to him, he moved into a dental practice on London Road, Reading.

It is here that he met my mother, Valerie, who visited his practice as a patient. Valerie was a local Reading girl, working as a nurse at St Mary's Hospital in Paddington, London at the time. Her parents, Norman and Audrey Wagnell, ran a town pub, The Traveller's Friend, in Friars Street.

Denys and Valerie married in 1950, and my brother and I were conceived almost immediately, during their honeymoon in Devon.

With twins on the way, my parents moved into a small detached house in Pitts Lane, in an eastern part of Reading called Woodley.

This was the backdrop for my first experience of wheeled transport: a grey Silver Cross twin pram, in which my brother and I would sit face to face. My mother used to walk us to a nearby railway bridge, to look down as the express steam trains rushed past, heading towards Reading General Station, and beyond to the West Country. Enveloped in smoke and steam, as the trains thundered through below us, we screamed with delight. Those visits to the railway bridge certainly developed my early fascination for all modes of transport.

I have fleeting memories of our days at Pitts Lane. The house, a small 'two-up two-down', had quite a generous garden, shaded by a massive oak tree. My

*My father (standing extreme right) with his parents, Percy and Ella, plus brothers. My mother, Valerie, seated in front of him with Roger and me.*

mother had daily help from a housekeeper, a local woman called Mrs Ruddles, or 'Ruggles' as we named her, and we also employed a gardener called 'Sparkey.' Sparkey had a long silver beard and always seemed to be pushing around a wheelbarrow with large sacks of potatoes. I guess these days the idea of employing a permanent housekeeper and gardener must seem like the height of luxury. For my father, at least, it was reminiscent of his childhood; in the household of a prominent town doctor like Percy Birtwhistle, a housekeeper, cook, gardener and even a chauffeur were considered the norm.

My father recollected that most years his dad used to travel from Lincolnshire to London, accompanied by his chauffeur, to visit the Motor Show. He'd depart Barton dreaming of purchasing a sporty model such as a Vauxhall. In the end, however, his conscience always got the better of him and he would buy a Morris, which to him seemed a more appropriate choice for a town doctor.

As with his father, Denys Birtwhistle remained faithful to the Morris brand, but not before a little indulgence, with his first car being a 1938 Morgan. He kept a faded black and white photograph of the car, with my mother seated in it, in his wallet until the day he died.

With my brother and I soon arriving, there was no justification for a small two-seater sports car, so the Morgan went. It was replaced by a black 1950 Morris Oxford, which was my very first experience of motorised transport. Those days, a Morris Oxford was a sensible choice but not particularly stylish. It was appropriate transportation, however, for a young dentist. With its design obviously borrowed from American models of that time, it had a split front screen, with a red leather interior. I have no recollection of how my brother and I travelled in the Oxford with no child seats or safety belts. I assume we just rolled around on the rear seat bank.

The summers were long and balmy in those postwar years, and life was positive. But things were soon to change. My mother was pregnant again, with a new family member due in the middle of 1954. They arrived in the form of my sister, Helen, the future 'Daddy's darling.' With Helen's birth, one thing became very clear: the house in Pitt's lane was too small. My father

*My grandparents on my mother's side: Norman and Audrey Wagnell.*

*The Silver Cross twin pram. Here with my grandmother, and my cousin Jennifer seated opposite Roger and me.*

*I was always in the driving seat.*

was now well established in the practice in London Road, and so we could afford the move to a bigger house. Reading had a couple of 'posh' locations, one of which was Caversham, or to use its full name, Caversham Heights. This would be the focus of their searching.

Located to the north of Reading, Caversham rises almost immediately once you have crossed the Thames over Caversham Bridge, hence the 'Heights.' I assume my parents were looking at Caversham because my mother had some connection to that part of the town. I know my parents were married in Caversham's St Peter's Church, and my mother's parents previously ran a tobacconist in the small parade of shops immediately to the left after passing over Caversham Bridge.

The Wagnell family were well known in Reading as the owners of a business in the centre of the town called The English Leather Company. The company shop was run by one of my grandfather's brothers or cousins. We used to visit when my mother took us into town for shopping. Located in Queen's Street, its proximity to the main Reading Transport bus station, gave it an added attraction when visiting the shop. The English Leather Company was a manufacturers of all types of leather bags and cases, and, according to my grandfather, supplier at one time to the late George V.

At an early age, our grandfather Norman had worked for the family business as a travelling salesman and rep, but must have had a big bust-up with his family, as by the time I was born, any links he had with the leather business were completely severed. He became a bit of a drifter, picking up jobs here and there. At the time of my birth, he and my grandmother were running The Traveller's Friend pub in Reading. Grandad, as we called him, could never hold down a job for very long and my grandparents were frequently switching between running pubs or some small shop or business. In the end, my dad, amongst others, had to help them out somehow. Nevertheless, both my maternal grandparents were very strong and proud individuals.

I will remember Grandad for two key things. Firstly, he wore a suit and tie every day of his life. Though, in the summer he might have loosened the tie if he was in a deckchair on the beach, and in the winter he would maybe wear a cardigan under the jacket. Secondly, he was meticulous about the cleanliness of the house. This trait I assume he got from keeping the pubs he ran, when he would get up at a very early hour to clean things. After he retired he kept this habit, which drove my grandmother to despair. He'd rise at 5:00am and start to dust and hoover, all while listening to the radio news at full volume. Later in life, when my grandparents moved into residential accommodation, this 'lifestyle' of his drew attention, and the volume at which he listened to the news at that early hour became a particular focus of many complaints.

*52 Kidmore Road, Reading.*

*My grandparents' pub in Friar Street, Reading: The Traveller's Friend.*

"Could he not perhaps listen a bit later in the morning?" the neighbours asked.

He reasoned, however, that when he listened to the news, he wanted it fresh and not hours late!

It was in the spring of 1955 that we moved from Woodley to our new house in Kidmore Road, Caversham: a 1920s detached 'mock Tudor,' with entrance porch and double bay window frontage.

We didn't leave all of Woodley behind though. Our dear housekeeper, Ruggles, despite having to travel by bus from the other side of Reading, would visit each day to continue helping my mother. She had grown very attached to our family and couldn't bear parting with us.

A great bond developed for many years between us and Mrs Ruddles. Her husband was a good handyman and would sometimes help with repairs in the house. For me, however, the best thing about the Ruddles family was their son John's amazing collection of Dinky Toys. Occasionally, if my father was driving Ruggles home, I would go along – maybe my brother came too, however, his passion for toy cars was nowhere near the obsession that I had. John Ruddles was meticulous with his collection, with every one neatly kept in their boxes. He must have been a Dinky nerd, since nobody else kept the boxes. My visits to him to view his collection had an almost religious quality for me. My personal Dinky collection was nowhere close to John's. It was simply mind boggling. His range of models included, for example, a complete collection of the Bedford delivery vans in all different liveries and, way out of my league, some Dinky Supertoys. Dinky Supertoys came in more robust boxes with blue and white striping – this classification was reserved for large trucks or earth moving equipment. They were expensive, and certainly not on my parent's shopping list, but John had them all: a Foden four-axle flatbed brewery truck and similar models. But the pinnacle was an American Euclid bulldozer with matching dumper truck, both in lemon yellow. This was truly another world for me, and of course not fair by any means.

The big move also brought with it the first day of school for my brother and I, at St Peter's Primary School, opposite the church of the same name in Caversham.

On this first day, I immediately fell in love with a girl called Elena. My only memory of her is that she had amazing curly blonde hair and was impossibly pretty. Of equal significance was a boy in our class called Paul Easby, who had a red Corgi model of a Jaguar Mk1. Corgi was a relatively new maker of model cars but had rapidly climbed to become the main rival to Dinky, and were judged by many schoolboys to be superior. One notable feature over the Dinky models was that they had clear plastic windows. This particular Jaguar model also featured a friction driven motor for the rear wheels. You could charge the flywheel by repeatedly pushing it on the floor, before letting it run off under its own power.

As far as model cars were concerned, there were two key camps of allegiance: Dinky or Corgi. There was also a fringe range of models called Spot-On, which were manufactured by the Tri-ang Toy Company, but Dinky and Corgi were the main protagonists. The smaller Matchbox models by Lesney Products should also not be forgotten, but they tended to act as a kind of consolation offering from distant aunts, or as a gift from my parents on rare non-celebratory days, such as a rainy weekend. I believe my first Matchbox model was an AA patrolman on a motorcycle with a side car, and certainly nothing that would have any real status in the classroom, especially when up against a Corgi Jaguar Mk1. Our family, however, were loyal to Dinky, despite the initial lack of plastic windows. I imagine this allegiance came from my father's childhood and his love of the Meccano metal building sets. The Meccano Group were a real toy manufacturing powerhouse, who also made Hornby OO scale model train sets, as well as the Dinky model cars and trucks.

Apart from Elena and the Jaguar Mk1, I can't recall too much from the days at St Peter's. Roger and I would attend this first school until the end of 1958, before finally being accepted to a preparatory school on the south side of town called Crosfields. I say 'finally' since both of us were already struggling academically, a trend that would unfortunately remain with both of us throughout our entire schooling.

Late 1958 saw mixed fortunes. My mother was expecting our younger brother Christopher, but there were complications and she had to stay in hospital. Added to this, I broke my left leg while going berserk in the garden of Paul Easby of Jaguar Mk1 fame. It was on November 5 – Bonfire Night. I tripped over a clothes line prop. There was a crack, and I lay there screaming. Paul was probably smirking to see a Dinky man go down.

Later, the X-ray at Reading's Battle Hospital revealed a severe fracture, and I left with plaster up to my hip.

As the Christmas holidays approached, to make things easier for my father, it was decided to split the family. Limited to getting around on crutches, I was sent to stay with my grandparents in the flat above the pub in Reading, while Helen and Roger went to my mother's elder sister, Jean, in Lincolnshire.

Staying with my grandmother for those three weeks before Christmas would mark me for years to come. Confined as I was with a leg in plaster, I was getting no exercise whatsoever, and my grandmother began feeding me non-stop, resulting in me inflating like a balloon. My father was staying at the house in Caversham and would call by in the evenings before visiting my mother in hospital, but he didn't seem to notice my dramatic gain in weight. It was only when he took me to see my mother one day, who apparently said I had got so fat that she didn't recognise me, that my father came around to the reality of things. The

*Me – in my St Peter's school uniform with a broken leg – my sister, Helen, and brother, Roger, in the garden at Kidmore Road.*

damage, however, had been done. 'Podgy Pete' had arrived, and from then on I was known as 'the fatter one of the two!'

My younger brother Christopher came into the world on December 22, 1958, and after Christmas the family was thankfully reunited once more.

The plaster on my leg was removed a couple of days before Roger and I would start at Crosfields. It was the only part of my body that had kept its original size. The leg was still too weak to support me and looked like a stick up against the other leg. Now with the extra weight to carry, things were not easy for me. I was still using crutches and the only way I could get up stairs was to sit on my backside and go up in reverse step by step.

I started the first day at Crosfields as the overweight twin brother with different sized legs – not a good image for the first day at a new school!

# Chapter 2

## *Crosfields*

**My father drove us to Crosfields that first day. He now had a new car – another Morris Oxford, though this time it was brand new. I had advised him on the colour, Forest Green, which was also available on the MGA; this was the closest we would get to that sports car. The interior trim was also green, leather, of course, but that was the norm. I remember the number plate to this day – LRD 331. Funny how young children remember things like that. Crosfields was on the south side of Reading in a part of town named Shinfield. The route passed by my father's practice, so it was no great detour for him to drop off me and my brother in the morning. After school, we would take the bus – which ran through town and up to Caversham, no change needed – from Shinfield Road. On that first day our mother met us at school and showed us how to catch the bus – understandable since we were only seven years old!**

Crosfields was, like many private schools, located in what was once a large country mansion, with annex buildings to accommodate extra classrooms. It had expansive grounds with plenty of room for sports fields: football in the winter, rugby the spring, and cricket in the summer. My father pulled in through the imposing gateway and along the gravel drive to the front entrance. There were a predominant number of Jaguars and Rovers in the car park, so I was anxious to get out of the car as quickly as possible, despite my handicap of crutches and school satchel, in the hope that our inferior Oxford would disappear.

Pleasantries were exchanged, and my father left me and my brother in the care of one of the senior school boys. Immediately things started to go wrong. The senior school boy flew off with my brother up a staircase, oblivious to the fact that I was on crutches. I flung my satchel around my neck and, being the only way I could negotiate a staircase, got on my backside and reversed up the stairs. By the time I had reached the top, 'senior school boy' and my brother were gone. This old mansion was like a rabbit warren and I was lost. Eventually, a teacher picked me up and we finally located the correct classroom. Red-faced, overweight, on crutches with a shrunken leg, I was greeted by hoots of laughter – welcome to boys' schools.

My brother and I were allocated to class 1A, and from then on were named Birtwhistle P and Birtwhistle R. The 1A and 1B classes combined to make up a total of around 40 pupils in the first year. That first day the combined classes were assembled in the school canteen for a joint dictation, to be used as a test of our literary abilities. The reading was from a scripture (religious) book about 'The Wise Shepherd.' After about two sentences I was already drifting away, my mind in the hills of the holy land. So what did I do? I started to illustrate the page with images of what was being read. The shepherd, crook in hand, surrounded by sheep grazing peacefully on the plains.

We handed in our sheets and the next day I faced the music: my first marks were 1 out of 10. Regretfully, that test set the standard of my academic level throughout my time at Crosfields, and it was only in art class that I excelled.

Weeks passed by, and physically I started to return to a degree of normality. The crutches went, my leg put on weight, and with the help of some sport, the rest of me shaped up, too. I remained, however, slightly overweight for quite a few years.

I eventually integrated well into the class, despite the scornful welcome of the first day, and new friendships started to form. A particularly strong friendship developed with a classmate called Michael Baylis. His father's business was the first American-style supermarket in Reading. Mr Baylis was an entrepreneur, as not only did he open the first local supermarket, he also introduced Green Shield Stamps to Reading. Green Shield Stamps were stamp-like vouchers you could collect when purchasing at Baylis'. You glued the stamps into an album, which, when filled, enabled you to exchange it for some kind of gift. Nothing too exciting mind you, a saucepan or a set of plates.

An added bonus to this friendship was the fact that his father was a car fanatic and had the money to indulge in this passion. He drove Jaguars and so the announcement of a new model meant that within weeks a new Jaguar was rolling up the drive of Crosfields. For me, the highlight must have been when Mr Baylis took delivery of the first Mk10 Jaguar in Reading.

Jaguar had two mainstream models: the Mk2, successful on the race track and also the car of choice for gangsters in the 'B'-rated films of the time, plus the rather sedate Mk9 limousine. The Mk9 was essentially the same body as the Mk7 and 8 preceding it, with just a few small differences. The Mk7, for instance, had a split front windscreen. They were quite conservative and featured a covered rear wheel. The announcement of the Mk10 was a revolution. Compared to the Mk9, this was a creature from outer space. Visually low and long, it oozed grace and sportiness, and above all it featured twin round headlamps, the latest design import from the States. Twin headlamps were a must on all modern car designs, but unfortunately never graced the products of the British Motor Corporation, which manufactured Morris and its siblings from Austin, MG, Riley, and Wolseley. The European outposts of Ford and GM (Vauxhall) introduced this feature, and even Rolls-Royce did, which was a true shock to the traditionalists, but not to BMC. No, a twin headlamped car never graced the Birtwhistle drive.

Back to the Mk10. The first day the Baylis' drove to Crosfields in that car, I made sure that my father took us to school early enough for me the witness the arrival. I waited nervously at the entrance gate until it came. Resplendent in dark blue, it just seemed to purr past me and up the drive, befitting of a Jaguar, the only discerning noise being the crunching of the gravel being moved under its weight.

These days it's hard to comprehend how the first sight of a new Jaguar could have such an impact on a small boy, but for me, with no exposure to the kind of media we have today, such occasions were truly out of this world. Very soon, another Jaguar would rock the automotive world in a way that I believe will never be repeated.

***

Back home, normality had returned, and after the dramas of Christmas the family was now intact. If Helen was the darling of my father, the new arrival, Christopher, developed a similar status with my mother. He had been born premature and his survival had been a very close call. My mother knew how hard she had fought to keep him, and a close, lasting bond developed between her and Christopher. Despite the special roles of Helen and Christopher, we were a close family and I felt very comfortable with life at home. In Caversham, I had developed a circle of friends apart from the ones I was cultivating at school. My parents had got to know a family across from our house: the Elliotts – Peter and Paula and their two children Tim and Anne (she would later change it to Annie). A relationship would evolve between our two families, strong enough to last a lifetime. Tim and Anne were slightly older and younger respectively than Roger and

*Two cowboys with Lone Star pistols.*

me, and Tim became our best friend in those days, with Anne joining Helen to play. Tim's father, Peter Elliott, had recently retired from his position as a Major in the British Army. Although 14 years had passed since the end of WW2, the knowledge of this conflict and the fact that 'we had won' was a mantra embedded in the minds of most young boys in the '50s and '60s and meant that playing war seemed completely normal. I took to this like a duck to water, and Tim's father was a welcome supplier of the entire military garb I needed. This included caps and berets, belts complete with various pouches and gun holsters, plus a collection of badges that my mother would sew on to my 'military' jacket. As far as 'killing hardware,' was concerned, Major Elliott was not quite as forthcoming, so we had to rely on our cowboy revolvers, which were usually chrome plated. The biggest manufacturer of toy guns was an American company called Lone Star. They also had a range of model cars, but these offerings were way inferior to Dinky or Corgi. Lone Star had cornered the toy gun market, and produced replicas of most weapons in the cowboy range, from a derringer to a full western-style rifle. It also supplied all the cowboy 'accessories' to go with its gun range – belts, holsters, spurs and, also highly prized, the Sheriff's badge in black and the Deputy's badge in red. Roger and I had to settle for being deputies since there could only be one Sheriff – another sacrifice for being a twin. Tim Elliott, of course, had a Sheriff's badge, so the pecking order had been established, and, understandably, he also had the best choice of his father's military attire. Tim's weaponry went one stage higher when he was gifted Lone Star's first military application, in the form of a German Luger hand gun, finished true to life, in matt black. By now, the Kidmore Road weapons race was on, and the following Christmas, Roger and I were given battery-powered 'Tommy Guns' capable, with a loud rat-a-tat, of firing red bullets. After two days of intensive battle, the batteries were removed. The weapons race, however, was not over. Never one to be beaten, Major Elliott bestowed a birthday gift on Tim that finally gave their side of the street the upper hand. It came in the form of what must be considered the pinnacle of 'Toy Warfare,' a life-size Browning sub-machine gun. This fully electric replica was about a metre long and came mounted on a tripod with fully operational bullet belt. We were beaten.

Never one to lose a battle, though, I answered the Browning's challenge with my own secret weapon: creativity. My father's work shed had all the tools and wood required to create an arsenal of weapons to cater for all the needs of street warfare. I spent hours sawing, drilling and assembling the most amazing weapons. My creations had a magic and power that contained all the fundamental value needed when playing – the use of imagination. These days, plastic guns have been replaced by virtual guns, and the days of children using their own imaginations to create games are disappearing, as is, sadly, childhood as I experienced it.

You may detect the irony in my description of those 'war days' but that was the mood of things in postwar Britain. Boys played soldiers and the girls were nurses. Our parents were proud of how Britain had survived the war, and the memories were still very fresh. As children, we only got the glossy side of war, and we watched the films of British heroism: an endless list that includes *The Dam Busters*, *Sink the Bismarck*, and *The Guns of Navarone*. My father had been drafted into the Navy at the end of war hostilities. He had not experienced front line action, except, that is, for duty as an air raid warden while studying in London during the blitz. My interest in war was mainly connected with tanks and planes, and I can't recall my father ever going into the details of the real horrors of the war. My mother had been a teenager in Reading, which also had its own dose of bomb damage, but again, it was a part of her life she didn't speak of in detail.

A few years later, a side of the war I barely knew about came to light when I discovered a volume hidden in a row of books on my father's bookcase. Written by Lord Russell of Liverpool, it was titled *The Scourge of the Swastika: A short history of Nazi war crimes*. Lord Russell was a lawyer and adviser at the Nuremburg war trials. The book was a well-documented and illustrated description of exactly what it said on the cover – primarily an account of the horrors of the Nazi concentration camps. The book had caused a big stir when first published and the authorities tried to have it banned, a sure guarantee for sales success. The contents, including the photo documentation, came as a big shock to me. My parents had talked about the Holocaust and the eradication of the Jews, but somehow the details were never discussed.

I don't believe that if my generation had been enlightened to the true horrors of WW2 it would have stopped us playing war. However, when I have related my childhood war games to German friends, they understandably find the notion of young boys running around in military garb unbelievable – yes, the wounds are deep. Later in life, when I lived in Germany, I would never experience children playing 'war games'. There will always be debates as to the wisdom of letting children run around with toy guns, but I truly don't believe it did me any harm. I think children need a safe outlet for their aggression, before they can do real harm. But that's a topic that could fill another book, so I'll leave it there.

Putting war games aside, at this stage in life I had an enthusiasm for many different interests. My twin, however, never seemed to share them with the same passion I had. He joined in but was very much a follower, and many of the activities I took up he didn't embrace with the same degree of interest. I was hungry to try everything: I became a Cub Scout, went train-spotting at Reading station, took up river fishing in the nearby Thames, just to name a few of my interests.

My academic life continued its meandering course, and regretfully both Roger and I were often at the lower end of the school's marks tables. I was, and still am, pretty bad at spelling, so maybe there was a trace of dyslexia in me, though it's not something my parents ever considered getting checked. In those days you tended to get labelled as being a bit 'thick,' and bad results or behaviour sometimes meant I would have to stand in the corner wearing the 'dunce's' pointed hat – an unthinkable act to bestow on a child these days. On the brighter side, however, where I did excel was in the art class. I could draw and paint and I loved every minute of it. My parents knew I had the talent, but only when visiting a school open day, where one of my paintings had won a first prize in a national school art contest, did they really become aware of my artistic potential.

The most used medium in art class was watercolour. The paints were made by mixing coloured powders with water, and the art room seemed to have a film of that powder dust settled over everything. I can't recall in detail what subjects we painted, but it was usually scenes involving animals or studies connected with the school surroundings. If allowed, however, my subject of choice would be some form of machinery. Cars, planes, ships, you name it, and the car that captured my imagination the most was the Jaguar E-Type.

In the late '50s, British representation in world motor sport was at a high point. In the Grand Prix class, the British Racing Green Vanwalls, with their 'bunch of bananas' exhausts, driven by Stirling Moss and Peter Collins, were winning over the Italian Ferraris and Maseratis. I would watch races on the television with the BBC's John Bolster commentating. Bolster was a cavalier kind of character, who sported a tweed deerstalker cap with matching hacking jacket. He always held a microphone that appeared to be permanently attached to his mouth, with his flamboyant moustache poking over the top of the BBC badge secured to it. Bolster covered all the major race events including the Le Mans 24-hour race. Here, the Aston Martins and, notably, the Jaguar C and D-Types were winning year after year. In particular, it was the D-Type Jaguar that was a car of unbelievable beauty. No wonder then, that when the road-going equivalent of it (in the form of the E-Type) came out, it was a sensation in the motoring world that I honestly don't believe will ever be repeated. This was a machine that's stance, lines, balance and proportion were just breathtaking. Perhaps the form of the Supermarine Spitfire aeroplane or The Mallard steam train had similar qualities, but for me they didn't quite match the E-Type. I'm often asked which car design is my favourite; without doubt, it's the E-Type. You need to put your mind back to the time when it was first shown at the 1960 Geneva Motor Show. The media exposure we know today just didn't exist. Here was a car that looked as if it had just come from the race track, a style of vehicle that most people would never have seen before, except, perhaps, in shorts reports on the television or pictured in fuzzy black and white photographs in the daily newspaper.

Malcolm Sayer, the Jaguar aerodynamicist and engineer credited with its design, took a pure form from the wind tunnel and made a sculpture on wheels of unparalleled beauty. This was not styled (a term sometimes used for automotive design and a term I never had any problems with) but a shape that drew all its influence from nature. The most efficient forms in the natural world, such as sharks or dolphins, have a functional beauty that the E-Type can be compared with. It will never be matched again.

My passion for any form of moving machines or mechanics continued unabated, and from Dinky toys I moved on to Meccano building sets, Hornby model trains, Airfix plastic kits and motorsport.

My first experience of a motor race was when my father took me and Roger to the Easter meeting at Goodwood in 1963. This was quite an important race weekend featuring all categories of race car, including a non-championship race for full-blown Grand Prix cars.

Goodwood was located within the grounds of a large country estate near Chichester, close to the south coast. The queue to get into the track on race day was endless; some things never change in Britain. While waiting, a Jaguar Mk2 came barging up the queue, resplendent in red, with a large sign on the front windscreen that read: 'competitor.' Yes, they drove to the track in those days.

Once we had parked the car, we took our place at the side of the track. There were no kind of barriers in those days to afford the spectators any sort of protection in the event of a car leaving the track, just a few hay bales strategically placed here and there. My father had brought a couple of small folding beach chairs and a picnic rug, so we settled down, our feet almost resting on the track.

This was a high-profile event, with drivers such as Graham Hill, Innes Ireland, and Roy Salvadori competing. Regretfully, Jim Clark, who later went on to be the 1963 World Champion in a Lotus, was not present. The race day was also significant as the anniversary of Stirling Moss' accident at the same meeting the previous year, an accident that was to end his international racing career.

There was a variety of races that day: saloon cars, including not only the Mk2 Jaguar we had seen entering the track, but also Mini-Coopers and huge American Ford Galaxies. The sports car race featured Jaguar E-Types battling against Ferrari 250 GTOs. Graham Hill won both the saloon race and the sports car race in the Jaguar Mk2 and E-Type respectively.

The highlight of the day was the non-championship Grand Prix race with not only Graham Hill and Innes Ireland, but also Jack Brabham and Bruce McLaren.

My father was an ardent Graham Hill fan. Hill's helmet had the same design as the caps worn by the London Rowing club – black with thick vertical pointed white stripes. I don't believe that my father had any connections to the rowing club during his years studying dentistry at London University, but he felt connected in some way, and from then on I was also a Graham Hill fan. Hill took the lead in the Grand Prix race in the works BRM in which he had won the World Championship the previous year. The BRM was quite a distinctive car, with its eight vertical exhaust pipes behind the driver. Unfortunately, Hill's luck was out and he retired, leaving the way open for Innes Ireland, driving a Lotus 23 BRM, to take the chequered flag.

Our days as a close-knit family were coming to an end, as at the age of 13, my parents judged a continued education at boarding school to be the best for Roger and me. Bringing up four children was quite a strain on my mother, so two of us being out of the house could only help. This was not a great shock to us: most of our class mates would also be going away to school, and coming from a middle-class family, this was quite normal.

I haven't dwelt so much on my days at Crosfields. My academic achievements really were nothing special, as was also the case with my brother. On the sports field, where I made it to the Rugby first 15, things were a bit better, but it became clear to our parents that neither of us would make the grade in the Common Entrance exam we would need to pass in order to earn a place in a public school. To make matters worse, we had both failed the eleven-plus exam needed to get a place at a grammar school, which ruled out that option, too; things weren't looking good.

By chance, one of my father's partners had sent his son to a newly founded boys' boarding school, on the Sussex coast near Chichester that required no Common Entrance exam. Earnley Boys' School, was named after the small coastal village where it was located and had only been in existence for two years. Everything about it was new; it sounded perfect.

We drove down to Earnley and after a meeting with the school director, Mr Betts, it was agreed Roger and I could commence at the school in the autumn of 1963.

Things wound to an end at Crosfields. Farewell parties were held at the homes of various school friends, plus a final leaving boys' party in the school gymnasium. Music started to play an increasing part in our lives and we all jumped around to the sound of the latest new group: The Beatles.

The summer of 1963 marked the end of an intact way of life in the Birtwhistle household. Most of our friends would also be leaving for new schools away from Reading. Of course, we would be together for holidays, but the family life we had all grown up with ended that late summer, and things would never be quite the same – Earnley was beckoning.

# Chapter 3

## *Earnley*

**The autumn term at Earnley Boys' School started in mid-September. New boys commenced the term a day early, and with that in mind we set off to drive the 60-odd miles from Reading to the south coast. This wasn't such a new journey for us since, as a family, we had spent a couple of summers on the south coast at Middleton-on-Sea, close to Bognor Regis.**

By now my father had another new car. The Morris Oxford was just too small for a family of six, so something larger was needed. True to form, he kept his allegiance to the British Motor Corporation and traded in the Oxford for an Austin Westminster A99. He would probably have liked to have remained loyal to the Morris label, but they no longer made a large model with a six-cylinder engine. The Westminster filled both these requirements, and though I was disappointed by its lack of twin headlamps, I was quite pleased with his choice, and for a BMC product it was quite flash. I also liked its horizontal chrome grille bar featuring the Austin badge in the centre. The engine was BMC's large three-litre six-cylinder, which also powered the new Austin Healey 3000. I could live with this car, and therefore had no concerns as we entered the car park of Earnley School.

Earnley Boys' School was the brainchild of John T Betts. Betts was a renowned lawyer and businessman whose expansive private home and estate, Earnley Place, was located in the small village of Earnley, south of Chichester on the Selsey peninsula, close to the seaside town of Bracklesham Bay. The centre of the estate was a large three-storey mansion, where Mr Betts resided. He was not married and shared the house with his personal staff, including a chauffeur and cook. Betts had served with distinction in the Royal Navy during the war, and through his success as a lawyer after the hostilities had built up a considerable personal wealth. Apart from his legal career, he was, amongst other things, a board member of Arsenal football club. In fact, he owned a large ocean-going yacht moored in Southampton named after the football club: Lanesra, or Arsenal spelt backwards. The Earnley estate also included a group of tropical greenhouses, manicured gardens and parkland. In the late 1950s, Mr Betts founded a sports club at Earnley Place for local boys. He had football fields laid out and a club house built on the estate grounds. This initial interest in serving the local community gave him the idea to expand, and through a trust he had founded, finance was secured to build a school on the Earnley Place estate, which opened in 1962.

No money had been spared in the design and layout of the school, and apart from the obvious dormitory and classroom wings, it featured an indoor swimming pool, indoor football gymnasium, language laboratory, library, and a 200-seat cinema. As well as the football fields on the estate, there was the latest in hard surface tennis courts, a squash court, and further sports fields located not far from the school's boundaries.

All of this made Earnley sound like the perfect solution for our future education. The school, however, had yet to prove its academic capabilities, having only been in existence for a year. Overall, I was very enthusiastic, though one fact saddened me: art was not a subject on the curriculum, and except for a carpentry shop the school had no facilities to cater for any kind of creative work, even as a pastime activity.

After unloading the two large trunks, containing our belongings, from the Westminster, one of which had to be secured to a roof-rack, we made our way into the school's entrance hall. Here, we were greeted with the predictable mayhem of parents and anxious new arrivals. A teacher gathered up me and my brother and without ceremony we were immediately whisked away from our parents, who we just managed to wave goodbye to, with no chance of even a kiss. Later, my father would tell us how saddened my mother was at

*A picnic by the Austin Westminster A99, on the way to Earnley School.*

such an abrupt parting of ways. In my mind, I think I was convinced that showing any form of emotion would immediately label us as being weak and so I wasn't too concerned, and perhaps having my brother with me helped matters. That security, however, would not last long, as we had been allocated to different dormitories. Little did I know, that apart from school holidays, this was the end of sharing a room with my twin brother.

The school had four dormitory wings, and though my brother and I were not to share dormitories, at least we were located in the same wing. The grouping of the dormitory wings was based upon age, ours being allocated to all the first- and second-year pupils. It was also the location of the school sick room and residence of the only female member of staff: the school matron. In fact, the school would provide a certain 'monastic' quality of life, since most of the teachers were single men, and aside from the aforementioned matron, there was a complete isolation from anything female for miles around. This masculine regime would instil a certain fear and shyness towards the opposite sex during my pubescent years to come.

The following day the 'old' boys arrived in dribs and drabs. The school had obviously gained the attention of parents from other lands and quite a few foreign pupils were arriving, including some black Africans, who were new to me. My dormitory began to fill, and tentative first introductions were made. So far, I had not done anything out of order, and there were just a few discussions as to why I had selected the cupboard I had. Yes, there was a definite hierarchy. I took on my usual strategy of trying to be everybody's friend. This, however, wouldn't prove so successful, since after delivering what I thought was a friendly comment in one boy's direction, I was pulled to the ground, with one arm yanked up behind my back, being threatened in no uncertain manner. Yes, there was a small selection of school bullies around, however, all in all, barriers were broken down and friendships started to develop.

Classes began and it was immediately apparent that the curriculum was fairly limited. As well as no art classes, there was no religion taught (no big issue for me personally), and no form of physics or chemistry. The school had no art room or science labs, and as I mentioned, the only creative facility was a woodwork shop. Things were very new still, so maybe more classrooms were in the planning. Our learning plan comprised Maths, English, and English Literature, with French and German taught in the new language laboratory. There was also History and Geography, which I loved, plus economics and public affairs. The school academic life was shared with daily sports, squash all year, football and hockey in the winter, with cricket, swimming, tennis and sailing in the summer.

The big bonus of the weekly curriculum for me, however, was Tuesdays, which were allocated to the Sea Cadets.

John Betts, having had a distinguished military career in the Navy, was keen to imbibe some of this military discipline into school life. Located on the south coast and with its close proximity to the big naval port of Portsmouth, Earnley was ideally suited for the Sea Cadet Corps. To this end, a new teacher with a naval background joined the staff: Lieutenant Commander Winn. For me, with my cub scouting and street warfare experience, this was a real highlight. I had enjoyed the cub scouts, but it is not a military organisation. The Sea Cadets, however, had a genuine link to the 'real' Navy and you also wore a correct military uniform. The whole of Tuesday would be taken up by the Cadets and conducted in full navy uniform. Instructors from the Royal Navy and Marines bases in Portsmouth also joined the regular staff for Sea Cadet day. I loved it.

The first day of Cadets was one to remember, with the entire school assembling in the central school courtyard. We were arranged in rows, the attending officers walking between the lines and screaming orders at us to look forward and maintain silence. Commander Winn, in full naval dress, stood proudly in front of us, chest extended with his campaign medals gleaming in the Sussex sun. When he addressed the new Cadets for the first time, you could have heard a pin drop as he spoke.

"My name –" pause "– is Lieutenant Commander Winn."

There was another short pause, but this was long enough for disaster to unfold. Josiah Prempeh, the son of African Ghanaian royalty and standing well above six-foot six, put his lips together to blow a raspberry of truly concert hall magnitude. Almost immediately Drill Sargent Fisher from the Royal Marines turned several shades of dark red, and screamed so loud, they could have heard him in Chichester.

"SIIIILEEENCCE IN THE RAAANKS!"

Order returned, and Commander Winn continued his address.

Despite the unruliness of that first parade, my appetite had been whetted, and I was determined to achieve something special in the Sea Cadets. A Royal Marines Bugler had started the day's proceedings with a crystal-clear rendition of the bugle call 'Sunrise,' and at the end of the day, just as moving, 'Sunset', to which the Union Jack was lowered from the school flagpole. This really caught my attention, and with the news that an important part of the school cadet force would be the creation of a band, I enrolled immediately as a bugler.

The Marine responsible for organising the school band turned out to be Sargent Fisher, the one who had screamed at Josiah during that first parade. He was a classic military professional, always immaculately turned out, with a keen eye for discipline, and I took an immediate liking to him. The aim of creating the band was to have a group of musicians capable of leading the entire Cadet parade at both the Admiral's inspection and school parents' day, by the end of next summer. The band would consist of a corps of buglers, drummers and flutes. Roger also took up the challenge and joined as a flute musician, while Sargent Fisher's 'friend' Josiah would play the bass drum. The anticipation of wearing a full naval cadet uniform and the chance of having my own bugle put me in a military seventh heaven. The bugles we were lent were not the silver variety as used in the Royal Marine bands but were made of brass, a material that needed to be polished with Brasso before each Cadet parade. I didn't mind this task, and kept mine in immaculate condition. Attached to the bugle was a navy-blue plaited cord with tassels, which enabling you to carry the bugle over the shoulder. I would practise in the evenings on the school squash court, where you could get the best echo and, of course, be out of earshot. I progressed well and soon started to develop the small muscle on the upper lip associated with brass instrument musicians. On cadet day, the buglers would stand in a row in the indoor football hall in front of Sgt Fisher, one after the other blowing their rendition of 'Sunset.' His favourite command was usually: "If you want any children in later life, keep your legs together when you blow that bugle, boy."

Or when threatening us with some disciplinary punishment, ending his sentence with: "I kid you not." For example: "If you don't stand still I'll have you running in full dress to the beach and back, I kid you not."

Over the course of the school year, under the watchful eye of Sargent Fisher, the band came together well. My own skills on the bugle impressed my mentor and I was given the honour of playing 'Sunrise' and 'Sunset' each week at the beginning and end of Cadet Day. To this day, many years later, when a Marine bugler plays 'Sunset' or 'The Last Post' on

Remembrance Sunday, I feel proud to know that I once did that, and pretty much just as well.

I loved Earnley because it offered so much, though not necessarily in the classroom. Opening a new school is a huge risk and only time would tell if Earnley could produce a high enough academic level to keep the Ministry of Education happy. I never got the feeling that the pupils at Earnley were over taxed. The academic day started at 9:00am and went on until lunch, then the afternoon was dedicated to sport. Afternoon classes commenced at 4:00pm and ended at 6:00pm, in time for supper, and there was a period from 7:00pm to 9:00pm set aside for revision. The academic day ended with prayers in the school cinema. The prayer meeting was usually conducted by one of the resident teachers, who delivered a small talk for us to reflect on, before all pupils stood to recite *The Lord's Prayer.*

J T Betts was obviously an atheist, since, other than the evening prayers, there was no religion taught at the school, and he never attended the Sunday service at the small village church. The Sunday service was mandatory for all pupils and was the only time when we were required to wear a suit, white shirt and tie. The church was known for being used by the nearby theological college in Chichester as a testing facility for their students – a kind of sermon skunkworks. This meant that each week we were subjected to some very questionable sermons. The daily evening prayer meeting took place in order to fulfil the requirements of the educational authorities when running a school.

After evening prayers, the pupils went to their respective dormitories and the day ended with lights out at 9:30pm for the juniors and 10:00pm for the seniors. Pupils at A-level age, from around 16 years old, resided in so-called 'studies,' where two to three boys would sleep and work, which was almost like living in your own flat.

The phrase 'rough and tumble' sums up the way many things ran at Earnley. The school facilities were amazing in what they offered, but putting things into practice was never as effective.

The school director J T Betts had achieved his goal of building this amazing educational facility, but when staffing it, he had run into problems getting adequately qualified personnel. Betts himself was not a teacher and there was little contact between him and the pupils. He resided in a distant office at the far end of the school building, close to his house. On the whole, he kept a good distance from the everyday running of the school, and over time he lost interest.

The sparse reports I could find on the Internet state that J T Betts was a generous benefactor of his boy's club, and was vital to the founding of the school and how it would later evolve. He died in 1975 and will be remembered for those achievements.

Nevertheless, I remember my time at Earnley for mainly positive things, and despite the apparent difficulty Betts had securing well-qualified staff, there were undoubtedly some teachers of note. In particular, I remember Mr Lymer, Mr Lowther and Mr Hill.

Mr Lymer was a junior teacher who I believe came straight from training college. He was from Liverpool, and was a passionate Beatles fan.

Mr Lowther was an experienced English teacher who drove a red Mini Cooper, with two King Charles spaniels residing on the back seat during class hours. He was slightly stocky with a pleasant demeanour and a twinkle in his eye. Single, he may well have been an actor in earlier life, for his lessons were always full of energy; it was in English Literature that he really came to life. He loved the prose of Shakespeare, and when reading the lines of our set book, *The Warden* by Anthony Trollope, he would melt into each character as if it was himself. Out of the classroom, he would take the lead in directing whatever school play or opera was planned for the end of the summer term. He lived alone close to the school and was certainly what you might describe as camp in his manner, revelling at the innuendos found in the books we were studying, even Trollope.

Mr Hill was the Maths and woodwork teacher. Also single, he lived locally and was also a hobby farmer in his spare time. This was mainly confined to raising chickens and looking after a few fields. Attending to his farm at the weekends, he, like Mr Lymer, resided in an apartment at the school during the week. Mr Hill had many interests. He was passionate about the film *The Sound of Music*, which he watched on a regular basis at cinemas along the south coast. His apartment adjoined one of the dormitories, which I was allocated to one term. Most nights after lights out you could hear the shrill tones of Julie Andrews resonating from his sitting room. I imagine she probably remained the only woman in his life.

Mr Hill drove a brand-new Singer Gazelle in two-tone green. The Gazelle was almost identical to the

Sunbeam Rapier, which was considered the sportier of the two and had rally pedigree, notably at the hands of the famous BBC commentator Raymond Baxter. Both were products of the Rootes Motor Group which also made Humber, Hillman and Commer commercial vehicles. Nevertheless, the Gazelle also seemed quite a sporty car at the time, and I was allowed to go for a drive with Mr Hill when it was new. I recall it had an impressive array of instruments with quite a few gauges and warning lamps, certainly a specification above that of my father's Westminster.

My interest for all things automotive continued during my Earnley years, and of all the teachers' modes of transport, it was that of Mr Lymer that inspired me the most. Soon after joining the school, he purchased a dark blue MG TF. The MG brand had established itself as an affordable sports car and was popular in prewar years. In the '50s, its reputation grew, and really put Britain on the map as a sports car manufacturing nation. It became a manufacturer to be reckoned with, even when up against the products of Germany and Italy. In particular, the MG TC had been a massive success in North America. The TF was the last of the classic-style MGs that came under the T nomenclature, featuring wheel fenders still separate from the central body panels. When released, it came into criticism for being too 'old school' – a bit of a last-ditch attempt to keep the T models alive. It's true that the prewar Magnette K3 and J2 models, both of which influenced the TC, were classic sports cars in the traditional way. The TC was followed by the more upright TD and for the first time featured steel wheels as opposed to wire wheels, in its basic specification – unheard of for a sports car. Despite its lack of love from some quarters of the MG fraternity, I personally believe that of all the T models the TF had the best proportion. Visually lower than the TD, it had beautiful flowing front wheel fenders, which, unlike the previous MGs, featured fully integrated headlamps. Wire wheels could be specified, but as with the TD, steel wheels were also a standard feature. Somehow, however, it worked OK on this design.

My fascination for the TF developed because at the age of around four, I received a tin model of the car made by Scalex. There was something magic about that toy car in shiny red with a tan interior. It had a deep effect on me, and ever since then I've hoped to own a TF one day. Apart from the MG, Scalex made quite a few model cars, including a Grand Prix Ferrari and V16 BRM. They featured a friction fly-wheel motor which could be charged, similar to Paul Easby's Corgi Mk 1 Jaguar, by pulling it backwards on the floor before letting it go. The models were quite large, at a scale of around 1:24, and were predecessors to the first slot car racers. Scalex were manufactured by Minimodels Ltd which developed the first slot-car racers based on its Scalex models and made under the name Scalextric (Scalex/electric). This venture, however, proved too big to bring to the market and it was taken over by Tri-ang which went on to develop Scalextric (still produced these days, albeit by the Hornby brand.)

Roger and I got our Scalextric set Christmas 1964.

# Chapter 4

## *Dancing classes*

**The thrill of returning home that Christmas was indescribable. My father arrived in the Westminster together with our sister, Helen, who was equally excited to see her two older brothers for the first time in what must have seemed an eternity for such a small child.**

Entering the house at Kidmore Road all the things I had so missed came back in a big rush. It started with the smell of the house, and there was something very reassuring about being back home. I rushed up to my bedroom, where everything was still in place: what to do first?

My mother broke the bad news to us over supper. She had enrolled me and Roger in ballroom dancing classes – two lessons a week for the duration of the holidays. This was a big blow to my plans.

My mother was a social climber, she was married to an up-and-coming Reading dentist (so was solid middle class), and moving in those circles it was expected that one should be able to perform on the dance floor. While those of our old friends who were not attending boarding school went to the local youth club, we had to put on our school suits twice a week and go to Miss Barret's ballroom dancing school.

Miss Barret was a well-known institution in Reading, having taught several generations of Reading children the skills of 'the parquet,' including my mother. Her studio was located in Southampton Street in the centre of the town. Roger and I were apprehensive about this new venture and I was particularly worried – there would be girls there, a definite case of stepping into the unknown. Up until now the opposite gender had not been a big part of my life, sure I had a sister and she had friends, but I had never really involved myself with them. They had been useful as hostages in our street warfare, but that was about as far as I had gone with regards to making any form of contact. Crosfields was an all-male institution and Earnley was like being in a prison or a monastery. At the age of nearly 14, it would be wrong to say that I wasn't interested in girls, but the thought of approaching them and having to join them on the dance floor, arms around their waists, was a daunting proposition for me.

Fortunately, a few of my old school friends were also attending Miss Barret's, including Michael Baylis, which helped me feel a bit more at ease. That security didn't last long though. Roger and I entered the dance studio, which was already buzzing with anticipation. Along each side of the room were lines of chairs; the boys sat on one side, with the girls opposite. Eye contact started straight away. Roger and I found our seats, eyeing the 'beings' on the far side of the room. Miss Barret took to the floor; she was a woman who looked as though she had been teaching dance classes for the last 200 years. Her grey hair was pinned up in a large bun, with wisps falling down her neck. She was completely confident in her stance.

"Ladies and gentlemen, to commence today's class we will do the waltz. Now, which lucky young prince wants to be my partner?"

A simple question with a simple answer – none of the 'princes' in that room wanted to be Miss Barret's partner.

"Come along, boys, I don't bite."

I, meanwhile, was avoiding all possible eye contact. Miss Barret made a bee-line past where I was sitting and pulled some other unfortunate fellow from his seat – I was, for the time being, safe.

Seated at an upright piano in the corner of the room was Miss Barret's assistant, who looked as though she had been there all those 200 years too. A signal was made and Miss Barret pushed her unwilling partner into action.

"One two three, one two three."

The two of them stumbled across the floor like a couple of drunken crabs. After this brief introduction, the moment I had been dreading arrived.

"Now gentlemen, choose a partner and ask her to dance."

Through the dizzy haze of uncertainty, I noticed that Roger had left his seat and was heading across the floor. I felt lost. It was like musical chairs and I needed to find a partner before Miss Barret spotted me. Boys outnumbered the girls, and this time I was to be soundly beaten by Roger. Slowly, I felt the shadow of Miss Barret envelope me.

It's all a long time ago and my performance on the dance floor with Miss Barret was almost certainly as toe crushing as her first partner that day. Soon I asked a girl to dance and barriers fell away, but my nervousness towards the opposite sex would remain for years to come.

Miss Barret's dancing classes became a regular part of our Christmas and Easter holidays for the next two years, culminating in a grand ball held in the works canteen of the Huntley and Palmers biscuit factory; now, it doesn't get more stylish than that!

School holidays also brought a welcome return of the television, another deprivation at Earnley that cut us off from the outside world. My father bought one of the first television sets in Reading – an Ekco. With a wooden outer case, it was almost cube-like in its form, such was the dimension of the tube screen those days. In the '50s there was only one channel, namely the BBC, further channels were expected, and the location to install a channel changer on the Ekco was covered by a circular plastic plate. That cover never got removed. The Ekco was unfortunately very reliable and remained a fixed installation in our sitting room for the next ten years, maybe more. When eventually the second channel did arrive (Rediffusion, later to be named ITV), the request to get the Ekco modified to receive the new channel was refused by my father. He argued that he didn't want us being corrupted by the advertisements featured for the first time on ITV.

The picture those days was black and white, with colour images not arriving until the '60s. Our first TV programmes for pre-primary school age were *Watch with Mother,* featuring a different half-hour programme each day at 2pm: *Picture Book*, airing on Mondays, featured a prim lady named Patricia Driscoll, who read from a large open book. It didn't take me long to see that the book, in fact, had no pictures in it at all; *Andy Pandy* was Tuesday's offering, and the week's first taste of questionable morals, with Andy's girlfriend, Looby Loo, confined to the wicker basket they lived in for most of the show. Her only significant role was to come out of the basket to sing "Here we go, Looby Loo," when Andy and his mate, Teddy Bear, had cleared off somewhere; Wednesday was *The Flower Pot Men*, featuring Bill and Ben, and more questionable morals. The two garden heroes with terrible speech impediments, lived in flower pots either side of their shared female partner, Little Weed, who would obediently keep her head bowed for most of the programme; Thursday was *Rag, Tag and Bobtail*, a tale about a group of animals living amongst the roots of a large tree; and finally on Friday, *Wooden Tops* – a loveable family whose bodies were made of – you guessed it – wood.

Thinking back, the amount of questionable connotations, of both racial or sexual nature, was quite amazing and not only on the TV. The Noddy books of Enid Blyton told the tales of a characterful young boy driving around in a yellow Morris convertible with red fenders, sporting a blue pointed hat with a bell at the top, whose best friend was a rather grumpy gnome called Big Ears. Try calling someone Big Ears these days. Even worse, their main enemies were a black street gang called the 'Gollywogs,' and young children of the day had a genuine fear of these villains.

The Gollywogs not only featured in the tales of Noddy, but were also a mascot for Robertson's jam. Each jar of jam came with a paper Gollywog, which, like a stamp, you could glue into a collecting album. When the album was full, you posted it to Robertson's and in return they sent you an enamel Gollywog badge. There was quite a large selection of characters, usually playing some kind of sport – a footballer or cricketer, for example. The more Gollywog badges you had on your lapel, the higher your street credibility.

Racial, let's say 'uncertainty,' was rife on the TV. An import from the US called *Amos and Andy* featured two black New Yorkers, who came across as complete buffoons. Finally, there was the Saturday night favourite of my parents': George Mitchell's *Black and White Minstrel Show*, which was a massive hit for the BBC.

*The Black and White Minstrel Show* was a head-on collision between a Mississippi Delta negro spiritual musical and the Moulin Rouge. A male dance choir dressed for the occasion often in white suits with tails, performed singing and dance routines with the BBC's female cancan dancing troop, The Toppers. Sounds pretty harmless – except for the fact that the choir members were all white males, wearing curly black wigs, their faces blacked out, and lips and eye lids painted white. Favourite songs were southern US classics, such as *Old Man River* and *The Camp Town Race*.

The lack of a channel changer on the Ekco meant that Roger and I would go across the road to the Elliott's to watch ITV. The Elliott's had waited until TV sets were available offering both BBC and ITV. Apparently, when Major Elliott first switched on the set, selected ITV and heard a voice announcing, "This is Channel Number 2," he assumed it was a perfume advert!

ITV's children's programme had all the US imports, such as *Popeye the Sailor*, *The Flintstones* and *Yogi Bear*, as well as – topping my list – the British space adventure *Thunderbirds*. The BBC couldn't compete with ITV, so during the week we were a permanent, late afternoon installation at the Elliott's. It was only with the arrival of Dr Who that the BBC began to gain some form of credibility.

They say you always remember what you were doing when President Kennedy was assassinated. Well, I was at the Elliott's watching *The Man from U.N.C.L.E.*

There may not have been a TV at Earnley but music was permitted. When I say music, I'm only referring to records, since apart from the Cadet band, music was another creative discipline lacking in the school's curriculum. As with TV, radios were also out-of-bounds. Some boys smuggled in small single transistor radios, which you could build as a kit. After lights out, using an earplug they listened to the first 'pirate radio' stations, such as Radio Caroline, which transmitted from old light boats or forts in the English Channel.

When joining Earnley in late 1963, Beatlemania was just about to get a grip on Britain. It's often been said that there will never be another mass hysteria the same

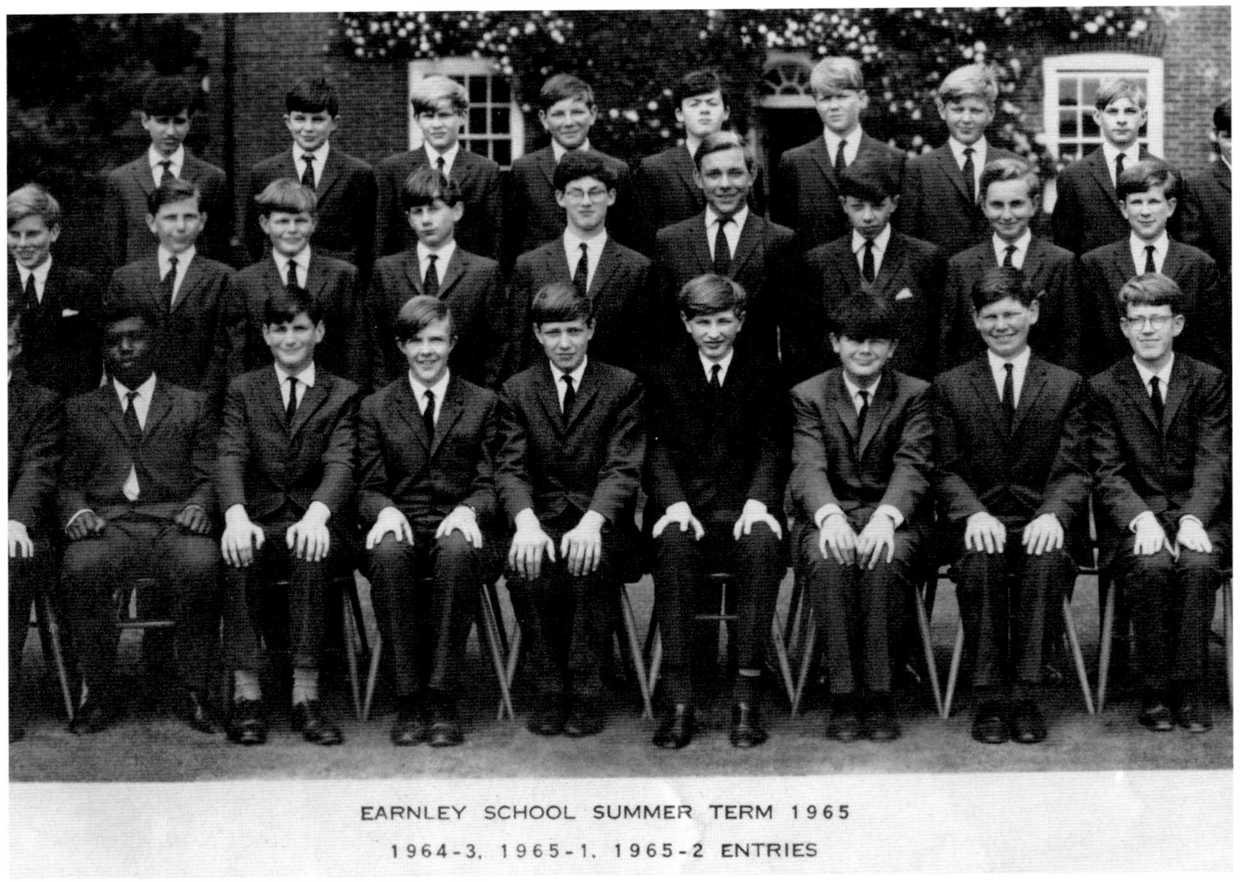


*Earnley School. I'm in the middle row, third from left.*

as that surrounding the Beatles, and I am in support of that opinion. It may sound a banal comparison to some, but I believe The Beatles were as significant to the popular music world as the Jaguar E-Type was to the automotive world – what a coincidence that they came to the world at almost the same time.

The first Beatle hit, *Love Me Do,* was released while I was still at Crosfields, and was soon to be followed by *Twist and Shout*. Up until then, my favourite popstars had been The Everly Brothers, Adam Faith, and, of course, Cliff Richard and the Shadows. None of these acts, however, had the air of expectancy that greeted the release of a new Beatles single, although Elvis Presley fans may prefer to argue. At Earnley, we all waited nervously for Mr Lymer, an ardent Beatles fan, to return from Chichester with the first copy of *I Feel Fine*. We all gathered in the communal room near the senior dormitories, where the school record player was located. The whining introduction of the song, with its electric feedback, started. Yes, this was truly another Beatles milestone.

Mr Lymer, a Liverpudlian, considered himself the school doyen of everything connected with the Beatles. His was a special relationship, almost as if he was their official school spokesman. Most of us could not afford to buy singles, let alone an LP. Mr Lymer would purchase every new Beatles release, and with it he knew he controlled a certain status and power amongst the pupils. Keep in his good books and he would invite you to his apartment to hear the latest release. I loved the Beatles, but not to the extent where I had to creep up to Mr Lymer. At the end of my time at Earnley, the Beatles had taken on an almost religious status for him and he chose to use the evening prayer meeting in the school cinema to give the whole school the first hearing of *Sargent Pepper's Lonely Hearts Club Band*. Seated on the stage in front of us, the chimes of *Lucy in the Sky with Diamonds* rang out. Mr Lymer didn't seem to be enjoying things, such was the significance of this record and the emotional weight on his shoulders. We listened to the whole first side before Mr Lymer got up in silence and with a stony and distant stare on his face left the cinema – there was no Lord's Prayer that day.

The summer of 1967, Roger and I got our O-level results. I scraped by with four passes in Maths, Geography, History and Economics (no idea how I managed Economics). Poor Roger, however, didn't get a single pass. My father was not pleased, despite my reasonable performance, and the topic of the questionable academic capabilities of Earnley raised its head. That year had seen the first A-level candidates leave the school, and the record was also pretty dismal. Nevertheless, I would move on, retaking the failed exams, and commence A-levels in History, Geography and Maths. The A-level Maths class only lasted a few weeks, a totally new dimension and way out of my bounds. However, other ideas had got into my head.

I was not the only pupil in the school who would have liked to have had art classes. The subject had been breached with the school director, Mr Betts, but he had a low opinion of art and wouldn't consider it as a serious subject. Fortunately, the school gardener and grounds man, whose name fails me, was a gifted artist, and proposed the idea of forming an Art Club. We would meet at a classroom in the evenings and when time allowed at weekends to draw and paint. The aim was to attempt passing an Art O-level.

After at least three years having drawn or painted practically nothing, my artistic skills had left me. My manual dexterity was still good, but I lacked the practise on the drawing board. The Art Club worked OK but there was little or no tuition, so I just muddled along as best I could and took the O-level exam in the summer of 1967. Regretfully, I failed.

By now, I had moved into a study room, which I shared with two other boys. As well as the three beds, it also housed desks for out of classroom studying, as was expected of A-level pupils. Here, I had the time to concentrate and improve my artistic skills, with the view of another attempt at the exam in the autumn of 1967. This time I succeeded.

Now with seven O-levels under my belt, my interest in studying further at Earnley began to diminish. The O-levels I had, including Art, were enough those days to get into an art college. That was my new goal.

During the Christmas and New Year holiday of 1967/68 I started to inform myself about local art colleges. I focused my search in the Reading area, since the big colleges in London required at least an A-level in art. The biggest local art college was attached to Reading Technical College: the Berkshire College of Art and Design. I picked up all the necessary paperwork and learnt about the courses. It turned out that the main college was not, in fact, in Reading but a few miles up the Thames in Maidenhead. The college had three main disciplines: Environmental Design with a focus on Interior and Exhibition Stand Design; Fashion

Design; and Printing. I now realised that I had no real idea what discipline of Art or Design I wanted to study, and a future career was not yet my key focus. Ideally, I would have liked to design cars or planes, but as my father wisely pointed out, "You will need an engineering degree for that." Of the courses offered at Maidenhead, Environmental Design seemed the most interesting. The application deadline for Maidenhead was spring the following year; I had up until then to get a portfolio together. I filled out the application forms and sent them off. Fully motivated, I returned to Earnley.

For the months of January and February I got my head down and spent all my time painting and drawing. In my heart I had already left Earnley.

One day in March 1968, I took a train from Chichester back to Reading and the following day my mother drove me to Maidenhead for my interview at the Berkshire College of Art. The principal of the college met me and enquired who I'd brought with me.

"Oh, then please feel free to join us," he told my mother.

The interview went well, maybe because the college principal spent most of the interview chatting up my mother, and we left with acceptance for the autumn entry.

The final summer term at Earnley dragged on, my interest in Earnley studies now gone, and, knowing that I would leave at the end of the term meant that the teaching staff showed no real interest either. I realised that for Art College I needed to lose my schoolboy look, so I found ways to avoid the monthly haircut and decided it was time to lose the puppy fat I had been carrying with me for years. My last contribution to the activities at Earnley was my involvement in the end of term opera, Gilbert and Sullivan's *H.M.S. Pinafore*. I loved singing and had always played a leading role in past performances. For this opera I would also have a lead role, but I also designed the stage set. It was a good way to say goodbye. My parents came to the last performance and after the encore, Mr Lowther said some words of thanks for all my contributions to the school over the years and how sad they were to see me go. There was no goodbye from Mr Betts. His enthusiasm for the school was long gone and a couple of years later he disbanded it entirely and turned it into a conference centre called The Earnley Concourse.

# Chapter 5

## *Berkshire College of Art*

**It was a late September morning in 1968 when I set off for my first day at art college: the beginning of a future lifetime in the creative world. I'd always wanted to draw and create, and Earnley had been a bit of a stumbling block on my way to where I wanted to be, but now I was focused and ready to begin that journey.**

My mother drove me to Maidenhead in our blue Austin 1100. The Westminster had been followed by a Wolseley 6/110 in a reddish brown colour. The Wolseley was pretty much the same car as the Westminster but with its own bespoke front grille, and was a better-appointed car. With a walnut wood dashboard, it was also one of BMC's top models, only bettered by the Austin Princess Vanden Plas. At 17, I had started driving lessons at a local driving school, and, as is permitted in Britain, could also practise in our car alongside my father, as long as it was fitted with the obligatory red-on-white L-plates, which we attached to the car's bumpers. The Wolseley must have weighed about two tonnes and was a bit of a tank. With its three-litre, six-cylinder engine, it had bags of torque and, quite frankly, was not the ideal mode of transport to learn to drive in. After a few nerve-wracking moments with me at the wheel, my father decided it would be a good idea to downsize; hence the Austin 1100. As a car, it was too small to take all the family, but the days of joint family holidays were over, so it didn't seem to be an issue. The 1100 was the best-selling car in Britain at the time, and was basically a 'blown-up' Mini, sharing many of the Mini's mechanical components. In design terms, it was the start of BMC's trend to base all its models on the design language of the Mini. The design of the Austin and Morris 1100 models worked, with a clear visual link to the smaller Mini. However, when this philosophy was applied to the mid-range cars a size up, in the shape of the Morris 1800 and its Austin and Wolseley derivatives, things got a bit uncomfortable. Not without reason were they nicknamed the 'Land Crab.' Undeterred, BMC continued to develop cars in this vein and, following the 1800 models, produced the Austin Maxi, which, despite being a hatchback, was nevertheless very conservative. At this period in the history of BMC cars, so-called 'Badge Engineering' reached its peak, resulting in some truly horrible models, such as the Mini-based Riley Elf, or the even worse 1300 Vanden Plas: a kind of mini Rolls-Royce for pensioners.

The 1100 pulled up outside the Berkshire College of Art's main building in the centre of Maidenhead. Apart from this site, the college had another location called the Annex, which adjoined the Maidenhead Technical College on the way out of town towards Reading. The first foundation year, as it was named, however, took place in the main building. Here, students learnt the basic hand work and fine art skills to equip them for future design work.

In the car boot, I had all the equipment required for my course, which we had brought with us. Two wooden drawing boards, a T-square for technical drawing, portfolio case, plus all the drawing pads, pencils, paint and brushes needed. At the age of 17, this was a big step in my life. Unlike when I had started at Earnley, this time I was able to give my mother a kiss, before struggling up the steps and into the college's entrance hall – into adulthood.

The Maidenhead college was similar to the many classic art college buildings, specifically designed for teaching art, that you can find all over Europe.

The entrance hall was dominated by a wide staircase. On both the ground and second floors were several spacious studios with tall windows that allowed light to flood the rooms, therefore producing the best conditions for creative work. I made my way in and was immediately directed up the stairs to a main studio, where all the new students were assembling. Nervously, I went in.

Before setting off on that first day, I was aware that other students I met would be assessing me straight away by my outfit. The look for male art students in the late '60s was fairly universal: an open-neck shirt with large 'donkey collars,' paisley neck scarf, bell-bottom jeans with spatters of paint, a cord jacket and beige desert boots. During the summer holidays I had let my hair grow, which, much to my father's displeasure, was now almost shoulder length. Entering the room of assembled students, I fitted in well. The space was quite expansive, with wooden floors and a scattering of large tables and drawing easels. Students were milling around in small groups. The first thing that struck me was the number of girls and immediately I felt myself blushing. The college had a Fashion Design course, which accounted for the predominant number of female students. They were mainly dressed in long gypsy style skirts with head scarves, however, one stuck out: an incredibly striking girl, with blonde hair down to her waist, wearing tight leather trousers and what looked like a chrome-plated bomber jacket with the words 'high voltage' written on the back. Her name, I would later learn, was Lucy. This was an entirely different dimension to the girls at Miss Barret's dancing classes: they might as well have been from another planet.

The college principal entered the room, accompanied by the entire teaching staff. He waited patiently for silence, and what I immediately noticed was how undisciplined some of the students were in comparison to what I was used to, several continuing to talk oblivious to the collection of tutors in the room. Eventually, silence prevailed and the principal addressed us with words of encouragement and the expectation for us to account for our acts as young adults now that we were no longer school children. After this we were instructed to go to our respective studios; my group, Environmental Design, remained in that room.

The curriculum for the first year would be a collection of weekly lectures aimed at developing all manner of artistic skills, both two and three dimensional. Subjects would include: still life, life and portrait drawing; communication; and sculpture and pottery. That first day we began with still life. The tutor looked exactly as you would expect an artist to look. Sporting a flat cap, obligatory scarf and a Gauloises cigarette in his mouth, he was not so dissimilar in attire to many of the students. He gathered us around and after a brief introduction instructed us to take a place at a table or an easel, remove a shoe and interpret on paper what we observed in it as a shape involving light and shade. What he didn't want was an illustration of a shoe. This was the beginning of my creative journey and I felt elated. For the first time in so many years I was being guided and encouraged to do what I loved above all in life, putting a pencil to paper. Being able to draw a three-dimensional shape, whether a shoe or anything else, and understanding light and shade, and perspective and balance are the key disciplines you need when designing. I left college that late afternoon for Maidenhead station feeling like a new person and I wanted more.

I commuted to college each day by train. Maidenhead was two stops up the main Great Western line to London, with the town of Twyford the only halt on the way. Trains left Reading General heading east past Reading power station and under the bridge at Woodley, where Roger and I used to watch the steam trains pass through from our Silver Cross pram. I would often look up at the bridge with affection.

Settling into art college life, I soon had a list of my favourite disciplines, and way at the top was Communication Art. This was a new subject for the college and dealt with the increasing importance of media design. The lecturer for this topic was a designer named Tony Rickaby, who worked in London for an advertising agency. Tony and I struck it off immediately, and under his guidance my most creative work during my first year was at his lectures.

Tony was into the pop art culture, both visual and written, and the big names of the genre at the time were artists Andy Warhol and Roy Lichtenstein, plus Tom Wolfe in literature.

My own first delves into pop art were during my last summer at Earnley. I had done a series of poster designs similar to the psychedelic themes appearing in pop culture of the time and notably on many record album covers. It was also at this time that the artwork of Roy Lichtenstein grabbed my attention. Much of Lichtenstein's work was based on images seen in US comics, and in particular 'war comics.' These small comic books could be bought at most newsagents, but were banned at Earnley, though that didn't stop a massive black market business in the school. I was an avid fan, especially of the *Air Ace* series, which featured illustrated tales of heroic British air combat in WW2. Here was a good source for my artwork and I produced a series of Lichtenstein-like posters. The posters depicted images such as a German Messerschmitt ME 109 going down in flames, with the pilot's screams of "*Achtung Spitfeur*" written in speech bubbles. Regretfully, none of that early artwork exists anymore, my mother having thrown it out when moving to a new house, along with all the other items reminiscent of our childhood.

Spurred on by my experiences at Earnley, and with Tony Rickaby's encouragement, I completed a series of pop art pieces during my first year at Maidenhead. One was a white oblong cube with the image of a launching Atlas moon rocket slowly changing into a Coca Cola bottle on the four sides of the cube. My masterpiece that year, however, was a sculpture (a construction) made of large plywood triangles and trapezoids. On the plywood panels I had depicted parts of drag-racing cars as bold graphic images. The sculpture came to life with the incorporation of buzzers and flashing lights. This work attracted the attention of the school staff and it was put up on display in the college entrance hall for several weeks.

When I think back to those creative experimentations, it often makes me wonder how I would have evolved as a fine artist. I guess at that age you are encouraged by your peers to be sensible and think about a future career in design. Maybe Tony Rickaby should have pushed me further … who knows?

The other disciplines in the college curriculum were, for me, more of a chore. Friday was pottery, which, during the first term, was spent manufacturing 'thumb pots' – or 'turd pots' as we disrespectfully called them. A ball of clay was modelled into a cup-like shape, and once several of these had been made, they were cemented together to form a cluster, given a tone pattern then glazed and fired in a kiln.

The first life drawing class was of a little more interest, given the prospect of being in the same room as a naked adult female for the first time. In the prude upbringing of the British middle class, a young boy would never have seen a naked woman, except maybe in photos, and those were hard to find. Pin-up magazines were never left hanging around in our house and I can't recall having seen any while at Earnley. My first recollection of seeing a picture of a naked female, was gracing the pages of a magazine called *Health and Efficiency* that a boy in the neighbourhood had managed to procure. *Health and Efficiency* was the magazine for naturist sun worshippers – in short, nudists. Usually on the top shelf at the newsagent, the magazine was quite strategic in the handling of its photos. Men were never pictured from the front, and most scenes were usually of naked couples playing volleyball or some other innocent outside activity. All the females pictured, arms stretched out with beach ball in hand, had their genital region airbrushed to a smooth surface between the top of their legs, so we were none the wiser as to the reality.

I was somewhat surprised then, when the first nude life model I was confronted with was quite a fulsome young lady with bright red hair not only on top but also 'below.'

The novelty was soon to pass, and life drawing became a weekly routine, which, at the time I found a bit boring, unaware of how important drawing a human figure is for the understanding of proportion, light and shade.

The first year at art college was approaching its conclusion and my foundation group was now equipped with the basic skills required to start studying design. However, before moving to the other side of Maidenhead to the Annex, three months of summer vacation lay ahead.

Having at last passed my driving test I was desperate to own my own car, but how to pay for it? A holiday job was needed.

I called my friend and fellow student Don, and we agreed to meet on the Monday at the yard of a construction company in Caversham called Sportsworks. I had got a tip from one of my father's friends that they were looking for students for the summer. Sportsworks, as their name implied, constructed sports grounds, but they also worked on some kinds of road construction.

We entered the office at the main entrance to their yard. Just outside I noticed a labourer was pouring a yellow, rather toxic looking fluid into a concrete mixer, whose mixing drum was clogged with hard concrete. We made inquiries about student work, only to be given a lecture from the guy at the desk, that they had just fired two students after a week's employment, who, as a goodbye 'gift,' had filled a concrete mixer with wet cement and left it to set.

This was not a good start, but after some persuasion, the guy at the desk, who we later learnt was the site manager, offered us terms for the summer: ten pounds a week plus overtime. I did my sums and figured out that £100 would be enough to buy a secondhand car – hopefully a Mini. We agreed.

All that summer Don and I turned up at around 6:00am, before getting into the back of a Morris Minor pick-up, to be driven off to a work site. As it turned out, most of the work was road construction, the majority of which was reparation work of jobs that Sportswork had bodged up the year before. For me, one key attraction of the job was that I could use a pneumatic road drill for the first time, as well as take control of a small dumper truck or tandem road roller. When I look back, how I didn't get injured or even killed at this job is a mystery to me. Health and safety didn't exist in those days.

The summer took its course and on July 20 Apollo 11 landed on the moon – "One giant leap for mankind" – and for me, having earned the £100 I needed, in my eyes, an equally big leap was the purchase of my first car.

With only two weeks before college started up again, I was avidly scanning the used car pages of Reading's *Evening Post*. It wouldn't be the end of the world if I couldn't find a Mini, and something like a Morris Minor was probably a more sensible choice, due to their simpler mechanical layout. Another alternative would have been a Ford Prefect or Anglia, but still my heart was set on a Mini.

And there it was, in the *Evening Post* on a Friday – '1959 Morris Mini – £100.' I made a quick phone call to the seller; yes, it was still available but I would need to hurry. Immediately, I got on the phone to a friend who was very knowledgeable about all things mechanical. He could drive me over to look at the car.

By the time we arrived at the address it was already dark, so not ideal for scrutinising a car. The present owner and seller, who seemed a normal kind of guy, had bought the car new, and being a 1959 model, it was one of the very first. In ivory white the car looked good, the body had been undersealed and there were no signs of major rust. Having said that, there were the tell-tale bubbles of paint blistering at the top end of the front wings – a typical rust area for Minis. This was the very basic Mini as Alex Issigonis had penned. It had sliding side windows and the door hinge mountings were located outside the body. The interior was very basic, too, with one single round speedometer dial in the centre of the front panel, which included just a fuel gauge and some warning lamps. The seats were in a grey and red vinyl, and it had red carpets.

I got in for a test drive. To start, you turned the ignition key located at the centre of the switch panel below the speedometer. On the floor between the seats, just by the handbrake, was a starter button to press. The gearlever was a long rod that went straight to the gearbox and not the short-linked version as on later models. All pretty basic, but it would do. We set off with the typical Mini whine coming from the gearbox, and everything seemed fine until I tried changing gear, which was met with a crunching of cogs.

"Double de-clutch," the owner shouted at me.

"What's that?" I asked. The synchromesh on the gears had either given up or this was normal.

"It's normal," the owner told me.

For those unfamiliar with double de-clutching, this is a method of changing gear, usually in cars with straight cut gears – so either competition cars or prewar models. Basically, you change gear by putting the gearlever into neutral and letting out the clutch between each gear change. This method allows the

*Now a student at the Berkshire College of Art, here I am with my 1959 Mini.*

next set of gears to match the speed of the previous gear.

It worked, and I had to believe the owner that this was normal.

Apart from crunching gears, the test drive proved uneventful and the car seemed to drive well. My friend who'd accompanied me gave it the thumbs-up and I parted with my hard-earned £100; the Mini was mine and I was ecstatic.

I've owned and run many cars in my life, but nothing will match the thrill of buying that Mini. They say you should always own at least one Mini in your life, as it's a car you will never forget, a statement I totally condone.

I was in love with it, and with a smile on my face I drove it to Maidenhead for the first day of my second year at the Berkshire College of Art.

The second-year students gathered together in the studio of the Annex building, which was located next to Maidenhead Polytechnic, right on the A4 as it entered Maidenhead from the west. The course principal for the Environmental Design course was a tutor named Leo Summers. He was a practising designer in the field of Exhibition Stand Design and was assisted by a younger tutor named Peter Wheeler. Both greeted us that first morning.

Leo Summers stood in front of us. He wore glasses, had shoulder length hair, sported a rather unkempt beard, and had a cigarette protruding out of his mouth, which was full of nicotine stained teeth. He welcomed us, and expanded on the fact that things would now get serious, we'd had enough of making pots and drawing naked women, it was time for 'real' design.

The Environmental course was divided into two disciplines, Exhibition Stand Design and Interior Design, both of which we had to learn the basics of in that year, before moving on to specialise in one of them for the final diploma. I had already made up my mind that I wanted to do Exhibition Stand Design, but accepted the fact we had to learn the rudimentary of both practices. The course was split further into learning the key skills required in all industrial design: technical drawing, perspective drawing, illustration and model making. On top of this, we were given basic design projects to develop our creative ideas.

Leo Summers only attended the college a couple of days a week, so it was left to Peter Wheeler to oversee the everyday running of the course. Like most of the college's tutors, he was also a practising designer, but was obviously not as busy as Leo Summers, since he was at the college almost every day. In addition to Leo and Peter, we also had visitations from more practising designers in the fields of both Exhibition and Interior Design.

The Annex would be my place of study for the next two years, until graduation in June 1971.

In those two years I would learn many manual skills that today have virtually died out. Technical drawing was done on transparent drawing paper pinned or clipped to a drawing board. The college didn't have any 'parallel motion' drawing boards as was familiar in the big design offices of the day, so we relied on T squares, curve and circle templates, and rulers as our drawing guides. Lines were drawn in ink with Rotring drawing pens of various thicknesses – 0.25mm to 1.00mm. For illustrations, we used white pencils and chalks on coloured Canson paper, and perspective views had to be calculated mathematically.

This was 'real' design, and I loved it. My favourite lecture session was Illustration, maybe because this was still a form of art. Once a week, a practising architectural illustrator named John Camb would visit the college. He had an office in Notting Hill Gate and worked mainly on illustrations for large public buildings, such as cinemas, or images of holiday hotels for tourist brochures. Most of the hotels illustrated by Camb were not finished, hence the need for realistic images.

I marvelled at John Camb's work. In particular I remember a large illustration of the Odeon cinema at Marble Arch. The picture was all the more stunning for the fact that it was a rainy night scene. The picture was somehow alive; John Camb was certainly a master of his trade.

Not only could we admire the work of John Camb himself, but he also brought in examples of illustration work from designer colleagues or friends. It was on one such occasion that he pulled out a board with an illustration that hit me like a bolt of lightning: it changed my life.

The image on the board was a wildly distorted rendition of the front nose and grille of an American Pontiac, it was simply signed 'Ken.' Despite its wide-angle perspective, the illustration was entirely believable, and, above all, what struck me the most was that it shone like it had been drawn in metal. I knew then, that this is what I wanted to do in life.

"Who's Ken?" I asked John. His full name was Ken

Greenley and he was working as a designer at the Luton-based car company Vauxhall. John had either studied with him or he had met him in designer circles in London.

Vauxhall was not a car company on my radar, although my uncle George up in Lincolnshire had owned a pink two-tone Vauxhall Cresta. The company was well known for promoting young artistic talent with its annual Vauxhall Craftsman Guild design competition. Young car designers could send their models to the Luton HQ for judgement. I never took part but I remembered seeing the entries in a magazine.

"How do you become a car designer?" I quizzed John Camb, and he immediately pointed me in the direction of one of the college's third-year students: John Saville. The third-year students worked in studios on the top floor of the Annex, and didn't like being disturbed by 'newcomers.' John Saville, however, had already come to my notice, since he also drove a Mini. His wasn't a Cooper, but it was pretty special nonetheless, with the latest Pirelli Cinturato radial tires mounted on Minilite style wheels and fitted with a wood rim steering wheel. Saville played the part: sporting Ray Ban pilot sunglasses, he wore a short bomber jacket and string-back driving gloves – he was the definitive 'boy racer.'

I went up to John Saville's desk and mentioned John Camb's advice to talk to him. Upon my enquiry, he wasn't too willing to free up any knowledge he had about becoming a car designer. This despite the fact his drawing board was covered with car illustrations. However, with more perseverance, he revealed that there was a new industry-sponsored course in car design at London's famous Royal College of Art (RCA). John was applying for the next year's entry and had an agreement with the college that if he completed his mandatory course work, he could also do car design in his extra time. He advised me to visit the RCA degree show that coming summer to see for myself.

That was all the information I needed. I had unlocked to door to becoming a car designer.

# Chapter 6

## *The path to the future*

**The Royal College of Art (or RCA) was founded in 1837 as the Government School of Art, a title that changed to its present status in 1896. It is still regarded as the world's most influential post-graduate Art and Design institution, and the pinnacle of design education.**

The yearly RCA summer degree show was a must visit for young Art and Design students. Furthermore, to go on to study at the college was the dream of most Art students.

In June of 1970, a group of us got the train from Maidenhead to London. Arriving at London Paddington, we took the tube to High Street Kensington, from where the RCA Darwin building is a 15 minute walk up Kensington High Street towards Hyde Park. On Kensington Gore, the Darwin building is located right next to The Royal Albert Hall, and opposite the Prince Albert Memorial in Hyde Park.

The yearly degree show features the work of all final-year students, covering almost all disciplines of Design, from Illustration, to Jewellery, to Textiles. The School of Industrial Design remains one of the biggest courses, and incorporates the Vehicle Design department as a special design discipline within it to this day.

In 1970, the two inaugural students of the Automotive Design course – as it was then called – presented their work. Upon entering the college doors, I headed straight to the Industrial Design show.

It was The Ford Motor Company that decided to set up a special course for Automotive Design at the RCA. Up until the '60s, the only academic facility in the world where one could study car design or automotive styling was ArtCenter in Pasadena, California.

In the years after the war, American automotive design or styling had developed to a polished art form, with a development process that had close links to the Hollywood film industry. Designs were sketched and illustrated by artists, and then creative sculptors, many who had learnt their craft making models for the film industry, sculpted models as full-size mock-ups, made of a wax clay material. ArtCenter's proximity to the home of the film industry, with its abundant source of creative personnel, explained its location.

In the '50s and '60s, the four key American automotive manufacturers – Ford, General Motors, Chrysler, and American Motors – all had large styling studios, as they were called, located in the automotive capital of the USA: Detroit.

Car design in the States was a glamorous job, and the automotive design centres were architect-designed cathedrals of style, oozing with the atmosphere of Hollywood.

The development of car design in Britain couldn't have been more different. Only the companies that were linked to American parent companies had design studios anywhere near the magnitude of the Detroit locations – meaning, primarily, Ford and the General Motors offspring Vauxhall Motors.

The traditional British companies were mostly located in the Midlands, and their design studios, if they existed at all, were part of the engineering drawing office. Jaguar or Standard Triumph, to name a couple, usually relied on a small team of artistically talented engineers or aerodynamicists, who worked with tool makers to create model prototypes. Jaguar's Malcolm Sayer, father of the Jaguar E-Type, is a good case in point. The British postwar designers would have drafted out the body forms on large seven-metre drawing tables. From these plans, wooden formers or moulds would have been made, which would then be used, along with metal beaters, to shape sheet metal in order to create the prototypes. The process would have varied from company to company, but there was little or no use of the American clay technique.

Another route was to commission an Italian styling house such as Farina (later Pininfarina), Bertone or Michelotti, to do the work. The Italians were the European masters of car design, working in gesso or plaster. The artists and sculptors of those design houses drew on a lineage of creativity in modelling forms that went all the way back to Michelangelo.

The British Motor Corporation (BMC) used a combination of in-house design-engineers and the Italians, notably Farina. The models linked to the Mini, such as the 1100 and 1800, drew their influence from the Mini's creator, Alex Issigonis, whereas the preceding, more geometrically-styled designs of the Morris Oxford, Austin Cambridge, and their Wolseley, MG and Riley derivatives, were the work of Farina, which incedently was also working for Peugeot – hence the similarity of some of their models of the time.

Soon, however, all the British companies started to set up their own design studios, mainly within the industrial Midlands, and all based on the American clay modelling process; but where to get the designers to fill those studios?

This brings me conveniently back to Ford's initiative to create Britain's equivalent of ArtCenter at the RCA.

The first students studying Automotive Design at the RCA, and graduating that summer in 1970, were an Englishman called Peter Stevens and a Scotsman named Dawson Sellar.

I made my way to their display and stood memorised by the artistic skills displayed by the work in front of me. Never more had I been so sure that this was what I needed to aim for, and I left London with a clear vision and goal.

Returning to college, it was obvious to me that I had to take the same work course as John Saville, who had by now been accepted for the RCA's autumn 1970 intake. I would developed a portfolio of car designs while trying to put forward the mandatory exhibition stand designs required for my final diploma. Examinations for the 1971 RCA entry were in the spring of that year, and with the 1970 summer vacations approaching, I needed to devote all that time to complete my diploma work.

The output for my final diploma was two proposals for exhibitions stands. This required the completion of a full set of technical construction drawings in all three elevations, plus perspective renderings and scale models of both stands. I had ten weeks.

Could I really spend the entire summer vacation doing college work while my friends were lying in the sun? I had to. For me, there was no choice so I got stuck in.

It was the longest summer of my life. In addition to my diploma work, I also started to learn how to illustrate cars, working in my bedroom in the evenings. The required tools to sketch cars were felt tip markers – or Magic Markers to use their trade name. I also needed drawing curves and ellipse guides to create the perspective renditions of the wheels.

John Camb had lent me the Ken Greenley Pontiac illustration. I tried to copy his illustration, in order

*My first automotive illustration for my application to the Royal College of Art.*

to learn the technique, before moving on to my own designs.

In my years as a designer, I have always told young designers not to be afraid of copying an expert's work. No matter how close you imitate, your own style will come through. Believe me, it works, and you save a lot of time.

The autumn term of college came soon enough, and with a smile of satisfaction on my face, I pulled into the Annex car park in my Mini, two boxed up exhibition stand models on the back seat, complete with technical drawings and perspective illustrations – I'd done it.

Over the following months leading up to the Christmas break, I fell into a focused work routine, the silence in the Annex studio only broken by the music playing on the in-house record player. It was a well-used piece of apparatus, but the choice of LPs was quite select. I recall the first year in the Annex there were only four albums to choose from: Leonard Cohen's *Songs of Leonard Cohen*, John Mayall's *Blues from Laurel Canyon*, The Beatles' *Abbey Road* and Pentangle's *Basket of Light*. By the final year, we'd moved on to *Led Zeppelin II*; Crosby, Stills, Nash and Young's *Déjà Vu;* and King Crimson's *In the Court of the Crimson King*. Pink Floyd was also well played, however, they weren't my favourite choice.

Being separate from the main college building down in the town made us feel aloof, but it also meant we had little contact with students from the foundation courses, and more importantly, the girls in the fashion course, so the highlight of our week was definitely when they visited the Annex to use its workshop. There were some beautiful girls on the fashion course but regretfully most were taken.

Luck, as it happens, did come my way that Christmas, although I don't know if 'luck' is the correct word for losing one's virginity.

At the tender age of 19, I was yet to sample the full delights of the opposite gender. My nervousness when in the company of girls – the result of attending all-boys schools until the age of 17 – didn't help my confidence, but things were soon to change, and quite out of the blue.

The college autumn term had come to an end, and as was the custom, most of the students assembled at a pub down in Maidenhead for a last festive drink before heading home for the holidays.

For a couple of hours the pub was buzzing, but then slowly people started say their goodbyes, and kisses were exchanged with the girls. Lady J, as I'll call her, one of the fashion girls of my year, made a beeline for me, and before I knew it, she had clamped her lips to mine, giving me what was way more than a normal goodbye kiss.

"Can you drive me home?"

*Ah*, I thought, *kiss for a ride*. As it would turn out the girl in question was certainly interested in a 'ride,' but not just me delivering her to her front door!

With my unexpected guest in the Mini, we set off on my usual route home along the A4, turning off towards Sonning. We had just passed the first bridge over the Thames when my companion suggested a walk along the river for some fresh air might be good. Obligingly, I pulled over. It was freezing and I wasn't in the mood for a walk – no, I still hadn't clicked – but we made our way along the tow path, her arm in mine. There was nobody in sight and suddenly she made her move, manoeuvring me off the path and behind some bushes. She pulled me down and I found myself lying on the ground with her above me – at last, it clicked.

In a hurry, she went straight for the zip of my jeans. I was all over the place, and still very much in a state of excited shock, but this definitely wasn't the first time for her, so she took control while I fumbled in my wallet to find the Durex I had stolen from my father's bedside cupboard. *How old was it?* I thought, trying to find the expiry date stamped in the plastic cover. Would it break? And then what? All sorts of things were going through my mind. Impatient of my fumbling around, Lady J took the small sachet, tearing it open with her teeth – "Don't rip it!" She withdrew the Durex and fitted it over me; things down there seemed to be working. She moved on top of me, and with some gentle persuasion I was where she wanted me. She started moving rhythmically above me, but my head was still filled with anxiety about the level of my performance.

I remember saying something like, "I don't understand it, things usually happen quicker than this," with absolutely no clue that my staying power was producing all the right results at the receiving end.

I concentrated and things took their course and I exhaled in relief. There was, however, no time for post-coital relaxation.

Lady J had straightened up in alarm. "A dog, A DOG!"

Our solitude had been broken, as a dog owner made her way down the tow path in our direction.

With Durex still attached, I frantically pulled up my jeans and we both sprang to our feet – it was close. Stumbling out on to the path, the dog owner clearly knew what she had disturbed as we passed her on our way back to the Mini.

The short journey back to Reading was contemplative. I dropped the young lady, as promised, at her house, and with nothing more than a quick kiss on the cheek and mischievous smile she thanked me for the ride, wishing me a Merry Christmas. Nothing more would develop out of this brief encounter, but like the assassination of Kennedy and the moon landing, you never forget your first time.

The deadline for the 1971 RCA application was not far away, and by early spring that year I had a good selection of vehicle designs mounted on boards with a protective sheet of clear acetate over them. I had developed a minimalistic technique for the illustrations (or renderings as they are called in the automotive world), and had also decided to vary the range of vehicle styles, adding some vans and trucks to the repertoire. I felt well prepared.

The RCA application forms arrived in the post, and after some careful reading I came across a small detail that could potentially jeopardise my chances of entry for that year. My heart sank. It clearly stated that the minimum age for college entry was 21; I would turn 20 that coming March.

I called the college and inquired if there was any way around this. They told me that it was unlikely I would be accepted that year, but nevertheless I should submit my work for referral to the 1972 intake. I didn't want to wait another year, but it looked like there was no alternative. The thought of filling a year by working at Sportswork was not appealing.

By 1971, in addition to Ford, Chrysler UK were also sponsoring the Automotive Design course. Chrysler UK was what was left of the old Rootes Motor Group, which manufactured the Hillman, Humber and Sunbeam cars. This meant there were now four places up for grabs.

The application process for the course involved three interview sessions. If the college and the sponsors approved the level of entry, then selected candidates would be invited for the first interviews at the college in London. Having passed that hurdle, a short list of prospective entrants would then be invited to the design centres of both Ford and Chrysler for further interviews.

I submitted my drawings in the hope of a positive reply. Within a couple of weeks a cream letter with the Royal College of Art's emblem on the front fell onto the door mat. Anxiously I opened the letter and read: "Dear Mr Birtwhistle, We are pleased to inform you that you have been accepted for an interview ..." and on it went. There was no mention of my age or a possible referral, so things were looking positive.

The first interview was critical, so I thought it best to take not only my automotive work but also my complete Maidenhead diploma work with me. I piled it all into the Mini and set off to London.

Those days, both driving into London and parking were no problem. I drove up Kensington Gore and, under the shadow of the Royal Albert Hall, found a parking spot right in front of the college. All my life I have prided myself on my punctuality and this day was no exception: I was way too early. I walked into the main entrance hall and asked a student for directions to the School of Industrial Design – it was on the third floor. Getting out of the lift I read a sign on the adjacent door: "Automotive Design Interview Sessions." I opened the door and went through. Seated at a desk was a generously proportioned lady with long black hair, thick black-rimmed spectacles and a cigarette in her mouth. Her name was Jane Howell, and she was the department secretary.

Beyond Jane Howell's desk were a row of seats, where other candidates were seated. They all looked quite a bit older than me, I thought. I moved over to join them. My interview wasn't for more than an hour, so I asked one of the other candidates if he could help with my 'luggage' down in the Mini. He reminded me a bit of Bluto from the Popeye cartoons: black hair, and beard, wearing a polo neck sweater with tweed sports jacket. Had this guy even studied Design? His name was Dave Evans, and he had indeed studied Design, in Newcastle. He agreed to help me, and we headed of down to my car. In the lift it immediately struck me that not only did Dave have a kind and jovial manner, but that he seemed to talk in some form of rhyme or riddle all the time. I stayed cool – he was a competitor and I couldn't afford to give away any secrets that may hinder my chances.

Jane called me. I could go through to the adjoining office; my interview was on.

I walked into the room, which was dominated by a long meeting table. On the far side of the table stood six gentlemen, all with their backs to some large windows that looked out towards Kensington High Street and beyond.

The interviewers were Professor Frank Height, head of the School of Industrial Design, and Nigel Chapman, resident tutor for Industrial Design (and, as I would later learn, a key driver in setting up the Automotive Design course). Also present were two visiting lecturers from the industry, Peter Ralph, best known for designing earth-moving equipment, and Carl Olsen, an American freelance designer who had previously worked in the States for GM. Finally, there were two representatives from the industry sponsors: from Chrysler, Rex Fleming, who, with his trim moustache, looked like an RAF officer, and an American from Ford, who could have been a bank clerk, wearing, as he was, a black suit with white shirt and tie; his hair 'Brylcreemed' into a neat side parting.

They welcomed me enthusiastically and after shaking hands, they sat down in unison, leaving me alone opposite them all.

Laying out my work, I commenced by going through my diploma work, I'd even brought the two exhibition stand models with me. They seemed politely interested, but it was clear they wanted to see my car designs. I moved on and took out the six finished illustrations I had brought with me, plus a sketch book. Immediately, I gained their attention.

"These are lovely renderings, Peter. Who taught you?"

I guess I must have mentioned the Pontiac sketches – Carl Olsen approved. Things were coming to a close, so now was the time: I mentioned the issue about my age. The Ford man responded immediately, his eyes directed at Professor Height.

"Peter, rules are made to be broken."

I inwardly sighed with relief: another hurdle seemingly overcome.

The mood of the room was warm, welcoming and friendly, and it was over. I left feeling I had accounted for myself well. I would hear from them within the week.

*My first attempt at a truck design, also for the RCA application.*

Outside, Jane gave me some words of encouragement."Look forward to seeing you in the autumn" – or something similar.

By now, Dave Evans had departed, so I had to find someone else to help me get my load to the lift.

Driving back to Reading, I felt I was in with a chance.

Back at college, my fellow students were all occupied finishing their diploma work. Some had employment possibilities in the pipeline, but for me a job was not an option I wanted to consider. It had to be the RCA.

Both letters from Ford and Chrysler arrived a day or two after my interview at the RCA, and both had the same message.

"Dear Mr Birtwhistle, Thank you for attending the interview at the RCA. We are pleased to invite you for a further interview at our design centre ..." etc.

I had been in bed when the letters arrived and my father, in a state of excitement, brought them for me to open – the chances for me of getting a place at the world famous Royal College of Art were starting to look ever bigger.

Almost immediately, I was checking a road atlas for the locations of the Ford and Chrysler UK design centres. Ford had a new development centre at Dunton, west of London, close to Basildon in Essex, which had been opened in 1968.

The Chrysler design centre was located at a site named Whitley, which in the 30s had been the location of the Armstrong Whitworth Aircraft Company, who had built the WWII Whitley Bomber named after the location.

The journey to Whitley via Oxford and Banbury, was fairly simple, but driving to Dunton would mean going around London on the North Circular Road, a trip I didn't want to tackle.

My father said I could use the 1100 to drive to Whitley, and I decided the train would be my best option for visiting the Ford development centre.

The interview with Ford was my first destination. Travelling via London, I took the train to Laindon, the station nearest the Ford development centre. From there I called Ford and they sent a car to pick me up.

The black Ford Zodiac that met me pulled into the front of the centre and dropped me off at the main entrance of a large, glass-fronted, five-storey, L-shaped building. Carrying my artwork in a black plastic portfolio, I walked through the entrance doors and introduced myself at the reception. I was directed to a row of seats; they would call for me.

The reception was a large marbled hall, where smart, besuited employees walked around in silence. I felt a bit out of place. As usual, I was early, and looking around I noticed another guy of my age looking nervous – no portfolio, I noticed. Just as I was thinking he looked a bit like Elton John, he suddenly got up, walked over to me, and asked if I was here for the RCA interview. He introduced himself: Martin Smith.

Getting straight to the point, he asked if he could look at my portfolio. I realised straight away that, like Dave Evans, this guy was an adversary, but I could hardly refuse to show him my work. I agreed but my body language must have shown that I wasn't too happy about this. I unzipped the portfolio and pulled out my mounted renderings. Martin looked at the first two and sighed.

"These are amazing." I felt relieved.

He was a likable guy and very modest. It turned out that he had studied Mechanical Engineering at Liverpool University but had always drawn cars in his spare time. He asked if I could give him some feedback on his work. To my surprise, he pulled out a small red folder not much bigger than a school exercise book. Was he serious?

In the folder were a selection of very simple but beautifully drawn designs. Some of them were just line drawings but they were very understandable, and it was quite clear to me that Martin Smith understood how to draw a car. He hadn't had the benefit of a design education, but it was very obvious that with some training he would have no problem achieving the level expected in the industry.

We were interrupted by a smartly dressed lady, who asked Martin to follow her.

"Will you be going to Chrysler next week?" he called. When I told him yes, he replied, "Good, I'll see you there," and they left. Nice guy, I thought.

After about an hour I saw Martin leaving. The same lady approached and asked me to follow her. She was wearing high heels with a thin seam up the back of her stockings. *Hmm*, I thought, this is a different world to art college.

I had hoped that I would get a glimpse at the studios, but no such luck. As is the case with most design departments, the administrative offices are all located outside of the main design area, and as I was shown into the office for my interview, I noticed some double doors with a key code. I couldn't help wondering what was behind them.

Getting up from behind a large desk was a blonde-haired gentleman of similar stature, his name was Ken Nelson. He was wearing a light grey three-piece suit with a dark red shirt and matching tie. On the wall behind his desk were a couple of massive renderings of some futuristic vehicles – I was awestruck. Ken Nelson was American, and a larger-than-life figure.

"Take a seat Peter."

He sat down as well and leaned so far back in his chair, for a moment I thought he may fall backwards. He pressed a button on an intercom on his desk.

"Peter Birtwhistle is here, can you come in?"

Almost immediately a door opened and the guy I recognised from the RCA interview entered, he still looked like a bank clerk. He was accompanied by a younger man, whom, because of his long hair, I presumed was a designer, but nevertheless was also wearing a three-piece suit.

Ken Nelson started the meeting explaining why the Ford Motor Company was the world's leading company for car design. Indeed, there was some truth in what he was saying. The Escort and, even more so, the new Capri were great-looking cars.

The Ford Capri was a sensation when it came out – a sporty looking four-seater coupé for the working man, 'The car you always promised yourself' – and it would be a massive success. The company had a very dynamic image, not only for its production cars, but also for its success on the race track, and Ken Nelson was justifiably proud.

At last I was able to present my work. There were a lot of questions asking why I had done a particular design the way I had, as well as "What could you bring to Ford if we decide to sponsor you?"; "Can you imagine working for us?"; and so it went on – I felt a bit uneasy and didn't warm to them, but then the interview was over, and the Zodiac drove me back to Laindon to make my journey home.

I was exhausted and slightly overpowered by the whole occasion. I had spent almost three years in the casual warmth of an art college studio, listening to John Mayall. At Ford, I was confronted with the smart coolness of a big business for the first time, and it made me feel anxious.

There was no time to be complacent. With my mind cleared of any doubts, I drove to Coventry the next week, to the Chrysler UK development centre in Whitley.

Approaching Coventry, I passed the factory buildings at Ryton, of what had formally been the Rootes Group's assembly line. Now Chrysler UK, they produced the Hillman Hunter and other derivatives such as the Sunbeam Rapier. The Hillman was quite a conservative choice, but its image had been boosted when in 1968 a Hillman Hunter – 'the most forgettable of cars,' as one report of the time described it – won the London to Sydney Marathon. It didn't help them, however.

In the '60s, Chrysler was looking to increase its presence in Europe or risk being left behind by Ford and General Motors. To start, they had already acquired a majority stake in the French Simca Company and the Spanish Barreiros Truck Company, and after failing to buy into British Leyland, went for second best, taking controlling shares in the British Rootes Group. Rootes would become Chrysler UK, and Simca, Chrysler France. The first new Chrysler UK vehicle designed and engineered at Whitley was the Hillman Avenger. The Avenger was the last car to carry the Hillman name, before the UK products took Chrysler. Only one car had the Chrysler nomenclature – the 180 – which was a disaster. The strategy proved unsuccessful and later all its European products would go on to be called Talbots. In the end, it proved to be a slow death and Talbot was swallowed up by the Citroën/Peugeot group.

At the time of my Chrsyler interview, the Avenger was new on the market and had given the old Rootes Group a new lease of life, so I felt positive as I drove through the gates of the Whitley Technical Centre.

The buildings had nowhere near the impact of the Ford centre in Dunton, but on the positive side they lacked the intimidating atmosphere I had felt there.

After parking the 1100, the security officer at the main gate directed me to the design centre, which was located in a row of low, hangar-style buildings, with the typical sloped glass roofing of a factory. I could clearly imagine the white-coated engineers of Armstrong Whitworth creating the Whitley Bomber here.

I went through the entrance door. There was a reception office with a small counter. Behind a sliding glass window sat a receptionist, who asked me to take a seat. As in Ken Nelson's office, there was some amazing artwork on the walls, but the atmosphere was more like a doctor's waiting room than a design centre.

It wasn't long before I heard a familiar voice and, looking up, I saw Martin Smith, whom I'd met the previous week at the Ford development centre. He waved in a friendly greeting and asked how things had gone at Ford. He was accompanied by Rex Fleming, whom I remembered from my first interview at the RCA. I was unsure how to handle this exchange and chose to be noncommittal, preferring to concentrate on Rex Fleming. I must have muttered something to Martin dismissing his question; I'm quite sure he thought I was an arrogant bastard.

Rex Fleming paid his farewell to Martin and turned around immediately, beckoning me to follow him up a long corridor, before showing me into an office. Seated at a desk was the design director of Chrysler UK, Roy Axe, who stood up and reached out his hand with a warm smile. He was tall with a stout figure, wearing thick-rimmed glasses, and he had a bald patch surrounded by curly dark hair. He reminded me of the BBC personality Eric Morcombe. Unlike Ken Nelson and his colleagues at Dunton, Roy Axe was undoubtedly British.

"Peter, welcome."

I felt instantly at ease.

Roy Axe was immediately warm and friendly and very enthusiastic about my work. This was a totally different atmosphere to the one I had experienced the week before. He didn't push the company line, preferring to concentrate on my work. The meeting ended, and with words of encouragement he bade me farewell, just short of offering me sponsorship.

Rex Fleming escorted me back to the front entrance.

"That was a good meeting, Peter, I hope we can do something."

As I pulled the 1100 out of the Whitley car park and pointed it in the direction of Oxford, I felt elated.

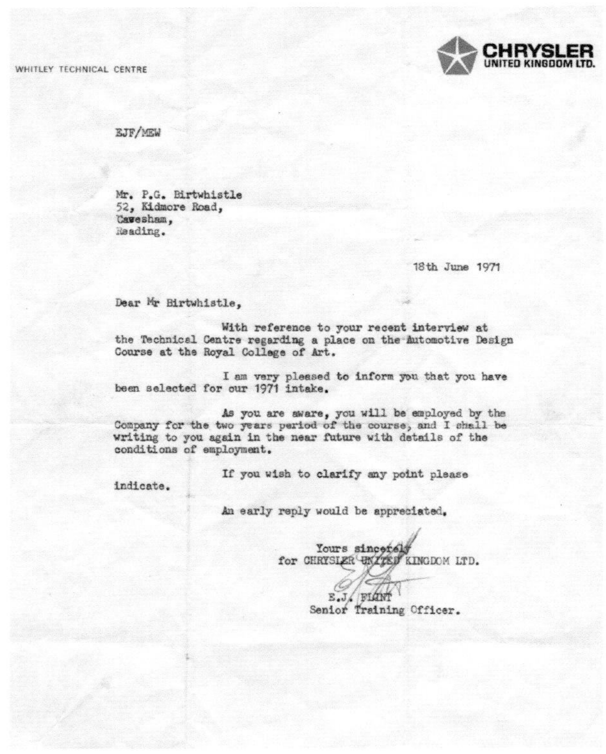

WHITLEY TECHNICAL CENTRE

CHRYSLER UNITED KINGDOM LTD.

EJF/MEW

Mr. P.G. Birtwhistle
52, Kidmore Road,
Caversham,
Reading.

18th June 1971

Dear Mr Birtwhistle,

With reference to your recent interview at the Technical Centre regarding a place on the Automotive Design Course at the Royal College of Art.

I am very pleased to inform you that you have been selected for our 1971 intake.

As you are aware, you will be employed by the Company for the two years period of the course, and I shall be writing to you again in the near future with details of the conditions of employment.

If you wish to clarify any point please indicate.

An early reply would be appreciated.

Yours sincerely
for CHRYSLER UNITED KINGDOM LTD.

E.J. FLINT
Senior Training Officer.

*I'd made it! My acceptance letter from Chrysler UK for entrance to the RCA.*

On the morning of June 20, 1971, a letter headed with the Chrysler 'pentastar' fell onto our door mat. As before, my father came in to my bedroom clasping the correspondence in his hand. I opened it.

I read: "Dear Mr Birtwhistle, With reference to your recent interview at the Technical Centre regarding a place on the Automotive Design course at the Royal College of Art. I am very pleased to inform you that you have been selected for our 1971 intake."

I read it, and without comment gave the letter back to my father. His face lit up, and giving me a light punch on the shoulder, he asked if he could take the letter with him to work – of course he could.

Up until I went to college, he'd had to face the fact that neither my brother nor I were top of the class. While many of the children of his colleagues and friends had been high academic achievers, he had to accept that we weren't on that same level, up until now. At that moment, I knew he was proud of me and wanted to tell the world about my achievement.

I laid back – I'd done it.

# Chapter 7

## *Royal College of Art*

**It was a crisp autumn morning of October 1971 when I walked apprehensively up Kensington Gore, a student at the world famous Royal College of Art, an employee of Chrysler UK, a trainee car designer.**

I entered the college's main doors opposite the Royal Albert Hall. Similar to Maidenhead, this first day was an introduction to the college for new students. Upon passing through the doors, I was directed down to the college assembly hall: The Gulbenkian Hall. It was packed, and I struggled to find a chair. Taking in all the different dialects, I noticed how many foreign students there were in attendance.

Silence fell and we were asked to stand. The tutors of the college entered on the stage, some dressed in their academic gowns. The college rector, Robin Darwin, after whom the college building was named, stepped forward and asked us to sit down. Addressing us, he spoke of what a privilege it was for us to be selected to study at the college and that we should use this opportunity wisely.

ROYAL COLLEGE OF ART

THIS IS TO CERTIFY THAT

Peter BIRTWHISTLE

is a full-time student of the Royal College of Art and is expected to remain in attendance until July, 1978..

BRIAN COOPER, REGISTRAR

October, 1974..

*It says it all.*

Looking around, I had no idea who the other Automotive Design students were, and as the introduction speeches concluded, I assumed I would soon find out. I made my way out of the hall and up to the fourth floor.

The Automotive course shared the same floor with the schools of Jewellery and Textile Design, although each faculty was separate. The main school of Industrial Design was located one floor below. This, combined with the fact that the Automotive Design students were company employees, created an invisible wall between the two disciplines. We were car stylists, as opposed to industrial designers.

I walked from the lift and through the doors of my new home for the next two years.

The Royal College of Art's Darwin building is in an enviable position, situated as it is opposite Kensington Gardens, with fine views out of the windows, which span from Kensington High Street up to Knightsbridge and beyond. Located at the front of the building, the Automotive Design studio was a truly amazing location to study. A single long room equipped with spacious drawing tables, there was plenty of room for the eight attending students.

The course's principle tutor, Nigel Chapman, was already in the studio to greet me and the other new entrants, two of whom I recognised from the earlier interviews – Dave Evans and Martin Smith. The four first-year students in the course were now all present, two from each sponsoring company. As it turned out, Dave Evans was the other student selected by Chrysler UK, while Martin Smith had been selected by Ford. Ford's other entrant was a Turkish student named Jean Nahum; apparently his family were connected with the Turkish car company Anadol. We greeted each other cordially, and were soon joined by visiting tutors Carl Olsen and Peter Ralph, who had both attended the first interviews at the college.

Nigel Chapman explained the curriculum and

impressed upon us that it was up to us to use our own initiative in developing the ideas we would tackle that first year. Students who had made it to the RCA were expected to be independent and not rely on a tutor always looking over their shoulders. The first year would focus on getting our design skills to the level expected in the automotive industry, before selecting our final master's project.

Finally, the senior professor of the School of Industrial Design, Frank Height, and the overall head of the department, Professor Misha Black, paid a brief welcoming visit.

Misha Black was an industrial designer of great note, and his portfolio included the City of Westminster street sign graphics, such as the famous Carnaby Street sign. He had also designed the Underground railway stock for the Victoria Line, as well as several of the first diesel locomotives. He was more of a figurehead for the college, and was a seldom visitor to the department, preferring to delegate most of his college responsibilities to other staff.

After the welcoming introductions were over, we spent the rest of the day sorting desk arrangements in anticipation of the arrival of the second-year students the following morning.

I returned to the basement flat in Earls Court that I had moved into. It had been a heady day – I don't believe I slept too well.

Returning to the college the next morning, the four second-year students arrived in the studio one after the other.

I already knew John Saville from Maidenhead. He was sponsored by Ford, as was a Scotsman, David Arbuckle. The other two second-years students, both sponsored by Chrysler UK, were Geoff Matthews, who, like Martin Smith, was an Engineering graduate, and Steven Ferrada.

At the age of 20, I was definitely the baby of the course, if not the whole college, and although Martin Smith was only 22, Dave Evans was closer to late 20s.

In all, it was a very mixed bunch.

John Saville was still the 'boy racer' I remembered from Maidenhead, all Ray Bans and string-back gloves. David Arbuckle was fairly conservative, a jovial character, and married. Geoff Matthews was also married and very much an engineer. He spoke in a deep, gravelly voice. Steven Ferrada had studied design, and I would soon learn that he was a very gifted artist. He could have been someone from the British nobility. Usually wearing a tight polo neck sweater, and occasionally smoking a pipe, he was a keen horseman, and on several occasions he actually rode to college.

In the first year, apart from myself, the only art college graduate was Dave Evans, who had studied Industrial Design at Newcastle. From Whitley Bay, and a proud Geordie, he possessed a questionable sense of humour, which in many cases went way over my head. Jean Nahum had studied Engineering and was a charming gentleman.Well-educated, he spoke fluent English and French, as well as his native Turkish. He was immaculately turned out in a well-cut suit, probably French or Italian – I don't believe he had any financial problems. Of the entire group I took warmest to Martin Smith. He was definitely the coolest of the group, a savvy dresser, and he oozed confidence. I regretted reacting so coolly to his warm approach at our meeting in the Chrysler technical centre.

The Ford Motor Company took the initiative to set up the Automotive Design course at the RCA because the traditional British Industrial Design colleges were not putting young designers into the field with the specialist skills required in the car industry. ArtCenter in Pasadena, California, was the key supplier of young talent for the American industry, so Ford figured that a similar style of design education was required in Europe.

So, what was the difference between Industrial Design and Automotive Design in those days?

Automotive Design in the '50s and '60s was known as 'styling.' It's a word that many Automotive Designers don't like to use, but to be truthful, styling is what car design is and was all about.

The core emotion when purchasing a car is usually how you feel about the way it looks. Sure, these days the image of the brand, build quality and, of course, the finance package offered are paramount, but deep down, it will most probably be the way a vehicle looks that secures the investment. Product Design is a whole other story, where the function of a product is usually the first consideration when buying. In postwar years, the Americans had also embraced styling for their domestic goods, the refrigerators of Electrolux being a good example. In Europe, however, you bought a washing machine based on what it did, not on how it looked.

In 1971, the famous mantra 'form follows function' was an invisible sign over the door leading into the

RCA School of Industrial Design. Products ended up looking the way they did based on their innovative content. Once that was established, the corners were rounded off.

Today, Industrial Design has caught up, and, in many areas, overtaken Automotive Design. Look at the products of Apple and Dyson – these are companies that now influence car designers, but in 1971 the ideology of looks first and function second was in its infancy in the field of Industrial Design.

No, the RCA graduates of Automotive Design in those days had to create emotion on wheels where function, questionably, took second place.

The design process associated with styling was nurtured in the North American automobile industry.

In the '50s, the American manufacturers changed their models at an amazing rate, with new designs hitting the showrooms on a yearly turnover. This meant that the cars had to have a fresh new look each time round to gain customer attention. Changing the look of the cars at this rate had nothing to do with rational design, being based purely on fashion or style. Likewise, the design process in the big American studios of the postwar era was a product of this philosophy. Put in a nut shell, it was 'show business.'

Designers were encouraged to create flamboyant and exaggerated illustrations or renderings of their designs, as a tool for selling them. The more wild and extreme the rendering, the higher your chances were of getting a proposal selected. Not only this, but the image on paper of the illustrated cars, had to have a realistic metallic or shiny finish – a really specialised technique.

These renderings, however, were just the start of a design process in the American car industry, and far removed from what you would find in a European design studio.

Once a design was selected it would move on to the three-dimensional modelling stage. The American car industry had developed a process where full-size models were sculpted in a wax clay material. Clay modellers would sculpt the forms with an assortment of tools and achieve a surface finish like metal. Once the model was finished it would be covered in a plastic painted film or foil known as Dynoc. The Dynoc would be painted according to which part of the model it was being applied to; windows, for example, would be painted a dark colour. For the final detail finish, front grilles or window frame mouldings would be finished in a chrome foil.

Prior to the modelling stage commencing, the designers would have to create the information needed for the model-makers to fabricate the forms and templates required to build up the first clay surfaces on the model armature.

The designers and draughtsmen would then draw all three dimensions of a car on upright draughting boards, covered with paper that had vertical and horizontal measuring grid lines printed on it. To do this they would, however, not use not a pencil, but rolls of flexible adhesive black crêpe tape of various widths, manufactured by the 3M company. Similar to the modelling clay, the tape was a 'spin-off' product from the film industry. Its usual application was for masking off film negatives, its key property being that it was light proof, a feature not required in the car design industry. It did, however, possess a further quality in that you could flex and bend it, enabling one to create lines on a surface as if you were drawing with it.

In the industry these full-scale drawings were known as 'tape drawings.'

The Automotive Design course at the RCA equipped the students with all the specialist knowledge they needed for future employment. This included rendering, tape drawing, and clay modelling. Learning these techniques was the key focus of the first year of study.

I soon settled into the daily routine of college life, which commenced most days with meeting Dave Evans and Martin Smith in the college canteen for a bacon roll before moving up to the fourth floor.

As well as a good canteen, the college also had a bar and a lunch time common room. Friday nights usually featured a disco in the bar, which I sometimes attended, preferring to leave for Reading by train on the Saturday morning. The RCA disco was well frequented by artists from all disciplines, including musicians from the pop/rock music industry. I remember that Brian Ferry and Eno from Roxy Music were regular visitors. Art colleges are a breeding place for musicians, so the appearance of such stars at the RCA bar was not unusual.

The platform for social exchange was well laid out at the college, but nevertheless I initially found it difficult to build relationships outside the walls of the Automotive Design department, with most departments tending to stick to themselves as groups. Another thing that I realised quite soon, was that most of the students were quite a bit older than me, in particular the girls; being in their mid-20s, they were not so interested in a smooth-faced 20-year-old.

Socialising within the Automotive Design group also proved a bit restricted. The second-year students were way too busy for nights out at the pub, and Geoff Matthews and Dave Arbuckle, both being married, had other priorities. John Saville was a guy I was never that close to at Maidenhead, and Steve Ferrada and I were just poles apart. In my year, Jean Nahum kept himself to his group of friends from Turkey. These guys were in another league, driving around in BMW 2002 Tiis and Renault Alpines. Dave Evans would usually join me for dinner in the canteen before returning to his apartment in the college hostel. Martin Smith had moved into a flat in Notting Hill with his Scottish girlfriend, Christine, so his time was limited. Nevertheless it was with him that I bonded with the quickest.

Martin was friendly with one of the previous year's graduates, who had gone on to take a position at the Ford studios in Dunton. Clive Potter shared a flat just off Kensington High Street with a guy named Dave Johns, a workshop foreman at a Fiat dealer called Radborne Racing. This was the catalyst for a weekly meeting in the pubs around Kensington that cemented us in a friendship that lasted way beyond the college years.

Clive was an amazing car designer, even more remarkable for the fact that he didn't even have a driving licence. He commuted from Kensington to Dunton each day, not willing to give up the 'London life.'

Other designers would often pop by Clive and Dave's flat, including Peter Stevens, one of the first Automotive graduates from the RCA. Peter was what you would describe these days as a true petrolhead. He was a regular visitor at the college and often showed up in some wild form of transport. Of these, the most outlandish machine was an open Ford Bronco, powered by a massive Holmann and Moody V8 engine. He had purchased it from the Ford competition department, having been driven in the famous Baja 1000 desert race.

Peter went on to hold senior positions at several companies, including Lotus, McLaren and Rover. Our paths crossed for many years to come.

Nigel Chapman had stressed at the beginning of my first year at the RCA, that it was up to the students to create their own projects, and it was here that I had real trouble getting into the swing of things. This was a remarkable fact, since it was my own initiative that had driven me to get where I was in the first place. Nevertheless, I began to struggle.

In the first year we primarily had to develop our presentation skills. As a platform for this, the students needed to select an original vehicle design, but what? Our visiting tutors, Peter Ralph and Carl Olsen, were both involved professionally in areas outside of the car industry and slightly tarnished by the 'product design brush.' Whatever we chose to design they were not satisfied with a 'normal' vehicle, so a conventional sports car or limousine were out of the question. Slowly, I felt the roof of the college pressing down on me.

Looking back now, the college's requirement for students to think outside of the box was the correct stategy, and with encouragement, particularly from Carl Olsen, I managed to pull myself away from this initial stumbling block.

All four first-year students came up with a selection of ideas way outside of the comfort zone of visiting designers from within the industry, so there was still some conflict between the styling philosophy of the car industry and the deep-rooted wish for fresh innovation within the college.

The sponsoring companies of the course sent designers to the college on a regular basis, and whether from Ford or Chrysler, they were all true masters of their craft. To this day, I don't believe there is another industry with the same focus on illustration skills as there is in the automotive industry.

The car design bible of the time was a magazine called *Style Auto*, which was published in Italy. It was more like a large format paperback than a magazine, and due to its specialised topic and small publishing numbers, was very expensive. Within its pages, you could marvel at the illustrations accompanying the development stories of various production cars or motor show concepts, and the top designers were superstars in our eyes.

General Motors had a particular reputation for having the best illustrators. In the US, Tom Semple had a beautiful minimalistic style, and at the GM European studios it was a Japanese man called Hideo Kodama at Opel. At Vauxhall, Chris Fields and Geoff Lawson came to mind. I would spend many hours studying the techniques of these designers. Some designers had become so accomplished that they had left the big studios to become freelance illustrators and designers. A true hero of the times, as he remains to this day, is Syd Mead.

Mead had studied at ArtCenter, and after graduating he joined the Ford studio in Detroit. He was a true visionary, and realised very soon that he would be too restricted working in the car industry. He left Ford and produced a series of paintings – the term 'illustration' would not do them justice. These images depicted future worlds and the transportation that went with them. They also featured fantasy animals and exotically dressed beings. He completed his first collection for publication in a book titled Innovations, sponsored by the US Steel Corporation. The book was a must for car designers. In the automotive design world Syd Mead became a legend and went on to work not only in all fields of transport design but also in the film industry, designing the sets and vehicles featured in such films as Blade Runner and Tron.

There are many stories in the industry about the creation of automotive illustrations, so magical were the styles. In one issue of *Style Auto,* there was a bird's eye view rendering of large American cars parked on a crazy-paving driveway. To the casual observer this was nothing special, but upon closer inspection the words 'fuck off' could be read in the shape and pattern of the stones. Some designers, including myself, would also use this form of messaging to disguise their signature, should the company wish for anonymity on the artwork.

The style of automotive illustrating or rendering that most visiting designers demonstrated used a special technique found only in the car industry. It involved a using a specific type of paper called Vincent Vellum.

Vincent Vellum was more like a thick tracing paper, being semi-transparent. It was this quality that enabled a designer to work using both sides of a sheet, and so build up different shades. Working with felt tip markers, the solid colours or shadows of the subject matter were drawn first. As an example, working on the vellum in black would produce a grey colour on its reverse side and in this manner tonal differences could be built up. Once the main solid colours were completed in marker, the designer then developed shaded areas and reflections by rubbing a chalk powder on the remaining untouched areas of the drawing. The chalk powder was created by scraping coloured chalks with a knife blade. Areas on the illustration could be masked off or protected by cutting out templates of paper. As a final touch, a background graphic behind the illustrated car could be added using a wash of marker ink called Flo-Master. Flo-Master was a petroleum-based ink used for refilling markers or in airbrushes; highly toxic, it is now banned from use.

Describing how to draw is never easy – it's so much easier to witness an artist at work, so I hope my description is half way understandable.

The visiting designers came at least once a week to demonstrate their rendering skills. Ford designers visited most frequently, as travelling from Dunton was easier for them and, like Clive Potter, many lived in London.

Notable visiting designers from Ford were Trevor Creed and Patrick le Quèment. Both these designers were the high-flyers at Ford, having worked on designs such as the first-generation Capri and the new Ford Escort. Both would turn up to the college in the 3-litre version of the Capri, usually attired in a suit with flared trousers and a tie of enormous proportion. They both exuded style and confidence.

In contrast, the visitors from Chrysler UK were true British motor industry Midlanders, with the Brummie accent to go with it. Not that the Chrysler designers were unable to hold their own against Ford, as the Whitley Technical Centre harboured a rich collection of talent, influenced by the US designers who would visit. Indeed, the designs from all the US manufacturers in the early '70s were legendary. Ford had established itself with the Mustang, Chrysler had its Hemi-powered Charger and Challenger, and AMC had the Javelin, but it was the competitors from General Motors – the Chevrolet Camaro and Pontiac Trans-Am – that really captured our attention the most. Indeed, GM was considered *the* address for aspiring automotive designers, such was the reputation of its Detroit-based design studios. This reputation also extended to its European outposts at Vauxhall and Opel. Strange then, that GM had no involvement in the RCA course.

From the fourth floor of the college, we were ideally placed to observe the designs of the cars driving along Kensington Gore, and the latest Camaro was certainly a highlight. RCA tutor, Carl Olsen, who had started his career at GM in Detroit, was naturally a big supporter of the brand, and there was very rarely a week when at some time or another he was not heard saying: "I remember when I was at GM ..."

In 1970, GM's UK division, Vauxhall, got attention when it showed a concept car right at the front door of the Italian coachbuilders at the Turin Motor Show. It was called the SRV – styling research vehicle.

The Turin Motor Show was recognised as a Mecca

for automotive design. Each year, the design houses of Pininfarina, Bertone, and Michelotti would vie for the title of best concept car.

In 1970, Vauxhall came to the Turin show with the SRV and really rocked the Italian car design establishment. The SRV was a design of great elegance and dynamics, and a featured article in the first copy of *Style Auto* in 1971 revealed the development artwork of the car, done by Englishman Chris Fields and Australian John Taylor, both designers working under the leadership of American Wayne Cherry. In particular, the sketches of Chris Fields were outstanding in their energy and pure simplicity; Vauxhall certainly looked like a great place to work, I took note.

That copy of *Style Auto* featuring the SRV was also notable for its feature telling the story of the development of the Lamborghini Countach, designed by Marcello Gandini, the in-house designer of Bertone. Like the SRV, this was also a design of unimaginable beauty and uniqueness. Yes, there was no doubt, we were privileged to be studying car design in the early '70s; the energy in the industry was just amazing.

Each of the first-year students began to develop their own individual style of presentation, and in particular Martin Smith started to show the potential I had seen in his sketches, becoming the year's benchmark.

Unlike the final year, where a single project would be developed, the first-year students would concentrate on a different topic for each term. Martin was using the opportunity to shake off his regimented Engineering background and produced a series of wild styling studies, choosing to ignore the advice of his peers to be 'sensible.' I, on the other hand, was more cautious in my approach and my first project was a delivery van with a modular engine system – the Industrial Design tutors loved it, but the visiting designers were not so enamoured.

Apart from the development of our two-dimensional skills, the Automotive course had a modelling studio, located opposite the rear entrance of the main building in Jay Mews. With large double doors, it must have originally been a stable for one of the large houses in Kensington Gore. The studio was equipped with an oven, which was required to warm up the special automotive modelling clay. In a soft, heated state it could be applied to a Styrofoam armature as the first stage in creating a scale model. The models were usually built in quarter scale and measured about a metre in length.

First-year students did'nt tackle quarter scale models but concentrated on smaller studies in order to develop an understanding of form and shape. However, as the year progressed we were roped in to help the second-year students complete their final degree models: a good insight into what we would face the following year.

Despite the attractions of London, my social life remained firmly rooted in Reading. I felt more comfortable in the territory I had developed while studying in Maidenhead. Establishing a similar situation in London was a vast obstacle to overcome when trying to develop friendships, and that first year I took the easy route by travelling home at the weekends.

I began to regret having sold the Mini, as I was now back to where I had been after I passed my driving test, having to beg to use the family car, which was now an Austin 1300.

As the year progressed, and with the summer vacation nearing, I needed some form of transportation in order to travel to Chrysler's technical centre in Whitley, where I would be working for the majority of the holidays – part of the course's industrial experience requirement.

It was clear, I had to get some wheels under me and made the decision to look around for a car. Setting my budget at around £200 I started to go through the classified adverts in the Reading *Evening Post.* Another Mini would have been great, but the demand was enormous, and sellers were asking high prices. Going down the list of makes, my attention was drawn to a car that hadn't been on my radar but was being offered by a local dealer at what seemed a good price for its age and mileage. I went over, and after a short test drive, I was the owner of a Triumph Herald 12/50. The Herald was well established on the market, but to be honest was not regarded as a very stylish mode of transport. For me, however, it offered several advantages. The mechanics were very simple, and with a front bonnet that tipped forward, including the front wheel fenders, it was very easy to work on. The 12/50 also featured a fully folding canvas sunroof; not a convertible, but pretty close. I was happy with my choice but still faced a barrage of negative comments upon announcing my purchase at college. Martin Smith immediately named it 'The Targa,' a title usually reserved for much sportier

*Me as a first-year student at the world-famous Royal College of Art.*

models, and even my father would refer to it as 'The Modest Herald.'

Despite the less-than-complimentary comments, it would do the job; there was no going back.

By now, all the first-year students had acquired some form of transport. Martin had inherited his father's Rover 90 and Dave Evans had set the bar very high by purchasing a secondhand Jaguar E-Type coupe. In his late 20s, he had obviously saved enough to afford such a car and was at an age where he could also pay the insurance. It was a very early model and it wasn't be long before Dave would be footing large bills to keep it running. He pulled up to Jay Mews the day he picked it up and I joined him on a short drive to Shepherd's Bush. Driving along, what immediately struck me was the whine of the gearbox, which sounded louder than the engine noise. At Shepherd's Bush tube station, Dave parked and let me take the wheel. It was a short turn, but I was there, behind the steering wheel of an E-Type, staring over that endless bonnet. To this day, that's the only time in my life I've driven that car, and believe me, it's very special.

Jean Nahum had driven from Turkey at the beginning of the year in a model produced by Anadol, to which his family had connections. It was an Anadol A1, a fibreglass car based on the mechanics of a Ford Escort. The A1 had been designed by the Ogle group and developed by the British Reliant Car Company, who specialised in fibreglass cars, such as its three-wheeler Robin and, in complete contrast, the very sporty Scimitar Coupé. Jean's Anadol was pretty special in so much that he had modified it to compete in motorsport rallies. It was fully stripped out with two bucket seats, seatbelt harnesses, and a roll cage. The engine was the famous Ford Kent motor, modified and sporting twin Weber carburettors. Jean's trip from Turkey had been a nightmare, since at the Greek border, custom officials had made him completely dismantle the car in a search of drugs. It had taken him two days to rebuild the car.

The first year at the Royal College of Art took its course and the famous end-of-term degree show was approaching. It was a pilgrimage for all art students. The pressure was on for the second-years and we all mucked in to help them finish their models. I remember, in particular, Geoff Matthews' model: an articulated truck with a cab positioned low in front of the driving wheels, similar to the ones seen at airports that collect the luggage off planes. To create a show, Geoff had a two-metre long trailer attached to the cab unit; it was pretty spectacular and I believe it earned him an honours.

In those days, the college had an open lower courtyard featuring two large trees, which I assume the college building had been built around. The courtyard separated the college entrance lobby, which was at the same level as the Royal Albert Hall across the road, and the lower part of the building that housed the canteen and library. For the degree show a large orange awning was assembled above the courtyard, similar in style to the famous Munich Olympic Stadium, which incedently had just been completed for the summer Olympics later that year. The Automotive Design students displayed their work in this area. It was a wonderful setting, except for the orange tint it gave to all displayed under it. It was a great feature for the show, but now no longer exists.

The '72 show was a good year, and on the days when it was open to the public, we would stand around bathing in the aura of being a student at that famous college. I felt proud to be there.

The degree show marked the end of the academic year. I packed the Herald and headed down the M4 back to Reading.

# Chapter 8

## *Whitley Technical Centre*

**There was no real time to catch up on old friends in Reading, as Dave Evans and I had to report to the Whitley Technical Centre the week after the Royal College of Art broke up for the long summer vacation. The autumn term commenced in October, and for most of the summer of 1972 I would be working alongside the professionals at Chrysler UK's design studios.**

Instead of driving directly to the Whitley Technical Centre, I had agreed to meet Dave Evans in Coventry. Chrysler had arranged for us to look at some accommodation in the town.

Dave was waiting for me at the address in his E-Type – a ground floor flat in a terrace house. We met the owner and after a quick look around decided to take it. It would suffice as a roof over my head during the week, as I intended to return to Reading each weekend. For Dave it was going to be a more permanent arrangement: the drive to his home in Northumberland, just for the weekend, was out of the question, and the E-Type was not proving to be such a reliable form of transport.

We moved in straight away and arrived for our first day at Whitley the next morning.

During my first visit to Whitley I had seen nothing more than the reception area and the interior of Roy Axe's office. Now, at last, I would see the inside of an automotive design studio; it was exciting beyond belief.

Rex Flemming, who it turned out was more responsible for the administrative side of the studios business, met Dave and myself, and after a brief introduction, we moved to the area of the design centre where we'd work on our summer project.

The Chrysler UK design centre was basically a converted factory building. At the front were a group of executive offices, and beyond were two main design studio areas. We walked into the studio where we would be working.

It was large and well lit. Mounted on the floor in the centre of the room were large steel surfaces or modelling plates, occupied by two clay models of large truck cabs. The studio in Whitley was not only responsible for the design of the Chrysler UK passenger vehicles but, as we learnt that day, the development of Chrysler's Spanish truck subsidiary, Barreiros. The two cabs stood alone with their front wheels, but no after-body was attached.

The commercial side of the Rootes Group was called Commer and included small vans right up to large lorries or trucks (to use the American expression for commercial vehicles). During the postwar years, the Commer truck was the popular choice for cement mixers or tipper trucks, and had a flat three-cylinder, two-stroke diesel engine with a sound like no other vehicle on the road. With the Chrysler takeover, the Commer name, like most of the Rootes brands, would disappear, to be replaced by the Dodge label, a fate that would also befall Barreiros.

For our time in the design centre, a dedicated corner of the studio was freed up for Dave and me to work in, with suitable drawing desks already in place. Standing by one of the truck models was Steven Ferrada, who Chrysler had employed after he graduated from the RCA, the other Chrysler-sponsored student, Geoff Matthews, having decided to join Ogle Design. A few introductions were made, and Rex Fleming asked us to follow him into the other studio where the Chrysler cars were designed.

Although the American Chrysler group had first invested in the British Rootes Group around 1964, it wasn't until the late '60s that it decided to concentrate on developing a new range of cars for Europe, based on the underpinnings of designs being developed by its French subsidiary, Simca. The last design to be badged as a Hillman was the Avenger, which was introduced to the market in 1970. However, from now on all future models in the UK would be named Chrysler, and design development was shared between the Whitley studio and the ex-Simca studio in Paris. The whole strategy was rather disorganised since Chrysler decided to remain faithful to the old Rootes buyers in the UK by keeping the Hillman name alive with the Hunter and Avenger models, and introduce parallel models named Chrysler. The first Chrysler-badged car in the UK – introduced just after the Avenger in 1970 – was called the 180, and, to confuse matters further, would be called the Chrysler-Simca 1609 in mainland Europe. The 180 was a large sedan with American flavoured styling, vaguely similar to some of the Plymouths of the time. It featured the flowing body shape named after the Coca-Cola bottle – 'coke bottle' design. This styling was a feature on many mainstream models of the time, including the Ford Cortina Mk 3 and Vauxhall's Viva and Victor.

As history tells, the whole idea of naming Rootes Group's future cars 'Chrysler' was not be successful. The name Chrysler meant nothing to the average British car buyer and poor sales reflected this. The original plan for the 180 was for it to be fitted with a new British-designed V6 engine, which was finished and ready for production. Had it had this feature as a top model it may have succeeded. However, Chrysler pulled the plug on that idea and despite all the tooling for the six-cylinder engine being complete, the car would only be offered with a Simca-derived four-cylinder engine. As a result, the model was a complete flop.

As Dave Evans and I walked into the Chrysler car studio at Whitley at the beginning of the summer of 1972, the cloudy times ahead weren't in anybody's minds and there was an exciting and positive air in the place.

The car studio was a long room with space for at least three full-scale models. When we entered, we were met by Chrysler UK's design director, Roy Axe, and his most senior design manager, an American named Curt Gwinn. Curt Gwinn, who is credited with designing the Hillman Avenger, was what you would describe as dapper in appearance, with collar length dark hair, and a neatly trimmed beard, wearing a fine polo neck shirt under his sports jacket. Nearby was a row of desks, behind which sat a small group of designers. Their names have faded in time, but it was here that Steven Ferrada, who had walked with us from the truck studio, was situated.

Roy Axe welcomed us warmly, and, just as I remembered him from our first encounter, was a larger than life character, very British in his manner, which I'm quite sure went down well with the American executives from Chrysler.

Introductions over, our attention was drawn to a model mounted on one of the modelling plates. Curt Gwinn took us over to it and proudly announced to us that this was in fact the new Chrysler 180 coupe. After the launch of the 180 it had been decided to improve the models appeal and produce a two-door coupe version of the car. The design in the studio, resplendent in metallic green, was a full-size model of the design. Work on the proposal was almost complete and the resulting car was handsome and had way more character than the 180 sedan it was based on. When I say 'based on,' this only refers to the mechanical running gear or platform. Seeing this, I felt encouraged.

Apart from the coupe model there was not much else to see. The studio walls were decorated with an array of spectacular illustrations, but what Chrysler was planning beyond the 180 was a mystery, though we would soon find out.

Curt Gwinn escorted us back to the truck studio and explained our work schedule for the summer.

Chrysler planned to develop a new range of small cars, all of which would feature transverse-mounted engines and front-wheel-drive, similar to the Mini and Austin 1100 models of British Leyland, or some of the cars manufactured by the French and Italians. In the early '70s the front-wheel-drive layout was still comparatively new. This sounded an exciting strategy, remembering that key competitors from Ford and General Motors (Vauxhall/Opel) had no front-wheel-drive cars in their line-ups.

Dave and I were to create exterior design proposals for a future hatchback car positioned below the Avenger model. During the summer we would develop sketches and renderings and finally have our designs produced as scale models – an exciting proposition.

It sounded great, and working in the environment of a car manufacturer's design studio, this was the real world of Automotive Design.

Dave and I returned to the flat in Coventry excited and full of anticipation at the task we had been set.

Living in the same apartment as Dave was also a new challenge. Over the first year at the RCA, we had got to know each other well, but the everyday relationship within the Automotive Design studio of the college was entirely different to sharing the confines of rented accommodation. The flat in Coventry was, however, spacious enough so we didn't tread on each other's toes.

The domestic side of my life was slowly developing. I knew how to handle a laundrette and iron my shirts, but cooking a meal of sorts was a skill still in its infancy. My mum (like all mums) was a great cook. In her kitchen it was always traditional British fare, but I guess that was the case in most households during the postwar years. Eating out was still a special occasion and usually confined to a steak house, like the Berni Inn, or a local Chinese. Indian restaurants were springing up all over the place but were still too exotic for my parents. Outside of London, the influence of continental Europe had not reached the English high street. Domestic and restaurant cooking in England was still Russian Roulette, and resulted in the negative reputation for eating in this country, which regretfully, up to this day, unjustifiably tarnishes the nation's culinary image.

In the Birtwhistle household the week would start on a Sunday with a traditional roast followed by one of my mother's amazing deserts – trifle, apple charlotte, or lemon freeze. Monday evening would be cold roast with bubble and squeak (which consisted of the remaining potatoes mashed together with whatever vegetables were left over and fried until they developed a crisp golden skin). During the rest of the week it would be the good old British standards, like toad-in-the-hole or shepherd's pie, and on Fridays I would always fetch fish and chips after my cub-scout meeting. My mother loved holding dinner parties on Saturdays, where she would try to stretch her cooking beyond the normal British fair. I was aware of her skills in the kitchen but did not pick up on the details until much later.

When in London, there was little point in me cooking, with the RCA canteen being as good as it was. Both Dave and I were regular visitors, with Martin Smith only joining us on Wednesday's for what we called 'Big Eaters' Night.' It was Dave who coined this title, which needs no real explanation other than the candidate who could pile up his plate the most, won.

Of our group at college, Martin was the most accomplished cook and it was he who gave me the rudimentary guidelines to create spaghetti bolognese or an Indian curry sauce.

Back in the flat in Coventry, the dishes placed at the table were still somewhat lacking in culinary flair. My special was fish fingers accompanied by frozen spinach with a slice of white bread and butter – fairly harmless, and certainly no match to Dave's own special brand of spaghetti bolognese, which had yet to reach the sophistication of Martin Smith's recipe.

The bolognese sauce was surprisingly simple: tinned mincemeat brought to the boil and mixed with Heinz tomato ketchup for taste and colour. For the spaghetti, Dave would follow the more traditional Italian route and boil it in a pan, but in order to add a bit of Britishness to this otherwise Mediterranean fair, he served the finished dish with Brussel sprouts that had lost all traces of their green colour due to the hour-long boiling they had been subjected to. I can only assume I had no problem eating this, but what I do remember is that Dave's spaghetti menu had a very cleansing effect on the digestive system.

This style of cooking remained throughout the summer and looking back, seemed totally appropriate for day-to-day life in Coventry.

Dave and I usually did the 20-minute drive to the Whitley Technical Centre in the Herald since his Jag was not too good in the mornings.

In the design centre we set about our task to design a small hatchback. These days you would describe such a vehicle as 'Golf size.' Back then, the Golf, however, was yet to be launched, so this was quite a revolutionary concept.

Commencing our design work, Rex Flemming and Curt Gwinn kept an eye on us both, with Steven Ferrada also feeling a certain responsibility due to the RCA link.

The process of developing a design until one was chosen to go ahead as a model was just as we had been instructed at college. Quick sketches would be drawn on A4 or A3 layout pads – usually in ball point pen or with Prismacolor coloured pencils. After this, more finished renderings would be developed on either Vincent Vellum or white drawing card. At college, Geoff Matthews had developed a style of illustration similar to the work of Syd Mead. He would take a large sheet of white card, usually about A1 size, and create a background wash by pouring Flo-Master ink over the surface. Using several different colours and sometimes adding water, he produced a marble effect. The result of this technique depended upon how flamboyant you were, and each designer would develop his personal signature. The real skill when applying this method, was to use the shapes created by the wash to develop a more identifiable background, such as a building or something natural, like a rock face or forest. Syd Mead was certainly an influence for many designers but the roots of this style of presentation came from the big studios in Detroit where Mead had also worked. There, automotive rendering reached levels of unbelievable drama, realism and finish. This style migrated to most of the studios in Europe, with only the Italians keeping a much more rigid form of presentation.

The summer took its course and our design proposals progressed from the illustration stage through to the mandatory tape drawings of all three elevations and commencement of models. The finishing deadline for our completed work was mid-August, when our period at the technical centre ended. Unlike the models produced at college, we worked in one-third scale, which meant that the resulting models were quite large, measuring around one and half metres long. A model this size is very heavy, and unlike the smaller scale of one-to-four, cannot be lifted without the help of a crane or fork lift. Fortunately, we got the help of the studio's clay modellers to finish off the models, and we both finished our work on time.

It wasn't all a regular nine-to-five day at the centre and occasionally the routine was interrupted. Rex Flemming took Dave and I on a tour of the production line in nearby Ryton. The factory was located a 10-minute drive to the south east of Whitley. I passed the factory each Friday on my way back to Reading.

Outside were rows of older Rootes models such as the Hillman Hunter and Singer Gazelle, which were basically the same car except for cosmetic differences, the Singer being a more up-market model. There were also rows of Hillman Imps, which, although manufactured in Glasgow, were stored at this location before distribution. The Imp was Rootes' answer to the

*Dave Evans' clay model (left) and mine in the Chrysler UK studio in Whitley.*

Mini and had borrowed its look from the American Chevrolet Corvair. It was rear-engined and had a sibling in Germany made by NSU called the Prinz. Both cars shared the same mechanical layout and were uncannily similar, although in no way related. Like the TT version of the Prinz, the Imp was quite successful on the race track. This, however, was not reflected in sales numbers and it remained a small seller compared to the Mini.

The Ryton factory was still assembling Hillman Hunters as we viewed the production line, both as limousines and estate cars – as they were known those days. At the commencement of the tour we were shown an estate model which had received the floor panels of a limousine model by mistake. Since the estate car was longer than the limousine, the result of this mismarriage was a gap in the floor of the luggage compartment. Displayed prominently, it was supposed to encourage the assembly workers to concentrate more on the job. I'm quite sure those responsible knew what they were doing – maybe an old April fool's joke – and despite the blunder, our tour leader was quite proud of this error. Also on display was the Hillman Hunter, which by great fortune had won the London to Sydney Marathon in 1968. The factory guide's chest expanded as he stood in front of the car, proudly announcing in a broad Birmingham accent:

"This gentlemen is known in-house as 'The wowat unter.'" (White Hunter)

The Ryton plant later ceased producing old Rootes models, before moving on to Talbot cars and later Peugeots before its eventual closure. Today, it's an industrial estate, housing Jaguar Land Rover's Classic works and SVO divisions, among other businesses.

Dave and I presented our models to Roy Axe and his staff, and they seemed happy; the chapter in Whitley closed for now. I didn't yet know if I would return next year.

Dave and I collected our belongings from the flat in Coventry. He got in the E-Type and headed back to Northumberland, while I once again pointed the Herald in the direction of Reading and a well-earned vacation prior to recommencing college work in October.

# Chapter 9

## *Southern Comfort*

**While Dave and I had spent our summer in Whitley, Martin Smith had been working in the Ford development centre in Dunton. He came to visit us one weekend, driving up to Coventry with his girlfriend, Christine, in the Rover 90 he had inherited from his father. Things between him and Christine seemed fine, so I was surprised when, upon my return to Reading, he called me to say they had decided to part ways, and to ask if I was interested in taking a couple of week's vacation with him. I had no plans so agreed spontaneously.**

Martin had an idea to travel through France to Switzerland and meet up with our Monday night pub friends, Dave Johns and Clive Potter, in the Swiss resort of Interlaken. From there, we could travel down to Turin and visit the famous Italian car design studios based in the city, before driving across to Monaco and the French Riviera, and finally return to the UK via Paris.

Because of our limited finances we would camp, but that was no real issue since mainland Europe was well distributed with campsites, which, unlike the ones in England, were better equipped, with more than just a field with a tap for water. It was agreed we'd travel in the Herald since it was uncertain whether his father's aging Rover could survive the journey. I had no problems, it was well maintained, what could go wrong?

The last week of August, Martin drove down to Reading from his home in Sheffield, and after an overnight stop we set off early for Dover and the ferry to Calais. Arriving in Calais we drove via Reims and on to Dijon, where we made our first halt for the night.

It was dark by the time Martin and I checked into a typical French tourist campsite on the outskirts of Dijon, and while Martin started to assemble the tent, I went off, torch in hand, to look for the toilets. I would soon learn that as far as lavatorial facilities were concerned, the French had their own way of doing things. Opening the door of a free cubicle I was confronted not with a normal toilet seat, but a porcelain platform with two foot plates and a dark hole disappearing into the ground between them. There was no light and in place of toilet paper a pile of old newspapers would have to suffice. Being a designer, I was easily able to figure out how I needed to arrange myself for this new experience. With one hand making sure none of my lower garments touched the surface of the porcelain, which quite clearly hadn't been cleaned that summer, and the other hand aiming the torch to make sure I was aligned with my target, I managed to keep my balance; I survived.

Tent erected, we cooked our first holiday meal of fried luncheon meat (Spam) with baked beans, together with French bread and beer – just great!

The drive from Dijon to Interlaken passed by Lake Geneva, and by the afternoon we had arrived at the campsite situated next to the lake in this beautiful Swiss resort. Dave and Clive had arranged to meet us there the same day, and they turned up not long after us. Dave had borrowed a new Fiat 127 from Radbourne's. It was finished in a light green, with an orange/brown cloth interior; a pretty cool car for its day. The Fiat group was making great cars in the early '70s and its recently launched, top of the range 130 coupe was just gorgeous. If Chrysler was to compete with the 127, it needed to make something very special; I took note.

We pitched our tents together under the shadow of the Alps. The lake at Interlaken is really an amazing location, and the first sight of those mountains was just breathtaking.

While we organised our tents, a friendly-looking elderly gentleman wearing a striped t-shirt and baseball cap wandered over to me – he could only be an American. A pensioner, he was touring Europe with his wife. We got into conversation and it wasn't too long before he spotted the yellow badge secured to the front grille of the Herald with the bold initials 'AA' on it. It indicated that I was a member of the British, Automobile Association, a roadside breakdown service. I figured that, travelling across Europe, it was a good precaution to join and take advantage of its European coverage, hence the badge.

My father had always been a member of the AA, and while travelling in the '50s the uniformed motorcyclists of the club – mounted on their sidecars similar to the low-status Matchbox model I had as a child – would salute members upon recognition of the badge on the front of the car. Those days were gone, but they still provided members with a sturdy enamel badge as an indication of membership.

Taking in the badge on the front of the Herald, my new American friend paused, and without any hesitation said:

"Well, that's just great, you know, some of my best friends are members of Alcoholics Anonymous."

By this time Martin, Dave and Clive had opened the first beer cans, so I thought it prudent not to comment on the kind American gentleman's observation and steered him away from our tents, concluding the discussion with a hand shake and best wishes for their remaining travels.

Back at the tent, things were just warming up and the conversation got louder, as did the crash of beer cans being randomly discarded. After the meal of grilled meat and sausages, Clive pulled a Duty-Free bag out of their tent, which contained two bottles of Southern Comfort, a particularly revolting, and to my taste, perfumed, Southern American bourbon.

We finished both bottles that night and to this day just the smell of Southern Comfort is enough to transport me back to that night in Interlaken and, even worse, the near-to-death hangover I had the next morning.

By the time any of us were halfway near being back to normal, I observed that the American couple had packed up and moved on, not, however, before taking time to put in a complaint at the campsite office about our night's activities. Later, we were accordingly warned by the site manager that there was to be no repeat of that night or we would have to leave, a point we were in full agreement of.

A day later we said our goodbyes to Dave and Clive, who would return to England – just as well, since I don't think we could have handled another boozy session with them.

Soon I would be threading the Herald out of Interlaken and pointing south. I would visit the town many times again in the future, but for now we were on our way to Turin, the traditional home of car design.

The route would take us via Martigny to the south-eastern end of Lake Geneva, up across the Alps, through the St Bernard's tunnel into Italy, and onward to Turin via Aosta.

Entering Italy, the contrast to Switzerland struck me immediately. Now at the beginning of September, Switzerland was lush mountain meadows and cows, while Italy was hot, unruly and dusty; I loved it.

We headed down the valley from Aosta, towards Turin, and decided to find a campsite just before entering the city. Leaving the A5 autostrada, we located a suitable site near the River Orco, which runs into the Po just north of Turin. The campsite had a restaurant and that night I was able to sample spaghetti bolognese without the help of ketchup or Brussel sprouts.

Prior to leaving England, Martin had written to three key players of the Italian 'carrozzeria' industry to ask if we could visit them and all had replied positively. Knowing Martin, he had made these arrangements long before he and Christine had split up, so having taken her place, I had lucked out on his foresight.

We would need two days to make all of our visits. Of the three locations, two were well known – Pininfarina and Bertone – the third, however, was the new kid on the block. Bertone's star designer, Georgetto Giugiaro, had quit the famous company to set up his own studio in the centre of Turin, and was the new sensation of the Italian car styling industry.

Pininfarina and Bertone had their design offices located close to their production facilities, where they built both bodies and finished vehicles for Fiat, Maserati and Alfa-Romeo. In addition, Pininfarina also built cars for Ferrari and Peugeot. Between the wars, both companies had developed real strengths as designers and manufacturers of exclusive cars of limited production, and after WWII their businesses had expanded to the extent that vehicle manufacture was now their major income. Nevertheless, both companies still offered an extensive design and engineering service for the industry, and not exclusive to the Italian companies. Pininfarina, for example, did a lot of work for Peugeot and BMC, being clever enough, in the true Italian manner, to sell the same design to both companies in the form of the Peugeot 404 and Morris Oxford.

As in the fashion industry, the design houses would choose international trade shows to demonstrate their design prowess, and in the '60s and '70s this was, without question, the annual Turin Motor Show, which, at the time, was located in two exhibition halls in the city, close to the banks of the River Po. To visit the Turin Motor Show was a must for car designers.

By the '70s, Bertone and Pininfarina were not the sole two styling houses in Turin. In 1971, Giugiaro was new on the scene; Ford had bought the Ghia studio and was producing outstanding designs; and Michelotti, famous for designing many Triumph models including the TR6, Spitfire, and my Herald, was also a force to be reckoned with.

As Martin and I motored towards Turin, the Italian styling industry was at its peak and the creations of this era took on a look that could almost be compared to modern architecture. Clean and chiselled surfaces were combined with bold graphic treatments, and the ideas seen in the concept vehicles produced by these studios would feed directly into production models. You only have to think of such cars as the previously mentioned Fiat 130 Coupe, the Fiat X1/9, and Alfasud Sprint Coupe, or the first iteration of the legendary Lamborghini Countach. The '70s were indeed a golden age for Turin. However, with the manufacturers' in-house design studios gaining momentum, Turin studios slowly diminished in design prowess.

Turin, when summer slowly turns to autumn, is foggy, and that early September morning in 1972, as we made our way into the city, was no exception. I was acutely aware that the Herald I was guiding into the city was returning to the place of its creation, a point Martin repeatedly pointed out with a degree of sarcasm. Our first port of call was Pininfarina and the address we had been given was an office located by the factory building in Grugliasco, not so far from the city centre. Today, Pininfarina has a bespoke design and engineering centre to the south east of the city in Cambiano, but at that time the design offices were situated near its sprawling factory location.

I pulled into the visitor's car park and we made our way to the reception. We were expected, and a smartly dressed Italian businessman greeted us, showing us to a group of chairs in the reception area.

Unfortunately, we couldn't visit the design office, but he called one of the designers to meet us. It was somewhat disappointing but understandable, and he had arranged for us to see some of the nearby production facility later on.

As promised, one of the design team came to join us and politely looked at the samples of our artwork we had brought with us. His English was non-existent and all we could get out of the gentleman who had met us was that he liked our work. The meeting was concluded, and we were asked to follow our host to the nearby assembly building.

Making our way through a side door we joined part of the production line where painted bodies were moving down. It looked similar to what I had seen in Ryton. I could make out three different models: the Fiat 124 Spider, Fiat 130 Coupé, and, very new, the Ferrari 365 GT4 2+2. Unfortunately, although designed by Pininfarina, the gorgeous Ferrari Daytona was not built by them, so it was nowhere to be seen.

That concluded our tour, which didn't go any further than a quick look. Nevertheless, it had still been worth it. All three of the cars we saw on that assembly line are now legends, and I'm pleased I was able to have witnessed a part of automotive history on that day.

From Pininfarina, we headed to the Bertone factory, located only a few streets from the Pininfarina plant, but we weren't expected until the afternoon so decided to drive into the city centre.

Located within sight of the Alps, Turin is the capital city of an area in Italy known as Piedmont. It has always suffered as a tourist attaction because of its reputation as an industrial centre, and arguably lacks some of the flair of Milan, Florence, Rome and Naples. Nevertheless, it is well worth a visit. It has a magnificent city centre featuring covered shopping arcades, with the famous Via Roma stretching up towards the main railway terminal. To its south, the rolling Piedmont countryside is home to wonderful towns such as Alba and Asti, where in local restaurants you can enjoy the best Italian cuisine, complemented by the fantastic local wine, including the famous Barbaresco and Barolo.

Many will swoon at the mere mention of Tuscany and understandably so, but the Piedmont is still, for me, the hidden jewel of Northern Italy.

That lunch time Martin and I would not be dining in true Piedmont style, instead satisfying ourselves with a Panini Prosciutto and beer in the Parco del Valentine on the banks of the River Po.

In Italy, everything stops for lunch, and we were expected at Bertone at 3:00pm. We drove back to the same industrial area in Turin where Pininfarina was located – Grugliasco – and pulled into the Bertone visitor's parking area, making our way to the reception.

Things here picked up, since standing proudly on display was the Lancia Stratos Zero show car, designed by Bertone's resident designer Marcello Gandini, who was also responsible for the Lamborghini Miura and Countach. After Giugiaro left Bertone to set up his own studio, Gandini had stamped his mark of authority within the Bertone design organisation, and he was one of the greats of the automotive design world.

The Lancia Stratos Zero was a landmark design; in a deep metallic red, it was a work of art on wheels, with the qualities of a modern sculpture or piece of architecture. Martin and I stood in awe as we took it in.

The welcome at Bertone was a duplicate of the morning. We were met by a smart Italian from Bertone's PR department wearing a dark suit, but something about his manner, however, seemed strange. We exchanged names and shook hands but then almost immediately he burst out emotionally in a broad Italian accent.

"Ees zo terrible, eez zo terrible." There were tears in his eyes – had he seen the Herald? My flippant thoughts were soon grounded as he told us news of what was taking place in Munich.

Neither Martin nor I had glanced at a newspaper since leaving England and were therefore pretty isolated as to recent world affairs. That year, Munich was hosting the Summer Olympics. Terrorists had taken the Israeli Olympic team hostage and the whole episode had ended in a bloodbath. The magnitude of this event had been lost on Martin and myself, unthinkable today, where we are bombarded with news 24/7.

Getting over his show of emotion at the happenings in Munich, our host managed to pull himself together but things didn't improve. Apparently, the entire design department was involved in the move to a new Bertone design centre outside the city, so a meeting with a designer was unfortunately impossible. However, he had organised a quick factory tour and a small surprise. We followed him, intrigued.

The factory visit proved very similar to that at Pininfarina, or Ryton, for that matter, since most car factories looked the same. Bertone were building bodies and assembling models for mainly Italian companies, specifically Maserati, Alfa Romeo and Fiat. That day, the production line was filled with the manufacture and assembly of the Fiat 850 Spider, a small rear-engine convertible they had been building since 1968. Also to be seen, but not on the assembly line, were some unpainted bodies of the Alfa Romeo Montreal coupé, another Gandini design, based on a show car produced for the Montreal World Fair, hence its name. Bertone only painted the finished Montreal body, fitting its interior before it was sent back to an Alfa Romeo factory for final assembly.

Our visit of the Bertone plant was more comprehensive than that of the morning, but it was in no way a complete tour. Part of the production line was being prepared for a new model and was out of bounds, which did, however, turn out to be the previously mentioned surprise. At one side of the production line was what appeared to be a covered car body on a rolling carrier. Our guide went over and proudly pulled back the wraps to reveal the bare metal body of a small sports car.

"This is our new baby," he said.

This presumably was the replacement for the Fiat 850 Spider, and although just a bare metal body, there was plenty to get excited about. A quick glance was enough to tell that it was most likely mid-engined, but there was no time for further inspection, as the cover was quickly replaced, and we were escorted away to the reception area. In November of 1972 all would be revealed, and we would then learn that we had been privileged to take a first glance at the new Fiat X1/9.

Bidding farewell to our still somewhat distressed host, and clutching a pile of Bertone brochures, we made our way back to the Herald. It had been a memorable day in Turin and we drove back to the campsite, still reflecting on the sad news from Munich.

The following day, the address of Ital Design, the company founded by Georgetto Giugiaro in 1968, took us to yet another industrial location quite near the Fiat Lingotto factory: a five-storey red brick building, located in a side street called Via Tapice. It was an unpretentious location behind Turin's main hospital. The only indication that this was the home of the rising star in Italian automotive design was a small sign at the entrance.

Upon entering, we introduced ourselves to a very attractive receptionist, and it was immediately clear that there was uncertainty concerning our visit, despite the correspondence Martin had received from them.

"Signor Giugiaro is a very busy man, I'm not sure if he has time."

She got up from her desk and escorted us to a small waiting room: she would see what she could do.

The room was like the inside of a greenhouse, if all the glass walls had been painted on the inside with white-wash. In fact 'room' is not the correct description, since judging by the noise and flashes of welding reflecting down through the open ceiling, it appeared to just be an area partitioned off from the company workshops. We sat down to wait. What was going on behind those painted windows? There was a small black line in one of the panes, which allowed a minute view as to what was hidden on the other side; we took a look. It was just about possible to make out the rear section of a car. In bare metal, it was low and probably a coupe. What you could clearly define was the end of the side window where it met the rear roof pillar. The body section was very geometric, as was the triangular rear pillar that ran backwards and out of view. This was obviously another example of the latest trend from all Italian design houses, where the forms took on an almost architectural signature. As it later turned out, we were looking at the body of the Lotus Esprit concept.

We heard footsteps and the receptionist came back.

"Signor Giugiaro has ten minutes for you – please follow me."

We did, and we were shown into what was presumably Señor Giugiaro's office. Our charming escort remained with us.

"Señor Giugiaro does not speak English, so I will translate."

No sooner had she said this, the man himself walked in, shaking our hands and quickly taking a seat at his desk. There was no doubt – he was in a hurry. He was a handsome man with thick swept-back black hair, wearing jeans with a brown polo neck sweater; he was every inch an Italian. Martin did the talking, explaining about the present stage in our design education and asking if he could give us some comments on our work.

We quickly laid out the illustrations we had with us and almost instantly he fired back at us: "Technique Americana! Technique Americana!" as he threw his hands up in the air.

There was no need for translation.

He went on: "These are nice drawings, but they mean little. I cannot tell from this work if you can design a car or not, there is so much more to designing a car than this style of pretty picture. In Italy we have a different approach to the Americans. I wish you both luck, I'm quite sure you will succeed." With that he got up shook our hands again and left the room.

Our Italian escort and translator watched him go before turning back to us, the smile leaving her face as he shut the door.

"Signor Giugiaro is a busy man."

We left the building and returned to the Herald. During the two days in Turin we had not seen or heard everything we were hoping for, but it had been a worthwhile experience. It was time, however, to leave the city of car design and head for France.

We crossed the French border at Ventimiglia, having made our way south from Turin to the Mediterranean coast at Savona. We were heading for St Tropez, where we planned a more restful end to our holiday, relaxing on the beach, stopping one night at Menton, just after the border, before continuing the next morning. The following day we briefly took in Monaco and my first look at the location of the famous Grand Prix track of the same name, before finally arriving at a campsite in the small village of Ramtuelle, located inland a tiny bit south of the port of St Tropez.

So much has been written about St Tropez, so I will only say that it's great if you have the bank account to go with it, which we clearly didn't. Nevertheless, we decided that even if we weren't among the famous and wealthy, we could at least observe them, and made our way to Tahiti Beach – the once noted haunt of Brigit Bardot. Parking behind the wall of high reeds separating the car park from the beach, we made our way down to this renowned location for sun worshippers.

The beach bars, with their rentable parasols and sun beds, were way out of our league and we carried on walking, soon to discover we were at the part of beach allocated for nudists. Immediately, my memories of the *Health and Efficiency* magazines of my youth came back, only here everything was on display and had not benefitting from the airbrush treatment.

True to form, people were running around trying to occupy themselves with some sort of ball game; no cricket, I noticed, and probably a good idea.

Martin and I agreed we would give it a go and laying our towels on the hot sand, removed all the garments we were wearing. It felt good, but neither Martin nor I were ready to start our own game of volleyball, remaining low and out of sight on our towels.

Nearby, I observed a large English group, obviously two families, who, from the slight nervousness of their conversation had probably met for the first time on holiday. Fathers and children were running around with beach racket bats in hand, encouraging others to join in, while the two pink mothers protected themselves in the shade of a large parasol, looking on approvingly. The sun got hot, hotter and hotter. Hats were put on, t-shirts, shoes, but as signal of solidarity to nudism, all middle areas remained uncovered.

For us, the parts of our bodies that this group were unwilling to cover needed some protection and we decided a beer in the shade was a good alternative to naked ball games.

As it turned out, for the next few days the weather changed for the worse and rain set in. After a couple of wet nights in the tent, we decided to call it a day and packing up, pulled out of Ramtuelle to head back home.

It had been a great break but the return to college soon beckoned.

# Chapter 10

## *London to Luton*

**With the commencement of the final year at the Royal College of Art approaching, I had nowhere to live. The previous year I had rented a basement flat in Earls Court but decided to end the arrangement at the beginning of the summer vacation, and Martin had suggested the idea of sharing a flat. I met him in London and we found a top floor flat in Holland Park Road.**

*A second-year student at the RCA. Very much a style icon.*

As was the case the previous year, the new student intake would start at college one day earlier. The new college year brought with it only two additional students for the Automotive Design course, and walking into the fourth-floor studio on that second day, both of them had already sorted out their desk arrangements, and were ready to meet us.

Lionel Chew and Peter Horbury had both studied at Newcastle College of Art and therefore knew Dave Evans, who had also been a student there. Introductions were made, and though both the new boys spoke with a Geordie accent, Peter Horbury's was much softer, while Lionel Chew could not disguise his origins, introducing himself as 'Layenal.' Both were slightly older than me, so at 21 I remained the course junior.

For us second-year students the final year meant selecting a design for our final degree show and assessment. The output was quite clear: full 2D development work, from sketches through to finished renderings, tape drawings and package to support the final scale model. 'Package' is a term that refers to the technical and ergonomic layout of the vehicle, so it needs to include the mechanical layout as well as the position of the occupants, and this in all three dimensions.

Martin was already quite clear on what he wanted to achieve, although he wasn't ready to divulge the subject. Jean Nahum, being from Turkey, was keen to create a vehicle suitable for his country's emerging car industry, which, since his family was involved in it, seemed a wise choice. Dave and I, however, were far away from selecting a direction: a situation that would prevail for several weeks. Whatever we selected, both the college and the industry observers – who would visit the final degree show next June – would be looking for innovation in form as well as function.

In Europe, the products of the car industry were still conceptually very conventional, and the future trends in terms of use and space arrangement were yet to evolve. It was up to us, as students at one of the world's elite design schools, to come up with these new provocative proposals that would shock the industry, but what?

We certainly weren't getting any clues from the forthcoming offerings of the British industry. I knew what Chrysler was planning, British Leyland was trying to pull themselves out of the quagmire of numerous iterations of the Mini formula, and while the Austin Maxi was at least a hatchback, it still had the typical Alex Issigonis signature that he had stamped on all his products. Ford and Vauxhall both followed routes that, except for Ford's Capri, produced safe, classic, front-engined, rear-wheel-drive sedans and estate cars, which were surprisingly similar in their execution and style – as seen in the respective Escort/Viva and Cortina/Victor models.

The British car industry was stuck in a rut, and as Martin and I had witnessed during our trip to Turin, in Europe, the Italians were still the masters in automotive design, with German companies such as BMW and Opel catching up fast.

Britain, however, still had one highlight, in the form of a vehicle that would start a trend that would change the industry forever. In 1970, the Rover Car Company introduced the Range Rover, a form of up-market off-road vehicle, not dissimilar to the Ford Bronco that Peter Stevens was driving around, albeit much more civilised.

The chief designer of Rover was a gentleman named David Bache. Although at this time Rover had no sponsoring commitment at the RCA, David Bache visited the department from time to time. He was exactly what you would expect of the head of the Rover design department, being very British and always wearing a well-cut Savile Row suit. His haircut was like his personal signature, being always immaculately combed back to resemble a Roman legionary's helmet.

It would be true to say that Bache played the key creative role in shaping all the postwar Rover models, right up to the final Rover created under his leadership, the SD1. It was rumoured that Bache had been so enamoured by the Ferrari Daytona, that he got one into the Rover studio and modelled an exact copy of its body section for the SD1. Not sure if anyone has checked if that was the case or not. Later in his career he led the newly formed British Leyland design department, with the Austin Metro and Maestro being the last cars developed under his leadership.

In the case of the Range Rover, it would be wrong to say it was purely a Bache creation, since the basic design layout was handled by Spen King. However, Bache did apparently play a role in refining the final design.

In Reading, I was lucky enough to sample the first model sold not just in Reading, but probably the whole of the South of England, and its buyer couldn't have been anyone else but the father of my school friend from my days at Crosfields: Michael Baylis.

His parents had moved to a large country house with acres of land in a small village in the Chilterns, and a Range Rover was tailor-made for this location. Michael called me excitedly and I drove over to sample a first ride. Upon arrival, Mike was already putting the Range Rover through its paces on a nearby meadow. By today's standards the first Range Rover was fairly basic, but nevertheless, seeing it that day, resplendent in its bright red body colour, it was like a creature from another planet compared to Rover's other off-road offering, the Land Rover. After the Mini and the Jaguar E-Type, it was clear to me that this was another British design that would surely go down in history as an automotive landmark.

I took my place beside Mike while he drove up and down the field, laughing as we had years ago in the playgrounds of Crosfields. We drove back up to his parent's impressive house and said our goodbyes. Michael Baylis was a wonderful bloke and a good friend – little did I know that that meeting would be the last time I'd see him. Those days, with no social media or mobile phones, friendships came and went so easily, and many of my Reading contacts slowly drifted out of my life.

1972 was winding to a close, and by now all the second-year students had decided on their final projects for the 1973 degree show. Martin Smith had chosen to develop a six-wheel chassis capable of taking two body styles, one of which was a low sporty body that was not dissimilar in its styling to the Pininfarina Modulo concept of 1968, while the other was a more practical SUV body, to answer the critics who might have labelled his project as being too styling orientated. Jean Nahum stayed faithful to his car for Turkey, while Dave Evans chose a sporty hatchback that he felt would please the powers that be at Chrysler UK.

I had decided to develop a multi-purpose car, featuring wide sliding doors on each side for ease of access, and removable upper rear side panels, which, when taken off, converted it into a kind of pick-up truck. It would feature a flat floor and have removable seats to make space for transporting large loads. It wouldn't be as stylish as the offerings from Martin or Dave, but bearing in mind we had yet to see any kind of multi-purpose vehicle (MPV) such as the Renault Espace (which first came to the market in 1984), I guess you could say it was a forward-thinking proposal, if maybe lacking the wow factor that Martin's design was bound to have.

1973 came soon enough and the pressure of completing our final year projects began to build up. The process of building our own clay models was very time-consuming, since it was expected that the students handled all parts of the model building, right up to the final painting and preparation of all the detail work.

As I previously mentioned, the Automotive Design department had a model building studio in Jay Mews, right opposite the rear entrance of the college, which was specifically equipped for clay modelling work. An old double garage, it was equipped with scale modelling tables and an oven for warming up the clay. There was no heating as such, maybe an electric fire, so in February, it wasn't the most comfortable of places to work.

I say that the students were expected to develop the models themselves, however, on occasions, clay modellers from both Ford and Chrysler visited for a day to give instructions on the basics of building up a model.

Developing a scale model was similar in many respects to the process of building a full-size model. Taking measurements from the scaled tape drawing, an armature would first be built using plywood and a high-density plastic foam material. The foam would have a shape that was approximately 2cm below the eventual outer surface of the final design. Once the armature was firmly secured to the modelling table, hot modelling clay would be taken from the oven and spread evenly onto the foam and left to cool and harden. The clay was hot enough that it needed to be handled with gloves, and care had to be taken when applying the clay so that no air bubbles developed. In time, once the model was finished, they would eventually come to the surface and blisters could appear under the final paint finish.

In fact, the best way to avoid a blistering model was to take the route Martin Smith chose: construct fibreglass models produced from casts he had taken from a clay model. This involved double the work, but Martin had started to develop his designs before the Christmas break and was pushing ahead much quicker than Dave Evans and me. For me, a fibreglass model

*Renderings of my final-year project.*

was out of the question, but I felt confident I could get the finish required with a painted clay model.

Once the first layer of clay had been applied to the armature, pieces of plywood (cut to the curvature of the planned shape of the car) would be secured to the top surface of the modelling table. Next, templates of the curvature of the side glass and body, of the front and rear of the car, and a larger template of the side view centre line of the complete car would be cut.

The first layer of clay would be applied slightly below the measurement of the final outer surfaces of the car, before the final layer of hot clay was spread on top. Immediately, the templates would be dragged along the side, front and rear of the model. The centre profile was also finished in the same way. The result of this process was a 'blocked-in' model. This was by no means the final finished design, but was a clean, basic starting point for sculpting in the final shape of the car.

The next stop was to take measured points from the tape drawing, in all three dimensions, and mark them on the clay using a measuring gauge, which was basically a metal right angle with an adjustable measuring needle attached to it. The paper on which the tape drawing was done was squared, with the lines every 2.5cm, making it possible in all elevations – plan, side and end views– to measure a point on a line of the drawing and then transfer that position using the measuring gauge to an exact point on the model. The points would be marked with white paint, then with a scraping tool, the clay would be cut away to marked point, and an exact surface would be formed.

It was a time-consuming process that these days is achieved with a milling machine working from CAD data, able to cut surfaces on the armature in a matter of seconds. In 1972, it would be at least another 20 years before that technological breakthrough came along.

By now, the final form of the model would almost be complete and subtle nuances would then be worked in freely, similar to a sculptor finalising his work.

On automotive models there is a final stage that creates a perfect surface, and this is known as 'highlighting'.

If you observe the metal surface of a car's body, it is not the physical surface you see but the light, shadows and reflections it creates. A good form (not restricted to just that of a car) should have a surface that pleases the human eye, in other words, it mustn't be too flat or fat. Flat surfaces tend to look cheap and weak, while fat surfaces make an object look ungainly or heavy. Nature is, as usual, the influence of this dimension, and we all know the shapes that please us. With automotive design, it's no different, which really makes you wonder why there are so many bad designs out there.

In order to achieve a good surface on the model, chrome foil, similar to normal household baking foil, is applied to the clay surface of the model. A white board with black parallel lines of tape is then held opposite the surface. It is then possible to see the black tape lines reflected in the surface, and thus judge how smooth the surface is. Ideally, the lines should remain parallel and change their flow over the surfaces in a smooth, even and progressive manner. Using this technique, it is possible to make minute adjustments to the clay surface by either scraping away or adding clay if required. It's a very fine art, working in fractions of a millimetre.

The importance of the highlighting process in automobile design should never be underestimated, as it's really what separates a quality design from a run of the mill product. In addition to the surface sophistication of a car, proportion and the quality of detailing are also of major importance before you even think of the visual signature. Today, so many manufacturers are in a hurry to meet deadlines that they just don't have the time to spend on maturing these important disciplines and it always shows in the final product. Product design also benefits from this attention to detail, Apple being a good example of how massaging an object until it's perfect in detail, touch and feel, can really single it out as coming from a quality manufacturer.

Spring and Easter approached and both artwork and models for the degree show would have to be completed by June. As with the previous year, the two first-year students Lionel Chew and Peter Horbury were pulled in to help. Both had fitted in well, the harmony of the course was good, and things were moving on well.

With my clay model finished, I had decided to paint it in a VW colour called Apple Green; it was a popular colour on the Beetle model. After applying a primer, I sanded down the surface ready for the final coat.

Getting a slot in the college spray booth was difficult since the School of Industrial Design were also painting their models. When I finally got in, I started laying on the bright green colour. It was looking good, and with the help of some students, I got the freshly painted model back to the fourth floor studio ready for the final detail work. As the degree show approached, I felt good too.

Soon, the familiar orange awning was assembled in the inner courtyard, where Automotive Design had its display. There was a buzz of last minute preparation throughout the college. Artwork was mounted, finishing touches were made to the models, and they looked good as they were positioned on their plinths for the final assessment.

The times for the final assessment, which would be carried out by the course professors and industry representatives, had been set. From Chrysler, it was Roy Axe who would attend. The first part of the morning on assessment day had been reserved for a special visitor; Her Majesty the Queen's sister, Princess Margaret, was to visit the college and review the work.

As a prelude for Her Royal Highness' visit, the students were briefed on how to greet and address Princess Margaret. It went roughly as follows: when Her Royal Highness approaches, men were to bow from the head and women should make a small curtsey. After this, you should then offer your hand, but rather than grip hers in a normal handshake, clasp your hand like a downward facing hook. Apparently, the royals shake so many hands each day that they need to avoid the pressure exerted by a normal handshake, since, with time, it could damage the nerves at the side of their hand. Once the exchange of hands is done we were to introduce ourselves by saying, "Good morning, Your Royal Highness, my name is ... It's a pleasure to meet you." After that it is up to her whether she wishes to continue any exchange or conversation, in which case you should subsequently refer to her as 'Ma'am.'

We waited as the royal entourage approached, and when it was my turn I duly addressed Princess Margaret as instructed. She asked me to explain my project, made a polite comment wishing me success, and moved on to the next candidate. I have no strong opinion one way or the other concerning the royal family, only that I don't envy them, but that encounter *was* still a special moment for me.

After the excitement of the morning's royal visit, the assessment exam in the afternoon seemed a formality. The jury took time to view the work as I explained my model and answered questions. It went well, and we concluded the day at The Queen's Arms, thankful it was all over. Now all we needed was an offer of employment.

The following day was the so-called 'private view,' an invitation only preview of the degree show, where industry representatives had the opportunity to meet the students and view their work before the college opened the doors to the public. For the students, this was more important than the assessment, since here you would hopefully meet the source of your future employment. Roy Axe had expressed himself satisfied with both my and Dave Evans' work, but without going as far as offering us employment, but maybe that wasn't his job. As I stood by my work, however, I noticed that no other representative from Chrysler UK had shown their face. The business for Chrysler in Europe had not been going so well, despite the reasonable success of the new Hillman Avenger. The Chrysler 180 was not a sales success and the Chrysler name was not working

*Completed clay model.*

*My completed model for the RCA degree show of 1973.*

for them. I believed I was in with a good chance of employment but needed to cover all my options.

These were the thoughts going through my head when I looked up to find a tall and distinguished-looking businessman, I guessed in his late 40s, wearing a double-breasted, pin-striped suit and sporting a full head of silver grey hair approaching me, smiling with an outstretched hand.

"Brian Adcock, Vauxhall Styling." He pronounced his name 'Braan.'

He was quick to point out that he had also studied at 'The Royal,' and that his boss at Vauxhall had asked him to visit the degree show to talk with any graduates who may be interested in joining Vauxhall Styling. (He called it styling, I noted.) It wasn't long before Carl Olsen came over and greeted Mr Adcock. Apparently, they knew each other since Carl's wife was the daughter of the recently retired Vauxhall design director David Jones. Carl put in a good pitch for me and after I had gone through my work, Mr Adcock got straight to the point and asked if I had time to attend an interview at Vauxhall's styling department in Luton. Without hesitation, I said yes.

"Splendid, we'll contact you concerning a date."

I was elated. Vauxhall and General Motors had a reputation in the industry for hiring the best designers, and there was a certain mystique about the work that went on in their studios. Suddenly I felt my allegiance to Chrysler slipping away. But I needed to temper those emotions – I wasn't there yet.

The remainder of the private viewing took its course, and for the rest of the week the public would see the show. My parents visited and viewed my work proudly. It was also good to see Leo Summers, my department head from the Berkshire College of Art, who had come to the show with a large group of students. Friends came by as well, and suddenly it was over. I had finished my education and was ready to face the working world.

College was not officially over, but for the second-year students the wind was out of our sails. I had yet to hear from Vauxhall. Martin Smith was being secretive about his career chances until he suddenly disappeared for a couple of days. Upon his return, he revealed that he had applied for a job at Porsche in Germany, and had been made an offer. He painted an amazing picture of his short visit: the head of design for Style Porsche, Tony Lapine, had driven with Martin in his special Porsche 911 to the Hockenheim race circuit, where the Porsche racing team was practicing. He could start in July. "Yes," he told us, "I will get a Porsche company car" – not true, as it turned out, and Martin knew it, but he liked to wind us up.

Dave Evans and I had yet to hear from Chrysler, but Dave said it was just a formality, and he was certain we would be going to Coventry.

I returned home to Reading where a letter from Vauxhall was waiting, asking if I could attend an interview the following week.

The Vauxhall development centre and nearby production facility were located to the south of the town, stretching from near the centre up towards Luton airport, which, positioned on a plateau, overlooks

*The degree show stand with finished renderings and model.*

the town. The engineering and styling buildings were located in Osborne Road. I'd set off early from Reading and arrived in good time. I pulled into the visitors car park and walked to the nearby entrance and reception desk. On my way, I noticed a row of cars parked in front of the main office building. In the number one slot was a dark blue Vauxhall Ventora, Vauxhall's flagship model. It was lowered and sitting on aluminium wheels that had a fine spoke design almost like a wire wheel; I'd never seen anything like them. I noticed the name on a sign in front of the car: E Taylor.

I walked up to the front desk and told them that Brian Adcock was expecting me. It occurred to me that apart from Mr Adcock I knew little about Vauxhall design – or styling as they preferred to call it. The copy of *Style Auto* with images from the development of the SRV concept car and Viva/Firenza models had signatures from Chris Fields and two others whose surnames were Taylor and Lawson. The designer, Ken Greenley, who had done the Pontiac sketches I had borrowed from John Camb while at Maidenhead, was also still working there. David Jones, the Englishman who had overseen the building of Vauxhall's new styling building had retired and his replacement was an American from the General Motors tech centre in Detroit.

I didn't have to wait long, and Brian Adcock came down a wide staircase that dominated the entrance lobby. He welcomed me enthusiastically and asked me to follow him. We climbed two storeys and went down a corridor to some double wooden doors without windows. There was a small sign that read "Vauxhall Styling." Mr Adcock pushed a card into a small slot and with a click the doors opened. On the other side, the corridor continued very much in the same style. We passed a number of offices and went through some further glass doors where a secretary was seated. There were a few more offices on the left and we entered one with Brian Adcock's name on a small plaque outside. I took a seat.

Mr Adcock explained the personnel structure of the department. He was responsible for the administrative side of things and in spite of his design background was no longer involved with the day-to-day design work in the studio. The incumbent design director was an American called Ed Taylor. I wouldn't be meeting him today, but instead his assistant, another American called Wayne Cherry. Mr Adcock picked up his phone, pressed a button and spoke:

"Wayne, I have Peter Birtwhistle here for you, can I bring him in?"

We got up and walked out of his office. Two doors down the corridor was another office, this one with Ed Taylor's name on the plaque. Mr Cherry would be using this office but had yet to arrive. Mr Adcock asked me to get my work prepared and, as he left, said Mr Cherry would join me soon. It was a spacious office, and looking through a large window with textile blinds hanging down, I could just make out a long outside terrace, which I realised was a model viewing yard. Unfortunately, there were no models to be seen. The office had no boards on its walls to hang my work on, so I decided to place them on the floor. No sooner had I done that the door opened and Wayne Cherry walked in with broad smile on his face.

"Peter, great to meet you."

He was a tall, good looking man who reminded me a bit of the Spaghetti Western cowboy start Clint Eastwood. Casually dressed, I felt immediately comfortable in his company.

"What wheels have you got?" he asked.

I was confused. He clicked immediately.

He tried again. "What car do you drive?"

I felt foolish for not understanding him the first time.

"A Triumph Herald."

"My God. Out of choice?"

Things hadn't started well. Would I be refused the job because of the Herald?

"I'm only joking," he added, the broad smile still on his face.

We got to business and, kneeling on the floor in front of my illustration boards, I took him through my work. He listened attentively and made some comments, and I started to feel at ease.

"Great, thanks, Peter."

When we returned to our feet, he said, "I like your work. When can you start?"

"First of August?"

"Fine. Welcome to Vauxhall! Sorry I have to go, but Brian will answer any questions. I'm looking forward to you joining the team." And he left.

Almost immediately, Brian Adcock bounced back into the office.

"I had a feeling Wayne would like you," he said. We returned to his office to discuss the terms of employment. They would send me a formal invitation of employment, to which I had to reply within a week, but I had already made up my mind: this was where I wanted to start my career in car design.

I returned to the Herald and made my way back to Reading.

There was one last meeting for all the graduating students before the college closed for the summer break: the graduation ceremony. I had passed the final exam and though I was graduating alongside my coursemates, there was regretfully one small difference: the academic level I had achieved at the Berkshire College of Art – SR Dip AD – was not recognised by the RCA, and for this reason they would only be awarding me a Diploma of Art and Design, rather than the Masters that Martin, Dave and Jean were getting. I had known this all along, and had agreed to it when I signed up for the course, but there had still been a small hope that in the end they might bend the rules. It wasn't to be.

In June 1973, in the Gulbenkian Hall, I walked up to the stage, with, as was the tradition, Professor Misha Black at my arm, to receive my Diploma from the rector of the college. My parents looked on proudly. The years of hard work had been worth it, and the long path of education had come to its conclusion. I was now ready to go to work.

Before we continue, now would be a good time for a dictionary of sorts, to explain all the specialist terms in automobile design used to describe different areas of a car:

*Standing with my proud mother after the 1973 RCA graduation ceremony.*

**DLO (daylight opening):** This refers to the side window area of the car.
**Valence:** The area below the front and rear bumpers.
**Belt line:** This is the line that separates the cabin of the car from the body, running along the lower edge of the side glass or DLO.
**Shoulder line:** This usually refers to a feature line running down the entire side of the car body at its widest point. It may be a hard crease or just a soft fold in the metal.
**DRG (down road graphics):** This is the look of the car when viewed straight from the front. It also applies to the impression you get of a car when it appears in your rear-view mirror. Prestige car manufacturers always play on the importance of this for the so-called 'overtaking prestige' on motorways.
**Green house:** The cabin area of a car: roof, front, side and rear screens.
**Plan shape:** The shape or curvature of the body when viewed from above.
**Top-hat:** The entire upper body of the car excluding the lower platform.
**Platform:** The entire underbody of the car onto which the bodyshell is mounted. It may also include the engine and suspension.
**Shut line:** A line in the body defined by all openings such as the doors, hood or trunk.
**Rocker:** This is the section of metal that runs under the side door opening between the wheel housings. It's basically what you look down at when you open the car's doors.
**Air duct:** Usually an opening where air is channelled into a part of the car.
**Fire wall:** This is the metal pressing that separates the engine compartment from the interior cabin of the car. The lower part of the windscreen is fixed to the top of the fire wall. That in turn is known as the cowl point.

Finally, it's worth noting the differences between jargon in the US and that in Britain:

Front hood (US) = Bonnet (GB)
Trunk (US) = Boot (GB)
Rear hatch (US) = A tail gate that also includes the rear screen (GB)
Wagon (US) = Estate car or shooting brake (GB)

# Chapter 11

## *Droop snoots and flutes*

**Sundon Park is a residential area to the north of Luton. I had called from a phone box about a room to let and the man on the other end (he sounded young), said I should come over to have a look. I drove from the centre of Luton, finding the address on a typical new housing estate. I pulled up outside a modern detached house with a garage. There was a red and yellow Vauxhall Viva van parked in front of the garage doors.**

I was met at the door by a slightly built guy with shoulder length hair and a fair beard – I guessed he was around my age. We shook hands, and he introduced himself as Martin Sellen. He motioned for me to come in. As we walked into the sitting room I immediately started to inform him about my new job. Smiling, he said that he also worked in the development centre at Vauxhall, in the engineering drawing office. We went upstairs to see the room, which was tiny. It would do, though. I only needed to sleep there. We went back down to the sitting room, and, as he seemed a nice enough guy, I said I'd like to take it. He agreed to telephone in a day or two to let me know. I didn't have to wait that long, since upon my return to Reading that evening, he'd already left a message for me to call him; I could move in.

I didn't feel like taking a vacation in the remaining weeks before I started at Vauxhall, so I spent most of the time meeting up with the few friends I still had contact with in the Reading area. While my younger brother, Chris, still lived at home, Roger and Helen had moved out of the house. Roger was finishing off his studies of Landscape Gardening and Helen was studying Nursing at the John Ratcliffe Hospital in Oxford. Chris was going on a canal boat trip with my parents and I was glad to have the house to myself, but I was counting down the days until I started my new job.

Finally, August was knocking at the door, so I put my hi-fi set and a suitcase in the Herald and set off for Luton. The drive was only about an hour and a half, so I knew that I would most likely still be coming back to Reading for the weekends.

I arrived at Sundon Park in the early evening. Martin (I could call him 'Mart') was watching television so I joined him for an hour or so. Mart wasn't a great conversationalist, so I decided to turn in; I would listen to some music and get good night's sleep.

The next morning, I left early. I had no idea how busy the roads in the area would be and did not want to arrive late on my first day. Turning into Vauxhall's engineering block car park, I encountered my first problem. A rather zealous uniformed man – he looked like a pensioner – waved me down.

"Where do you think you are going?" he asked. It turned out that I would need a works pass to enter the parking area. I had no alternative but to turn around and drive up a street opposite. Everything was limited parking and so I kept driving. Eventually, the limited area stopped, but it was at least a 20-minute walk back to the centre; this would need sorting out.

I made my way back down to Osbourne Road and walked up to the main reception, Ed Taylor's (Vauxhall's design chief at the time) blue Ventora, with its distinctive wire-look alloys, was already in its parking spot. The receptionist called styling, and said someone would come and meet me.

It wasn't long before a guy, who I assumed to be a designer, came up to me. With long blonde hair, he was wearing a light yellow summer suit with a bright yellow tie to match.

"Hello, Peter. Ken Greenley."

We shook hands, and I nervously began to explain all the things I already knew about him: the Pontiac drawings of his I'd seen at art college etc. We made our way up the stairs and into the styling department. I followed Ken and we passed through the double wooden doors that separated the outside world from the inner secrets of Vauxhall Styling. He directed me into the first office we came to on the right. A sign by the door read, "Administration." I was going to need a door pass. There were three guys sitting at desks: Kelvin Holland, Derek Baxter and Mel Armstrong.

Mel Armstrong greeted me by saying, "Congratulations on your fifth place."

He was referring to the place I had won in the *Daily Telegraph Magazine*'s styling competition.

Each year, the colour magazine supplement that came with the Friday edition of the *Daily Telegraph*, ran a design competition to design a car. As a prize, the winning contestant would have their design built. That year, the design was to use the Austin Maxi as a base. The competition was open to anyone who wished to submit drawings, including designers from the industry. It had occurred to me that the design I was working on for my degree at the RCA was very close in size to the Austin Maxi, so I submitted some illustrations of my work. As it turned out, my entry was selected in the top six, taking the fifth place that Mel Armstrong was referring to.

Vauxhall designers had done well in the competition, with Chris Fields (who had since left the company) winning the competition, and Mel Armstrong taking third place.

The results had been announced earlier in the year, before I had signed up to join Vauxhall, and Chrysler UK, which I was then an employee of, had rushed a photographer to the college to take pictures of me for the company magazine. Despite his third place in the competition, Mel Armstrong was no longer working on the front line of Vauxhall Styling, instead working in

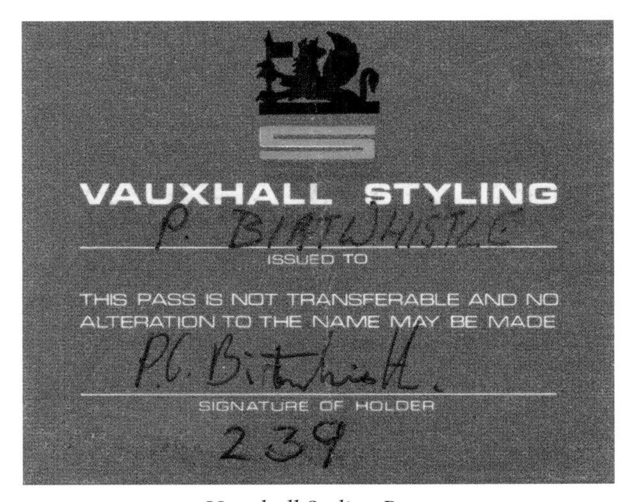


*Vauxhall Styling Pass.*

administration, but having achieved third place in the competition, he'd obviously not lost his touch. Anyway, now was not the time to discuss the politics within the department, and Ken Greenley was keen to get me into the main studio.

Having picked up my door pass and company ID, we moved on to another set of doors located close to the row of executive office where I had attended my first interview with Wayne Cherry.

The Vauxhall development centre had been built in the early '60s. It comprised of three key areas: a front four-storey office building where the engineering department was located and where my new landlord, Mart, worked. On the ground floor behind the offices was the so-called 'running shop,' where prototypes were constructed and worked on, and it was above this area that the styling department was located. The main styling area had two key areas each side of a central corridor. To the left when entering, were the large design studios, including a presentation hall which faced out onto the outside viewing area, and on the right were the workshop and fabrication areas, including a trim shop for making seats and interior models, plus a paint booth.

It was an Englishman named David Jones who was the driving force behind setting up the new Vauxhall Styling centre. Also a graduate of the Royal College of Art, he was a contemporary of the renowned sculptor Henry Moore. Jones had made his way up in the Vauxhall design organisation, and was design director from 1937 to 1971, the longest a person in the car industry has held such an office. He will go down in history as the designer of many of the key postwar Vauxhall models, but it was perhaps a commercial vehicle that was his most famous design: the Bedford CA delivery van, with its sliding entry doors. Not only was the CA a huge sales success as a delivery van, it was also a favourite vehicle for ambulances, camper vans or dormobiles.

Stories have it that Jones always considered the occupation of car design as a true romantic art form, and to create the appropriate atmosphere within the confines of the new Vauxhall Styling department, he had two Dalmatian dogs purchased. These two ownerless creatures would be able to walk gracefully around the studios – well, that was the idea anyway. In addition to this, he arranged for a pigeon coop with white doves to be installed in the outside viewing area, with the hope that, like the Dalmatians, they would add grace and style to the department's ambiance. Needless to say, the idea didn't quite go to plan, and with no one specific to train them, the Dalmatians ran around out of control while the doves left trails of droppings all over the viewing yard. Neither dogs or doves remained long in the department.

As a result of a car crash in 1967, which had a lasting effect on his health, Jones finally retired from Vauxhall in 1971, making way for Ed Taylor, an American and GM veteran.

As Ken Greenley opened the main doors into the design area, I had yet to meet Ed Taylor, but that was the last thing on my mind as we entered the main corridor, which stretched the entire length of the department. If David Jones hadn't quite achieved the atmosphere he was looking for with the Dalmatians, then the impact of this corridor was still full of drama. It must have measured about 100 metres in length. I immediately took in the amazing renderings that were mounted to the walls. Visionary concepts created, I assumed, by the in-house designers, including those of Ken Greenley and the other designer I had yet to meet, who signed his work with a blocked stylised 'Lawson', a name I remembered from the *Style Auto* magazine.

We walked on. At the far end of the corridor was a large backlit two-metre-square monolith, featuring the Vauxhall griffin symbol. In front of this was the SRV concept vehicle. Compared with the Midlands charm of the Chrysler design department in Whitley where I had spent the previous summer, I felt I had now entered a temple of style. The impact of it all was out of this world.

At a little over half way up the corridor we turned left through a door into a light-flooded hall. It was the main exterior design studio. Again, I was taken aback by the scale of Vauxhall's operation in comparison to that of Chrysler. The vast room was about 20 metres wide and long enough to accommodate at least three full-size clay modelling surfaces in line. This meant, when considering the width of the studio, it was capable handling at least six vehicle models at a time. At the far side of the room were floor-to-ceiling windows, at least ten metres in height, which looked out on to the external viewing yard. If the light from outside was not enough, the ceiling was a surface of translucent lighting panels.

I was bombarded with the plethora of things to take in at once. Each end of the studio had twin sets of movable boards the size of a car, three-fold in depth, which could be moved upwards and out of sight. Some were covered in full-size, side-view, airbrush illustrations of a car, the others had engineering layouts of car bodies on them. The designers were seated at desks that ran parallel to the studio walls and on the wall opposite the viewing yard were further boards covered with evenly spaced renderings.

I noticed that the modelling areas were not metal plate surfaces in the studio floor, like those I remembered from Chrysler, but instead the clay models were secured to floor-mounted jacks and surrounded by a kind of wide steel rail, also fixed to the floor. On the rails were steel towers with measuring gauges secured to them. Several clay models, in various stages of build, filled the room, but the focus of attention in the studio were two fibreglass models placed to one side.

Around these models were a large group of design staff including both Wayne Cherry, and, as I would soon learn, Ed Taylor. Ken Greenley explained that the two models were the first hard styling models of Vauxhall's new Viva segment car, and featured full interiors. It was hard to make out the design of the vehicles since they were surrounded by both designers and fabrication staff. Observing the group it was easy to separate the two, since all the designers appeared to be wearing suits and ties, and the fabricators were clothed in various forms of overalls. The clay modellers, I would later learn, wore a uniform of brown trousers with beige coloured tunics.

Despite the obvious distractions, introductions were made. Wayne Cherry came over with Ed Taylor, who, slightly shorter than Cherry, was wearing a double-breasted light suit. He had a Beatle-style haircut with a moustache that made him resemble the film star Burt Reynolds. Other designers came over to say hello; John Sowden and John Heffernan were from the interior design studio, which was run by yet another John – John Hartnell. The exterior team comprised Ken Greenley and Geoff Lawson. So this was the face behind that signature, I thought. Geoff Lawson was also an RCA graduate and had studied Furniture Design, as he had attended the college prior to the Automotive Design course being founded. He was friendly guy, and, as I soon learnt, my first allocated desk would be next to him.

More designers came up: Dave Garrard and David Reibscheid, an Israeli who had graduated from the famous ArtCenter. They were both working in the commercial vehicle studio. American designer Karl Fischer also said hello, together with some of the studio engineers: Eric Tooley, Charley Nurse, John Boughton, Morris Franklin, plus a few more.

There was no doubt, with a total staff of around 200, this was a big operation, and it had to be. At this time, Vauxhall had two main car lines plus their derivatives, including the Viva/Firenza and the Victor/Ventora range. As well as this, it had its full commercial division of Bedford, featuring large trucks down to the CF and Viva vans. The Bedford truck division was then the biggest in Europe, with huge contracts, one being with the British Army.

The crowd began to disperse from around the two models, leaving Ken Greenley and John Heffernan to explain them to me. There was a four-door limousine in metallic blue, and a coupé in a striking Candy Apple Red. These were the intended replacements for the Viva model and its coupé derivative, the Firenza, although they appeared a bit bigger in size compared to the models they were replacing. Apparently, the underpinnings of these cars were not based on those of the Vauxhall Viva model, but on those of General Motor's German partner Opel's models, the Ascona and Manta coupé, which explained the increase in size. The so-called 'top-hat' or body of both cars, would, however, be individual to Vauxhall.

I was impressed; both cars looked great and it was noticeable that neither featured a grille opening between

*Planned replacement for the Vauxhall Viva based on the Opel Ascona body.*

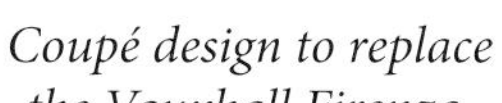

*Coupé design to replace the Vauxhall Firenza.*

the headlamps, with air being taken in below the front bumper instead. This was a front-end look usually associated with exotic sports cars; indeed the blue sedan model had a front design that looked uncomfortably close to the Ferrari 365 GT4 that I had seen on the production line at Pininfarina one summer ago. As it turned out, Wayne Cherry was a Ferrari fanatic and had recently purchased a Ferrari 275 GTB. The Ferrari influence was clear to see.

The coupé also featured the grille-less front design but was much softer in its character. There was no visible front or rear bumper, with the whole front of the car being a plastic cap, and the lower area below the air intake featured a shovel-like feature that was called an air dam or spoiler, a detail only seen, up until now, in the motorsport world. The upper area of the car featured a wide middle pillar behind the side door, and a rearward sloping flying buttress pillar similar to the treatment on some of GM's American models such as the Pontiac Le Mans.

Having inspected the models, Ken Greenley, who, as it turned out, was the most senior designer in the large exterior design studio, took me over to the desk they had arranged for me, next to Geoff Lawson. Immediately, Geoff suggested we go over to the material store to get some sketch pads, plus all the drawing equipment I would need. At the store, I felt like a kid in a toy shop, since everything an artist or designer could possibly need was there. I selected some pads and felt-tip pens, plus a selection of Prismacolor coloured pencils and pastels; I was set.

I had hardly been a minute back at my desk when Wayne Cherry came up and said he'd like to take me round the studio and show me what was going on.

Talking as he walked, he explained that in Europe General Motors was going through a transition period. Up until that time the company had been producing cars in England under the Vauxhall name, predominantly in right-hand-drive, and therefore exclusive to markets where you drove on the left, so basically all the previous British colonies. The cars from GM's subsidiary in Germany, Opel, had nothing in common with Vauxhall, which just didn't make any business sense, and was a situation that GM now wanted to change.

We moved from the large studio and through a door into the viewing auditorium. This was a large presentation hall equipped with turntables, and used primarily for design reviews. Placed on one of the turntables was a red car smaller in dimension to the Viva model. It had a similar front-end treatment to the two fibreglass models I had just seen, in that it also featured no upper grille or air intake between its headlamps. A two-door, it had a sloping rear end which I recognised as the new hatchback trend. In other words, there was no conventional boot but a rear door that swung upwards, allowing access to the whole rear luggage area behind the rear seat. The rear seats could be folded down flat, increasing the load capacity much in the same way as an estate car. Wayne (everyone was on first name terms, including the executives) explained that this would be the future layout for all small cars, and that Vauxhall wanted to be one of the first on the British market with this feature. It was, of course, nothing new in mainland Europe, where the Italians and the French had pioneered the hatchback trend.

Looking at the back of the car, I noticed the badge: it was called Chevette. In addition to the fresh layout of the car, Wayne explained that this new model was the first of a range of new Vauxhalls that would have some commonality with the Opel products, and in this case it was to be based on the underpinnings of the next Opel Kadett. It was what he described as a 'world car,' where derivatives would be built by GM globally, thus cutting down on development costs.

I took in the design, and thought it wasn't bad, but looking at the front end I noticed they must have run out of money, since the headlamps were not flush with the body, but set back in square plastic frames. Wayne explained that they had to watch the costs carefully on this car and that the front screen, side glass and doors were the same as the Opel model it was based on. The focus of my attention, the headlamps, were in fact borrowed from the Vauxhall Viva. This, he stressed, was the only future model in the Vauxhall range to share common external parts with Opel and Vauxhall. I took his word for it.

We carried on through the department into the interior design studio, where John Sowden and John Heffernan were working. The studio also included an area dedicated to colour and trim material development.

Moving across the main corridor we entered the fabrication workshops. Wayne explained that Vauxhall built all its own hard material prototypes in house, which included the manufacture and covering of all the interior components such as seats. Leaving the workshop area, we moved back into the final design

*Design model of the planned replacement for the Victor. (Vauxhall Heritage)*

studio, which was responsible for Bedford truck and commercial vehicle design.

The studio was dominated by a hard model of a massive articulated truck unit, sitting on three axles: front steering wheels and twin rear drive axles. It was pretty squared off in shape, with a very dominant black grille, but looked modern. This was a truly huge truck, with a cab that featured a rear area compartment with full sleeping accommodation for the driver. It was simply called the TM and would be Bedford's new flagship. Looking around, I noticed other smaller derivatives of this design being worked on, so Vauxhall was obviously keen to strengthen Bedford's presence in the truck market.

Wayne moved back into the main studio, where my new desk was located, but he hadn't finished. At the far end was a large clay model being worked on by several clay modellers. I had taken it in earlier but did not wish to step too far on my first day by going over to look at it. We walked over. Wayne explained that the model was the first idea for Vauxhall's new large car, intended to replace the Victor range. Although in clay, I could detect all the same design features I had seen on the other new models, but most notable was the fact that it was a fastback design, something unheard of for this class of car. As of yet it had no name, but Wayne was quite clear: it wouldn't be called Victor. Until a name was chosen, it was referred to by its engineering programme number V78 – the V standing for the body class or size and the 78 for its launch year. This meant that the model had another five years of development before it reached the showroom. Another feature about the V78 that Wayne pointed out was that it would be the first of a new range of GM cars to feature a Wankel rotary motor. Over the next decade GM planned to replace all its vehicles with Wankel-powered models, a trend that could also be observed in other companies such as Citroën, Audi NSU, and Mercedes.

By now, my head was so full of information and I could hardly think straight. There was, however, one last surprise. Looking out of the vast glass windows that separated the studio from the viewing yard, I could see a car being driven into sight. It was silver and from where I was standing, the front end looked vaguely similar to a Ferrari Daytona or Renault Alpine A310. We walked out into the sunlight and towards the car, which by now had come to a halt. The driver's door opened, and a grinning Geoff Lawson got out. Approaching, I could now see that behind the sloping front end was the coupé body of the Vauxhall Firenza. The car was transformed, and it looked amazing.

The Vauxhall Firenza was originally intended to be Vauxhall's answer to Ford's massively successful Capri. Geoff Lawson's original sketches of the Firenza featured in the 1971 Style Auto magazine, featured not only a sloping rear coupe body but also a smooth aerodynamic front-end, different to the convention Viva model. Designer sketches from development programmes used in publications are very often done after the design has been completed, so I guess those Lawson sketches may not have reflected the true story. However, it did appear that, originally, Vauxhall wanted a bespoke front for the Firenza, as well as coupé rear end. As was often the case, cost considerations got in the way and the final model retained the same front end as the Viva model. In design and image terms, the Firenza couldn't compete with the Ford Capri, with its legendary 3-litre top-of-the-range model. Vauxhall introduced a performance version of the Firenza called the 2300 Sport SL but it was no performance match for the Capri.

Seeing this model in the viewing yard, it looked as if Vauxhall had got the message and were eager to change the Firenza's image.

Sitting on wide alloy wheel rims, which had a vague resemblance to competition Minilite wheels, the rear of the car featured a small 'duck tail' spoiler mounted on the trunk lid, similar to the one on Porsche's wild 911 Carrera model. As it turned out, the car, which was named the Firenza HP (high-performance), was only planned as a limited production model – a kind of Firenza swan song, if you like – before the replacement for the Viva range was introduced a couple of years later.

Geoff was justifiably proud of the design, which he'd

*Designers at Vauxhall were encouraged to do 'fantasy' renderings for the entrance lobby. This was my first attempt.*

worked on under Wayne's direction. Despite its official name being Firenza HP, the car had already attained a nickname within Vauxhall Styling, owing to the sloping nature of the front nose. They called it the 'droop snoot,' a title that would stick with it even after the production launch.

I was quick to settle in at Vauxhall, and felt very welcome. There was a great team spirit throughout the studio, and being seated next to Geoff Lawson meant I had one of the best automotive designers and artists in the business to learn from.

Apart from the V78 Victor replacement, the development programme for the next mid-sized car was still very much a work in progress. The two hard models I had seen on my first day were still quite far away from production feasibility, and Geoff Lawson had the task of leading that programme; he asked me to help him with sketch input. This, then, would be my first work as a professional automotive designer.

Over the coming months, I would get to know most of the team and their individual characteristics. Most of the modelling team were local Lutonians who had learnt their craft either as tool builders within Vauxhall, or had been former makers in Luton's hat industry. The relationship between the designers and modellers could be compared to that of army officers and their foot soldiers – a common respect, but with very different training and backgrounds.

*Some of the Vauxhall Styling team. Extreme left, John Heffernan, fourth from right, Ken Greenley, with me left, behind Ken.*

Of the designers, John Heffernan, Ken Greenley and Geoff Lawson were a trio that I would soon count as friends during the working day. John and Ken shared a flat together in Harrow, while Geoff preferred to live locally in Harpenden. However, all three were slightly older than me, so socially I felt I needed to find contacts closer to my age.

Within Vauxhall, the designers, or stylists as they were known, were treated with a certain distance, not to say suspicion. We were artists who painted pretty pictures and it was the engineers who really designed the cars; we simply illustrated their ideas. Geoff thrived on this reputation and would do anything to upset the barrel.

He drove a white Sunbeam Tiger, which was the Ford V8-engined version of the Sunbeam Alpine. Each morning he would arrive at the works car park in spectacular fashion, driving through the gate and by the security guard who had stopped me that first day, with little regard for the speed limit. As a result, Brian Adcock was always having to deal with complaints. On top of this Geoff was dating one of the best looking girls in the engineering department, which only added to the jealously he stirred up.

The designers were poncy, overpaid, and had the choice of the best girls, which could be why, back at the house in Sundon Park, Mart was so uncommunicative.

Usually, I'd get home and Mart would already be making his supper, which he would put on a tray and eat in front of the TV – and there he'd stay for the rest of the evening. Meanwhile, another guy had moved in to share the house, though I can't recall the three of us going out to the pub or anything similar. Over time, Mart warmed up, but it was hard work and, as at the RCA, I lived for the Friday afternoon drive back to Reading and the weekend.

However, in the October of 1973, driving anywhere came to an abrupt halt. OAPEC, the organisation of oil-exporting Arabic countries, had proclaimed an oil embargo in response to the US decision to re-supply the Israeli military in the Yom Kippur war, and any US allies who also supported Israel in the conflict would also be penalised, which included Britain.

Almost immediately, fuel was in short supply and with it careful consideration about using a car. Massive queues formed in front of petrol stations, and the notion of driving back to Reading for the weekend had to be put on hold until things improved.

On the positive side, Mart and I shared the drive to Vauxhall and back each day, so a degree of communication eventually developed between us.

The 1973 energy crisis had a wide-ranging effect on not only world politics but also global industry, and in particular the car industry.

In the Vauxhall design studio there were soon mutterings of gloom and despair doing the rounds, but I was so low down the ladder that I didn't let it concern me unduly. However, in the higher echelons of General Motors, product strategies were being reconsidered, which would have a direct fall-out on the future cars Vauxhall developed.

I continued to assist Geoff Lawson with ideas for the next Viva, if that's what it was to be called. I pinned up my ideas next to his, but it was very clear that my contribution was being used as a means of developing my skills further, to a level where I could be relied on to design a project of my own.

Fortunately, I didn't have to wait too long before Wayne and Geoff asked me to handle design work for a special version of the Ventora, a model that was still in production.

Since the demise of the larger Cresta and Viscount models, Vauxhall did not produce a large car in the luxury segment, which was then dominated in Britain by Jaguar and Rover. In order to address this situation, Vauxhall was in talks with Panther, a small car company owned by an entrepreneur named Bob Jankel, with various ideas for cooperation. Panther produced a replica of the Jaguar SS 100 called the J72. Jankel wanted to produce a smaller roadster and had contacted Vauxhall in connection with supplying engines for this car. During negotiations, Wayne Cherry had become good friends with Jankel, and an idea formed for Panther to produce a luxury version of the Ventora. It would have a longer wheelbase allowing appropriate legroom for rear passengers, and would be powered by the V8 engine used in a Holden model (Holden was GM's Australian division.) Wayne was keen to push this idea and got approval from the Vauxhall board to make a design model for management review.

Wayne thought that this project was an ideal platform for me to cut my designer teeth and asked Geoff to oversee my work on the project. Although the interior design modifications were confined to the application of luxury materials such as wood and leather, the exterior included a new front nose treatment and some adjustment around the rear roof pillar to compensate for the extra cabin length. I set to work.

In order to develop a convincing design of the exterior, I needed to do a full-size tape drawing, and based on that a full-size airbrush rendering. I had done neither – it was clear I was being thrown in at the deep end.

An airbrush, for those unaware, is a pen-like tool running off compressed air that is used to spray coloured ink onto a surface. It's basically the same as when spraying paint, but on a smaller scale. When applied to a flat surface it allows the user to build up colour in gradation, thus enabling the creation of shades and shadows.

The key reason for the airbrush was to convince Wayne and his managers that, in a realistic illustration, I had a good enough design to go forward into a clay model.

Producing a full-size tape drawing, which would then be used as a guide for an airbrush illustration, was one of the key tools in the industry that could not be tackled at the RCA – there just wasn't enough space available. These lifelike illustrations, usually of the side view of a car, were once again the product of the big American studios, and there were legends about the lengths designers would go to in order to create a realistic vision of their design. Rumour has it that one designer set the illustration of his car in front of an autumnal background, even going to the extent of covering the studio floor in front of his work with real leaves. Yes, car design in the States was show business.

As I planned my design, I was not thinking about leaves or fancy backgrounds, with Geoff Lawson wisely advising me to keep things clean and simple and to worry about complex backgrounds at a later date. But before all that, I had to do my tape drawing.

When planning a side view tape drawing, and the airbrush illustration that follows, one has to consider that when looking at a car, you do not take in its full dimensions. Viewing from the side, you will focus on not the theoretical centre line that runs down its length, but instead on the parts of the vehicle body nearest to you – in other words, the front and rear corners and the outer edge of the roof and wheelhouse wings, which are also nearest to you. Simply said, the dimensions of a car when looking at it in real life are shorter and lower that the true external measurements of the car's body. This is basically because of the laws of perspective: everything beyond the points closest to your eye will start to diminish in size. The human eye takes this in when it looks at all objects and will send appropriate signals to the brain. Not only will the eye judge a car to be lower than it really is, it will also focus on the wheels and increase their dimensions. When illustrating a car, especially from the side, automotive designers will always make their cars lower and place them on sometimes disproportionally large wheels, this is because the human mind expects and likes it that way.

When Giugiaro looked at the illustrations that Martin Smith and I showed him in Turin during our summer vacation, and he cried "Technique Americano," he was referring to this trait in our artwork. The cars were low and sitting on massive wheels. The Italian design studios of the day handled their illustrations in a much more accurate manner. A side view would include the full length and height of the middle of a car, and the wheels and tyres would also be true to life, always appearing small. Even in the side view of a car, perspective will always play its part. Believe me, the images of the Italian studios, despite their honesty, just didn't look right.

This form of artistic 'adjustment' that only the Italians declined to use, is known in the car industry as 'cheating,' and it applies today just as much as it did then.

In order for me to commence my tape drawing, I first needed a copy of the technical drawing of the side view of the Ventora, or so-called 'master draft,' to base it on.

The master draft was hand drawn on a plastic drawing folio, which was rolled out on a large, flat drawing table. Using large aluminium curves and long strips of plastic, known as 'splines,' the body draughtsmen would create what was known as a wire frame, using measurements taken manually with the measuring gauges from the final approved clay model. It's called a wire frame because a line is drawn at 100mm (probably six inches in those days) intervals, representing the surface measurement in each so-called XYZ dimension, almost like the rib cage of a skeleton. The end result, in simple terms, looked like chicken wire. Draughtsmen would draw the surface contours using a plastic ink, and needless to say it was a priceless documentation of the car's body.

The three dimensions when measuring a car body are recorded as XYZ lines. This is what is known as the Cartesian principal, and dates back to technical drawing from the 17th century. Lines recorded for the car's length are referred to as X lines, Y lines are the width and Z lines the height.

When measuring a car body, a vertical line is established, usually through the centre line of the front wheel; this is called the 0 line. Lines measured for the X (length) dimension that are behind the 0 line are in plus, and those in front of it are minus. So, for example, taking a point on the body surface that is 1230mm behind the centre of the front wheel is known as X 1230.

The same principal applies to lines measured for the Z (height) and Y (width) dimensions, thus enabling the body engineers to record the entire surface of a clay model as a series of 3D points, which they compiled in a book.

To maintain accuracy during the modelling stage, the clay models would have a set-up point embedded in the clay on a small metal plate that could always be referred to: for example, whenever the model was returned to the modelling plate after a design review.

Getting back to my drawings, I secured the technical drawing of the Ventora to one of the large boards at the end of each studio. On top of this, we fixed a sheet of vellum drawing paper, on which I would create my full-size tape drawing.

The process of doing the airbrush is not dissimilar to doing an illustration with markers and chalk, only it's done in life-size dimensions.

The outline of the car will be drawn first in tape. Once this is satisfactory, a further sheet of vellum will be put over it and on this the airbrush illustration will be done.

By masking parts of the image that had been created by the tape drawing, separate areas could be painted in gradations using the airbrush, until the final illustration was complete. Finishing touches would then be added in tape, with photos of the wheels applied to give further realism.

Due to the size of the illustration, it required quite a bit of physical dexterity, since you had to literally run up and down the image in order to create an even covering of ink.

Creating a full-size tape drawing is a true art form that today, with the use of computer images and full-size projection, has in many automotive design studios all but died out. The lengths to which designers used to go in order to create an unbelievable level of realism in their work is a now sadly a piece of automotive design history.

Having completed my first full-size airbrush drawing, both Geoff and Wayne Cherry gave me a nod of approval.

Various designers from the other studios gathered round. Smiles could be seen and pats on the back were distributed; I'd done a good job!

Looking back over my career, I went on to do numerous such works, sadly none of which have survived.

Wayne Cherry invited Vauxhall's then president, Bob Price, to look at the design in the studio. Price was a larger than life figure who had a good relationship with Wayne, and was supportive of his enthusiasm to liven up the Vauxhall range with special models. Nevertheless, his key job in Luton was to rationalise the product development in conjunction with Opel, as we later found out.

One has to question if Bob Price really thought there was a good business case for a long-wheelbase luxury version of the Vauxhall Ventora as a weapon against Jaguar, but at that point in time he gave the green light to move forward and agreed for a model to be built.

In terms of creating a model, clay work was confined to the front area of the car, so specifically the bonnet and front grille, plus the lower front valance. The rest of the car was based on the stretched body of an existing FE Ventora model. Other details on the car, such as lamps and bumpers, were also treated differently to the standard model.

A Ventora body had been lengthened and the various panels that were modified had been removed

*My first full-size airbrush renderings for the Ventora V8.*

and replaced by foam armatures on which the new clay was applied.

Since I had no experience working on a full-size clay model, Geoff Lawson took the lead and I supported him with further sketch work. Despite the limitations on what could be changed on the design, it didn't stop Wayne using this as a platform to develop what he considered to be a future key design element for all Vauxhalls, harking back to its glorious past – the fluted bonnet.

Perhaps the most famous of all Vauxhalls was the historic Prince Henry model of 1911. The Prince Henry was considered a very sporty car of the time and was instantly recognisable due to its distinctive sculpted 'flutes' at the top of each side of its front bonnet. It was a design signature that Vauxhall kept right up to the postwar models of the '50s, but by the time cars such as the Viva and Velox had hit the market in the late '60s, the famous Vauxhall flutes had all but disappeared.

Wayne Cherry saw this feature as a key part of Vauxhall's heritage, and wanted to reintroduce it as a design feature on all new models.

With the shadow of Opel design influence hovering over Vauxhall, it was important that the company sent out a clear design message of its own. To this end, the historic flute went on to grace the bonnets of all future models, and it started on the Ventora V8.

Geoff did a good job executing the design, and a final hard model of the car was completed. That, however, is as far as it went. Vauxhall had other priorities, and a car in this class just didn't make any sense. Nevertheless, to this day, I still have a photograph of myself and Geoff Lawson, together with the Vauxhall Styling archive chief, Derrick Baxter, seated in the back of the model, all looking suitably important.

Looking back at those times at Vauxhall Styling, the winds of change were blowing hard against this proud company in Luton, but nevertheless, under Wayne Cherry's leadership, there was a passion and wish for the company to carve its own way despite the odds stacked against it. Wayne Cherry wanted Vauxhall design to send out a strong message, even if in some cases it was just confined to some flutes on the bonnet. In 1974, he found himself with a world-class team to give him that support. Still, there was a hard fight ahead.

*Myself, Derek Baxter and Geoff Lawson seated in the back of the Ventora V8 prototype.*

*The styling model of the Ventora V8. It wouldn't go any further than this. (Vauxhall Heritage)*

# Chapter 12

## *The ghoster supreme*

**The 1973/74 energy crisis essentially halted the development of the Ventora V8, but it didn't end there. GM was making radical changes globally to streamline its organisation, and this had an immediate fall-out on the future models developed by Vauxhall.**

The two models I saw on that first day were known as the 'U' car programme and were based on the same platform as the Opel Ascona and Manta Coupé. The original plan was for Vauxhall to have its own top-hat on the Opel platform; in other words, completely different bodies to the Opel development.

News soon started to trickle into the department that Vauxhall would now take the same body as Opel and be confined to just badging the cars as Vauxhalls using the new model name 'Cavalier.'

As Geoff Lawson ironically put it: "Looks like our future job will just be putting chickens on the front of Opels." He used the studio's in-house term, 'chicken', with reference to the famous Vauxhall griffin badge.

Soon, the clay models in the studio would be remodelled to take the exact same form as the Opel sister car. As with Opel, the new Cavalier would have both a sedan and coupé version, plus a new derivative known as a sports hatch, which took the shape of a kind of fastback wagon.

Wayne Cherry, always looking for the positive side of what was a pretty disappointing situation, persuaded senior management that there needed to be more differentiation between the Vauxhall and Opel models, this despite the fact that the two brands shared very few markets. It was therefore agreed that the Vauxhall Cavalier sedan would take the same front-end design as the coupé version based on the Opel Manta. This fitted in with his strategy to have a common Vauxhall look, where all cars had a sheet metal panel between the headlamps, a look pioneered by the Chevette and 'droop snoot' Firenza. It was a token gesture, and it really began to look as if all future Vauxhall models would become rebadged Opels, a thought that worried me when considering my future as a designer.

Still, at that point in time things weren't all bad, and the model planned to fall above the new Cavalier, the V78, was still being worked on as an individual Vauxhall body design.

As with the U cars, the 'V'-bodied cars were also based on an Opel platform, in this case that of the Opel Record. The original plan for it to have a Wankel rotary engine had also been thrown out, as a result of the energy crisis. Wankel motors were super smooth and compact, but they drank petrol, so it was an easy decision for GM, as well as many other world manufacturers, to turn their backs on the Wankel engine, leaving only the German NSU and Japanese Mazda brands to continue with this engine in their products.

The sedan version of the V78 was still going ahead as a fastback hatch design, as opposed to the Opel model which was a normal 'three box' sedan design. What, however, was of real interest was that Opel not only based its Record model on this platform, but also planned a larger upmarket car to replace the aging Diplomat model, which would also have a coupé derivative. Once again, Wayne Cherry saw an opportunity for Vauxhall.

With little design work to be done on the Cavalier models, Wayne asked me to start looking at ideas for a V Coupé.

This was the opportunity I had been waiting for: a complete car to design from the wheels up, and I got to it with the first sketches.

Wayne wanted to develop ideas for a full-size clay model as his intention was to show the design to GM senior management when they visited Europe during the latter half of 1974, for the London Earls Court Motor Show and the Frankfurt Motor Show, that followed soon after. The executives visiting Europe

*Proposal for a Ventora-based coupé.*

*Full-size airbrush of Ventora coupé proposal – later to be called Monza Coupé.*

included the chairman of GM, plus all his immediate reports, but of most interest was GM's legendary design vice president, Bill Mitchell.

In automotive history, Bill Mitchell will be remembered as not only the father of some of General Motors' most significant designs, but also as a great showman.

His mentor had been Harley Earl, who, in the postwar years, had built the General Motors design centre (located at the famous Tech Centre in Detroit) into a true temple of American style. It was Earl who had driven GM's designers to create the winged dream machines that drew their design inspiration from America's space and aircraft industry.

Many believe Bill Mitchell was responsible for the finned giants that graced the driveways of America in the early '60s, and he undoubtedly directed their creation under Earl. Once he succeeded him in 1958, and took over overall leadership of General Motors styling, he would go on to develop a much more sophisticated design language, resulting in some true landmark designs.

Mitchell was the father of the famous 'Coke bottle' body language featured in the Chevrolet Corvair, and his masterpiece, the 1963 Buick Riviera. Probably of most significance, however, was his creation of America's first true sports car: the Corvette Stingray.

In GM's global design organisation, Mitchell was revered and held in awe, and his yearly tour of the design outpost of his design empire were events of unbelievable pomp and show. In the autumn of 1974, I would experience my first Mitchell review.

Preparation for the Mitchell review began months ahead of his arrival in Europe, and the lengths gone to in order to keep him satisfied were probably unmatched throughout the industry.

All the main studios had to be redecorated, full-size airbrush illustrations of future design proposals

*Bill Mitchell, GM Design president.*

completed, and there needed to be at least one finished, full-size clay model of a visionary concept – in this case the V Coupé. In addition to all this, models of all the future production programme were prepared.

Yes, a Bill Mitchell review was an occasion of military precision.

Everyone was on high alert as the final week approached before the Mitchell review, the whole department going into overtime mode. This was a time of the year that the modellers anticipated with great eagerness, since there was the potential to earn at least double their normal pay by working long hours.

For overtime, there were two in-house expressions that referred to the length of overtime being worked. If the work went on until midnight, it was called a 'ghoster,' this being rewarded by an extra half hour's pay for each extra hour worked beyond the normal eight-hour day. However, the icing on the cake would be the last night before the big review, when generally, most staff would work beyond midnight, and for every hour after that point, they would be on double pay. This was called a 'ghoster supreme.' The preparation for the Mitchell review that year was the same as always, and a 'ghoster supreme' was heading everyone's way.

The night before the review, I experienced my first ghoster supreme. All the studios were fully lit, music blaring from a car radio, while half asleep staff walked around in a trance. It was all pretty surreal.

Meanwhile, Bill Mitchell had already arrived in London the previous day. For his arrival, the truck studio had prepared a special surprise for him, in the form of special Bedford CF van. Fitted with a unique interior it had been painted in two-tone silver and was fitted with Wolfrace wheels. The CF was Bedford's bread and butter product, a delivery vehicle that was also popular as a base for Dormobile mobile homes. Its styling was similar to a Chevrolet van that was the favourite of the so-called 'trucker' cult in the US, which in England had yet to catch on.

For this special Mitchell van, the rear loading area, which had no windows, had been panelled off, and the side walls and ceiling covered in red leather. The floor was fitted with a deep pile red carpet, and in the middle of this were a couple of swivel captain's seats. The whole compartment was subtly lit, and to be perfectly honest, looked like a brothel on wheels. This was the vehicle to pick Bill Mitchell up at the airport and bring him to his hotel – it seemed a risk.

The designer responsible for Mitchell's personal transport was David Reibscheid, an ArtCenter graduate working in the truck studio.

On the day of Mitchell's arrival, Dave set off to Heathrow wearing a suit and chauffeur's cap, with Wayne Cherry joining him for the trip. After meeting Mitchell at the airport, Cherry and Mitchell would then travel together in the back compartment to his hotel.

Upon arrival, Mitchell came through customs, walking with the aid of a stick. Apparently, he'd had an accident on one of his motorcycles. Wayne escorted him to the waiting van but the step into the rear compartment was too high for him to move his injured leg, and so a taxi had to be ordered. Dave Reibscheid would drive the empty van back to Luton alone.

I must have only got a couple of hours sleep before returning to the studio early for Wayne's pre-review briefing.

Wayne called the designers together and we gathered around in a circle. Wayne quickly informed the team about Mitchell's accident, which had happened the previous week at the GM testing ground – nothing serious but he needed a stick to support himself. We were not to comment on this unless he brought it up himself.

Bill Mitchell was well known for his love of

motorcycles. He had a big collection, all of which he'd had customised by a small design team at the GM Tech Centre. The customisation went beyond new paint schemes and design modifications, with Mitchell also including custom crash helmets and leathers. One of his famous bikes was based on the BMW flat twin, and called the 'Red Baron' after the WWI German air ace, Baron von Richthofen. The BMW was painted red and had a black German Iron Cross painted on the fuel tank. To match this, Mitchell had red leathers and a crash helmet modelled on the typical helmet worn by the German military officers. Yes, he was certainly a showman.

Once, one of GM's visiting executives, an Australian named Leo Pruneau told us that he had seen Mitchell's leathers all laid out in the tech centre trim shop. I remember him saying in his broad Australian accent:

"They looked like a row of stretched hippo skins."

Wayne finished his briefing by encouraging us to laugh at all of Mitchell's jokes. The show could begin.

Brian Adcock met Mitchell's car at the front of the main reception. Brian put on his best British manner, which the Americans loved. If Vauxhall couldn't employ a member of the British royal family, then Brian was the next best thing.

Brian took Mitchell to Ed Taylor's office where they were joined by Wayne Cherry before commencing the studio tour.

Ken Greenley, Geoff Lawson and I waited in the main exterior studio, next to the models.

The entourage arrived. Bill Mitchell was wearing a pale yellow suit with white and black patent spats, and was carrying the black walking stick Wayne had warned us about – it had some kind of silver motive at the top – maybe a skull.

Mitchell was about 5ft 5in tall and carried a bit of weight round the middle. He was bald, and his face and head were slightly flushed.

Walking towards us, he got straight to the point.

"That Benelli Six was a pig and it kicked like a mule. I told my guys to scrape that piece of Italian shit off the tarmac and send it right back where it came from."

Wayne had a wide grin on his face. We had the all clear to talk about his crash. Mitchell went on: he was renowned for his chatting. The Benelli Six (or Sei to use its Italian name) was Benelli's answer to the Honda CB 1000 six. With a six-cylinder motor it was spectacular, but obviously the handling left something to be desired, or at least that's how Mitchell put it. He had lost control on the banked corner of GM's test track. Fortunate to have not hurt himself too badly, he now had a great platform to talk about his 'legendary' skills on two wheels. We listened in awe, and Wayne was happy.

The great man was in a good mood and continued:

"You know, guys, I can't imagine a more exciting job in the world." He paused. "Unless I was a woman, that is!" Fortunately, there were no female members of staff present, so this was a cue for more laughter: the review was going well.

Models were moved outside, and Bill's comments and recommendations noted, with occasional swipes at what the competitors were doing. Wayne started to relax. The ghoster supreme had been worthwhile.

Much has been written about Bill Mitchell, his showmanship, and his one-liners, but you don't get to command the design department of the world's biggest car company on jokes alone. He was an outstanding designer with a level of personality and charisma that is sadly missing in the ranks of today's design leaders. It was a privilege to be in the company of this great man.

# Chapter 13

## *The Continent beckons*

**Bill Mitchell was not only passionate about motorcycles, he was a great motorsport fan. The Corvette Stingray open prototypes of the '60s (which he created with GM's legendary Corvette designer, Larry Shinoda) were raced in North America for a while, before, regretfully, GM got cold feet about motorsport and official corporate involvement was stopped.**

Nevertheless, the company managed to disguise some motorsport support – probably the most famous example being the support of the Chaparral cars of Jim Hall, which continued to get help from Chevrolet engineering and design. To what extent GM styling helped Jim Hall is not so clear, but you only have to look at some of his early racing cars to see some Corvette influence. Also of note were the wire-look aluminium wheels on the Chaparrals. These were a true first, and in the '80s they were copied by many aluminium wheel companies. As it turned out, GM Design's indirect support of all forms of motorsport led to some interesting design spin-offs, to be imitated by many companies.

At Vauxhall, motorsport was, in line with the corporate policy, not directly supported by the company. Nevertheless, it got around this, by forming a team known as Dealer Team Vauxhall – DTV for short. The team was run by a man called Bill Blydenstein and his driver was a colourful character called Gerry Marshall. Marshall was a legend at British club racing level and very approachable, always seen drinking a pint of beer with fans after a day's racing – imagine that these days! That he liked beer was obvious from his generous stature, and with his physical frame, touring car racing was the only style of motorsport that would accommodate his build. He was, nevertheless, blindingly quick in race car.

The DTV cars ran mainly at British club racing level, hardly competing outside of Britain, and were very successful. Blydenstein had started with modified Viva GTs before then moving up to a wildly modified Firenza Coupé, which was nicknamed 'Old Nail.'

Wayne Cherry was a great motorsport fan and was supportive of giving Blydenstein some design help on his race cars, which the Australian John Taylor (JT) handled. It was in a special studio at the end of the viewing yard where this work took place.

Blydenstein's Old Nail was reasonably competitive, but with the new super saloon racing series being introduced for 1974, the four-cylinder Vauxhall engine just didn't have the grunt needed to keep up with some of the American V8-powered cars. Super Saloons was a so-called 'silhouette' class, which meant the fibreglass bodies had to have a centre line profile the same as the production cars they were based on, and a motor positioned roughly in the same place as the production car they were based on. The engine you chose was very much open, which meant, for example, that you had VW Beetles with rear-mounted V8 motors.

Based on these regulations, Wayne, JT and Blydenstein came up with the idea to create a super saloon Vauxhall based on the yet-to-be-released Ventora V8 that I had been working on. The car would have a fibreglass body that loosely resembled the Ventora and be fitted with a racing version of the Holden V8 motor planned for the production car.

The clay model development of the car started in 1973 and was handled in JT's special studio under a veil of complete secrecy. Slowly, however, rumours started to float around the department that something wild was being created, and eventually I was allowed to go and have a look at the model.

The studio was quite cramped, with just about enough room for one car, making it difficult to judge the clay model from the side. However, just seeing it from the front, when walking into the studio, was enough to tell that this was a pretty wild machine. It must have been at least a foot wider than the production car it was based on, although still recognisable as a Ventora, and was fitted with massive split rim racing wheels and slick tyres.

The plan was to have the racing car finished for the 1974 Super Sport season.

Once all clay modelling on the body was finished, it was spirited away to Blydenstein's workshop for tool casting and manufacture.

JT would continue to support the development of the Ventora super saloon, which by now had been christened 'Big Bertha,' and I still recall his Australian drawl when leaving to visit Blydenstein's – "I'm just poppin' over to Blydie's."

The mechanics of Big Bertha were quite sophisticated compared to the other offerings in the Super Saloon class, Blydenstein having pulled in the help of racing car designer Frank Costin, who came up with a very complicated De Dion rear axle for the car, featuring in-board brakes.

There were six races for the first 1974 Super Saloon season, of which Gerry Marshall won three in Big Bertha. Luck, however, didn't go his way and he managed to wreck the car quite comprehensively at Silverstone. The damage was terminal, and picking up the pieces, Blydenstein decided to use what remained of the Ventora and build a new super saloon based on the Firenza HP droop snoot.

Once again, JT got involved and created a car that was to be even wilder in appearance than Big Bertha. The Firenza super saloon was only recognisable as a Firenza from its roof shape and front and rear profile, the rest could have belonged to a slab-sided sports racing car such as a Lola or a Chevron. The new car was logically named 'Baby Bertha.'

Big Bertha had always been a bit of a handful for Marshall on the track, resulting in its demise in the

*Big Bertha.*

Armco barriers of Silverstone, when it lost its brakes. However, with Baby Bertha Blydenstein had got his calculations right and it was an immediate success on the track.

In 1976, DTV and Blydenstein made the decision to enter World Championship Rallying and the super saloon involvement was wound up, but Baby Bertha still makes guest appearances at historic car meetings to this day – she was pretty special.

For World Championship Rallying a vehicle much closer to production would be needed. It would be my next design project.

In 1976, Wayne Cherry, who was now heading up the Vauxhall studio, and Geoff Lawson, were summoned to work in Detroit for the summer. In their place came a GM senior manager, an Australian by the name of Leo Pruneau, and a young American designer named Clark Lincoln.

Pruneau was no stranger to Vauxhall and had worked in Luton in the mid-'60s, on the 'coke bottle' style Viva HB, a design that had drawn its influence from the first Chevrolet Camaro. Pruneau had been working in the states on the Camaro and had been told to go to Luton to 'fix' the Viva being developed by David Jones. After that he returned to Australia to head up the Holden Studio.

Clark Lincoln was one of GM's high flyers and a summer placement in the Vauxhall studio was a kind of bonus for him. He was a very likable guy and naturally a very talented artist and designer. We would join as a team and work on a special version of the Vauxhall Chevette that would become the 2300 HS, the 'HS' short for homologation special.

Vauxhall president, Bob Price, was keen to develop Vauxhall's image across Europe and deemed the World Rallying Championship a good platform to promote the Vauxhall name.

As with the British Super Saloons Championship, the DTV organisation, under Bill Blydenstein, would put together a team to take part in the international rallying championship running a Group B rally car based on the Chevette.

The rules for Group B competition cars were somewhat vague with regards to the number required to be produced in order to go rallying, but there was a gentleman's agreement with the motorsport governing body, the FIA, that a company would need to show an 'intent' to produce a number of cars before the final production number was completed. With the Chevette HS this rule would apply.

The car would be based on the three-door hatch, but instead of the standard 1256cc engine, the Vauxhall slant four engine used in the Firenza 2300 Sport SL, fitted with a twin-cam head would be shoe-horned into the Chevette's engine room. Suspension and brakes would be modified to cope with the increased power output, and visual tweaks would be made to the exterior and interior design, in order for it to stand apart from the more mundane mainstream model.

Clark had been given the task of developing the design and I would assist him. As far as exterior modifications were concerned, we were restricted to modelling a new front nose and rear tailgate spoiler. The car would be fitted with the same alloy wheels used on the Firenza droop snoot, and it was agreed that all models would be finished in silver with a contrasting graphic stripe treatment.

Although the design modifications were limited, it was clear that we needed to get the maximum impact from what could be applied to the front nose of the car. In the USA, there was a race series called IMSA that featured RSR Porsche 911s, BMW 3.0 coupés, and a variety of modified 'Americana.' Clark suggested we take some design influence from these cars, most of which featured very boxy front spoilers. Instead of having a front lower spoiler like the 'shovel' treatment of the Firenza droop snoot, we decided to create a more box-like profile, similar to the IMSA racers, featuring side scoops for brake cooling. At the rear of the car, a 'duck tail' spoiler was mounted to the rear hatch ... and so the Chevette 2300 HS was born.

Due to the urgency of getting cars produced to go rallying, the design was signed off quickly, and a styling model fabricated for senior management approval. The final car was silver, and had multiple red pin-stripe graphics running along the top of the front fender and into the belt line. The interior was finished in black with red tartan cloth inserts in the seats and doors.

At the end of the summer of 1976, Clark returned to Detroit and I was left to complete the final design development. Probably of most interest to automotive historians is the nature of how the production line for the car was prepared for FIA approval.

Up to this time, Vauxhall had produced only a handful of prototypes, and not yet the number required for Group B homologation. This is where the phrase 'production intent' came into play.

For the visiting FIA officials, the factory area where the Chevette 2300 HSs would be built was not quite finalised, and only three finished cars were at the end of the production line, with a couple of models behind them in final assembly. Alongside the cars were a scattering of parts, wheels, interior panels, seats, etc. I went up to the site and applied the red pin-stripes to the three finished cars, and we were all set to welcome the FIA officials. At a signal, when the delegation entered the production hall, the three finished cars were driven out of the assembly area and the incomplete

*Chevette HS 2300 styling model.*

models moved along the line. The officials were satisfied that Vauxhall could produce the required number of cars and the car was homologated.

This procedure was no different to all the other manufacturers producing competition specials, and in the latter half of 1977, Vauxhall commenced the full production run of the 2300 HS, all hand-built and very much sought-after today.

The Chevette 2300 HS's story wasn't quite over for me. The first rally cars made by Blydenstein featured rather squared-off wheelarch extensions, in order to cover their widened track. Blydenstein had developed these features without help from Vauxhall Styling, and it showed. By now, Wayne Cherry had returned to Luton and declared himself not too impressed with the look of the cars competing in the field, and a more elegant solution to the widened fenders needed to be found. A further development of the car would be required, and a small production run – named the 2300 HSR – solved the fender issue. I started to get some ideas together.

At this time, many of the new Opel models on the drawing board featured a style of wheelhouse known as teardrop or box fenders. The first of these was the new European GM 'small car': the Opel Corsa and Vauxhall Nova.

The teardrop fender is a physical way of widening the body side over the wheels in a horizontal swelling shape, similar to the form of a teardrop, rather than the predictable way of simply widening the body in line with the circumference of the wheelhouse.

There is much discussion as to who first developed this design feature, but it seems to have come from modified touring cars in American motorsport; notably those from the tuner Greenwood. He ran converted Chevrolet Monza Coupés and Corvettes with this feature, and since he no doubt had help from GM Design, you can probably trace this design feature to GM. Take a look at a 1971 Pontiac Le Mans Sports and you'll see what I am talking about.

Greenwood made road versions of his Corvette, which featured not only the teardrop fender but also a hoop-like rear spoiler: another feature that later turned up on a very special sports car. I'll get to that story later.

Wayne thought the teardrop fender treatment was a good solution for the HSR, and I completed a full-size

*Full-size airbrush for Chevette HSR.*

rendering of the proposal, with the boxy front spoiler also being smoothed out on the way. As it turned out, I only planted the seed, and did not actually get to work on the final model of the car.

By the beginning of 1977, it became very clear that the days when Vauxhall designed its own models were over. The Cavalier was a mixture of Opel Ascona and Manta body parts, and the Victor replacement, the Carlton, was the last car from Vauxhall to feature its own bespoke front-end treatment. At the same time, Vauxhall was planning to release a larger car above the Carlton, the Royal, in both sedan and coupé forms, which were re-badged versions of the Opel Senator and Monza Coupé. This decision had killed off long ago the idea of the 'V' Coupé that I had worked on and had been shown to Bill Mitchell at his design review back in 1974.

In the early part of that year I began to work on Vauxhall front ends for the next Chevette, which were based on the Opel Kadett model. In my heart, however, I realised being a designer at Vauxhall had its limitations, and a future modifying Opel designs was not for me. It was time to look for a new job, but where?

The omnipresence of Opel-based projects in the studio gained momentum, and it was only too clear there would be no more stand-alone Vauxhall products to leave the factories of Luton and Ellesmere Port.

Wayne endeavoured to motivate the team with a range of small projects intended to keep spirits up. He had maintained his close contact with Bob Jankel and the Panther Company, which led to Geoff Lawson getting involved in a wild concept of note. An office in the corner of the main studio was sealed off from prying eyes, but Geoff soon let me in on the secret. Spurred on by the trend in the F1 world for six-wheeled cars, notably the Tyrell, Jankel decided a six-wheeled super car may have appeal. Geoff created a scale model in the studio office for this unofficial project, and it was quickly delivered in secrecy to Jankel, who then replicated it as a full-size model. Geoff and Wayne visited Panther to supervise the first prototype and later, to an astonished motoring world, the Panther 6 was revealed. In the end, only two Panther 6s were built. With power from an eight-litre turbocharged Cadillac motor, wild claims were made about its straight-line performance, but since the front twin-wheeled arrangement was taken from a Duple motor coach, it presumably had questionable handling.

No record of Wayne and Geoff's involvement in that car was ever recorded, but I assure you it came to life in the Vauxhall studio.

I got myself involved in a similarly questionable project, namely a plan to try to lift the sales of Bedford's ancient HA Viva van of Royal Mail and Postman Pat fame. Scotland's main Vauxhall outlet proposed a special version which was intended to catch on to the new craze of custom vans. We prepared a metallic brown HA Viva van, sitting on chrome Cosmic wheels, with fake wood foil decoration on the lower body side. The interior was finished in a Burberry cloth – yes, we were way ahead of our times. This whole package was sold as the Bedford Scotsman. I began to feel that this was not the destiny for an RCA graduate.

As luck would have it, my close friend Martin Smith called me. Having started his career after graduating at Porsche, he had returned to England for a short period to work at Ogle Design, before going back to Germany to join the relatively little-known south German company Audi.

*Carlton five-door hatch full-size airbrush.*

Apparently, the Audi Design department was a small operation, and relied to a certain extent on outside design houses such as Giugiaro's Ital Design. Indeed, the latest VW Passat, penned by Giugiaro, shared most of its components with the new Audi 80. Things were very much on the move at Audi, and the new five-cylinder engined Audi 100 – developed under the leadership of the new R&D chief, Ferdinand Piëch, and the similarly new design head, Hartmut Warkuss – was creating quite a stir in the industry. Yes, there was a fresh wind of change in this company.

Audi had head-hunted Martin, and his arrival at the almost 'cottage industry' design operation at Audi's HQ in Ingolstadt had evidently impressed enough to motivate Warkuss into looking at expanding his operation. He therefore asked Martin to contact anyone who may wish to join the company – hence his call to me.

*Carlton three-door wagon proposal.*

I looked at a map of Germany, taking quite a while before I could locate Ingolstadt, located 40 miles north of Munich. For me, Germany still had the feel of a country that was recreating itself after the war, even though that had ended 30 years previously. On the map, Ingolstadt looked as though it was in the middle of nowhere, and just neighbouring the border of the Iron Curtain. Did I want to live there?

Well, things had to change in my life, so I let Martin know I would be interested to meet and talk.

It wasn't long before he got back to me. He and Warkuss would fly to London, where we could meet at their hotel.

By now, I had managed to get the attention of Vauxhall's interior designer, John Heffernan, and persuaded him to also meet with Warkuss.

I drove down to London. Martin and Warkuss met me in the hotel bar. My first impressions of Hartmut Warkuss were that he looked very Germanic; almost Nordic, with bright, piercing eyes. That would be about the only thing I learnt about him, since he spoke no word of English, or at least,

*Middle and left: Sketches for Opel-based Astra five-door hatch.*

what he may have understood or spoke, he didn't wish to disclose, and he let Martin do all the talking. Things went well, or at least I think they did. The meeting ended, and I left them to meet John, who had been waiting in the hotel lobby.

A few days later, Martin called me. Warkuss had liked both my and John's work, and wanted us to visit Ingolstadt for a second meeting.

I flew separately to Munich and Martin picked me up in Warkuss' Audi 100. We drove from the Munich airport in Riem through the centre of Munich, up the Leopold Straße, and on to the autobahn in the direction of Ingolstadt.

It was planned for me to meet Warkuss once more the following morning and, on the afternoon of the same day, John, who had decided to travel later. Both of us attended the Audi Design summer party at a nearby lake in order to get to know the complete design team.

As we blasted up the autobahn towards Ingolstadt, I felt that this was a country I could take to: it felt modern and invigorating.

The next morning, Martin picked me up from my hotel in the centre of Ingolstadt, and we drove to Audi Technical Centre located to the north of the town. We passed the production facility, dominated by a large blue neon sign reading 'Audi NSU.' The design centre was located to the far north side of the factory complex, and, having signed in at the security gates, we drove on past a six-storey block that housed the engineering offices – similar to Vauxhall, I thought. In the distance, I noticed a compound full of bright green and blue NSU RO80s, which the Audi NSU group still manufactured at its factory in the town of Neckersulm, north of Stuttgart.

The design department was an unpretentious single-storey affair that reminded me of the entrance of the Chrysler UK design centre in Whitley. Martin and I walked through some glass doors and straight into Hartmut Warkuss' office to the right.

"Ah, guten tag, Herr Birtwhistle."

Warkuss smiled and motioned for me to sit down. As in London, the meeting was conducted in German.

*Sketch for Opel-based Astra wagon.*

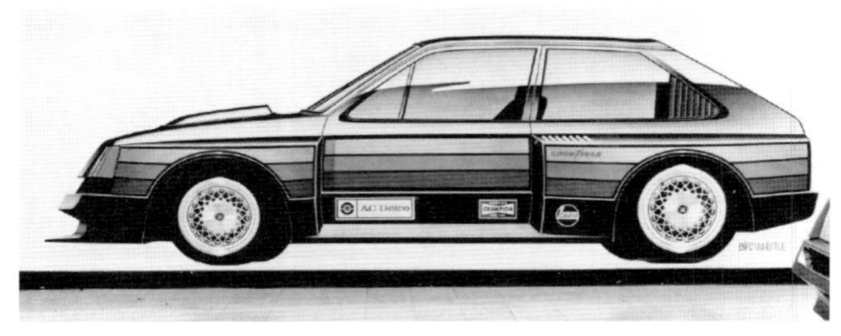


*Dreaming of an Astra racer! Full-size airbrush. (Vauxhall Heritage)*

*Sketch of Astra-based coupé.*

*Polaroid image of Astra clay model. Note the Opel rear bodywork.*

Warkuss got straight to the point, with Martin translating, and asked if I'd like to work for Audi.

I didn't need to give him an answer right then, but he wanted me to spend a day in Ingolstadt and join the team that evening to make sure I felt welcome – it was a nice gesture. He explained that Audi was a small company with big plans, and that a strong design department was essential. Things were just beginning, but he could promise an exciting future with the company. The meeting concluded, and Martin drove me back into town, agreeing to meet me later and drive to the evening's party.

I spent the afternoon walking around the town and soaking it in.

Ingolstadt is situated halfway between Nuremberg to the north and Munich to the south, and is located on the River Danube. The centre of the town is dominated by its theatre and adjoining castle: home of the German military museum. Another landmark is its red brick cathedral, which can be seen from miles around. Apart from Audi, the town's main industry was refining oil at the refineries located to the east of the town.

It was a pretty market town that appeared not to have been damaged in the war, and predictably spotlessly clean. Fashionably-dressed people were enjoying the sun in street cafés, and it felt like being on holiday. The Arndale Centre in Luton seemed like another planet, and I knew I was ready for this change.

By now, John Heffernan had also met with Warkuss, and, talking to him in the hotel lobby as we waited for Martin to pick us up, his impressions were the same as mine: we had to move on from the slow death at Vauxhall.

The party was great, the beer flowed and the usual jokes were made about the non-goal at Wembley in 1966. It was a friendly team, but as with Warkuss, hardly anybody spoke any English, and as I left Munich for London the next day, I knew I faced two difficult tasks – handing in my notice to Wayne, and learning German.

Wayne Cherry sank back in his chair, and, for one minute I thought his head may disappear below the typical tall collar of the shirt he was wearing.

I think the only thing that may have changed my mind about leaving Vauxhall was if Wayne had offered me the opportunity to go and work for Opel, but it wasn't forthcoming.

He avoided any discussion about why I wanted to leave, preferring to go into a long dialogue about his career at GM, how he'd always dreamt of working for them, and why he would never leave the company.

It was a short meeting. We shook hands and he wished me well. I turned and left his office.

If you look for Osborne Road, Luton, on Google Earth you can still make out the Vauxhall Styling yard, with its single round turntable, on the back roof of what is now an administrative building for Vauxhall Motors.

*'The last straw' Special version HA van.*

*Vauxhall Styling team, Christmas 1973. Ed Taylor (left) and Wayne Cherry, seated in the front of the Vauxhall Prince Henry. Brian Adcock is on the rear seat (right). That's me circled, standing behind Geoff Lawson. (Vauxhall Heritage)*

It's all history now.

Wayne Cherry strove to the bitter end to keep that great studio in Luton going, but it wasn't to be. He soon moved to head up the Opel studio in Germany, taking John Taylor with him, and the two of them later moved back to where they had begun their careers at the GM technical centre, where Wayne later fulfilled his ambition to become vice president of GM Design.

My good friend Geoff Lawson went on to take up the position of chief designer for Jaguar, the company ironically having bought the Chrysler UK facility in Whitley to set up a new design operation. We stayed in touch.

The other Vauxhall Styling department designers, modellers and engineers found positions with suppliers, or, as in the case of Ken Greenley, started their own companies.

To this day, I can't think of a better place for me to have cut my teeth in the discipline of car design than the Vauxhall Styling centre. It represented the pinnacle of the traditional art of automotive styling: a true craft that has long since disappeared.

*Vauxhall artwork.*

# Chapter 14

## *Audi NSU AG*

**My father had already said goodbye to me the previous evening as, by the time I was ready to depart the following day, he was already at work.**

As with my previous moves, my luggage consisted of a couple of suitcases and my hi-fi system, which fitted comfortably into the trunk of the blue Opel Manta I had acquired. Since commencing work at Vauxhall, I had run several cars after the illustrious Triumph Herald. I took advantage of Vauxhall's ex-fleet car offer and bought a Firenza 1800, which was then replaced by a Yamaha XS500 motorcycle and a Porsche 912 that I bought from the father of Dawson Sellar, a designer at Porsche I had got to know through Martin Smith. I deemed the Porsche 912, which was using about as much oil as petrol, too much of a risk to travel in across Europe, and thought another vehicle made in Germany would be less of a problem should things go wrong. It had been a clean swap with the previous owner of the Manta and it seemed a good deal. Today, I would have known better.

The Manta packed, I went to my mother, who was standing somewhat forlornly on the drive of the house in Kidmore Road. I hugged her and gave her a kiss.

"Don't worry; it's just for three years."

I started the motor, and, with a quick wave, set off for Dover.

I had decided to take a night ferry to Ostend and travel via Brussels to the German border at Aachen. From there, it was a straightforward drive via Cologne, Frankfurt and Nuremberg to Ingolstadt.

When I left the ferry in Ostend it was four in the morning and still dark. The brightly-illuminated motorway to Brussels was traffic-free, as was the centre of the Belgian capital, which I passed through unnoticed. The ring road around the city was still to be finished a few years later.

I figured I could get to Ingolstadt by late afternoon and pressed on, the Manta purring away. I felt happy with my new set of wheels; it had been a good choice.

Passing the border into Germany, it was a glorious day, and the flatness of Belgium changed into autobahns lined by dark dense forests. What a contrast to Britain, I thought, and motored on.

A couple of things struck me. Although the war had ended over 30 years ago, there was still an amazing military presence in the form of truck convoys of American forces. The last vehicle on these seemingly endless lines of military vehicles was usually a Jeep, with a large sign on the back reading "*Konvoi*." Were we expecting another war?

It was also apparent that there was a distinct hierarchy on the autobahn in terms of lane 'ownership', with the outside or overtaking lane being the clear territory of Mercedes and BMW. Any vehicle getting in their way was flashed vigorously, and almost pushed out of the way. I decided it wise not to venture there.

German trucks also had a different design to those commonly seen in the UK. Many of them were from the company MAN, and had a more traditional cab-front design similar to the old postwar Bedfords, with the engine compartment stretching forward of the driver's cab. They also appeared to be air-cooled with a large fan dominating the front grille. Yes, there were so many new experiences to take in.

I continued, passing the city of Cologne with its impressive cathedral just visible on the skyline, and onwards south, towards Nuremberg, eventually arriving in Ingolstadt mid-afternoon.

Martin had agreed to let me use his spare room until I found a flat. I didn't want to over-extend his hospitality, so finding accommodation was number one on my list.

September 1, 1977, driving behind Martin's silver Golf GTi, we left the centre of Ingolstadt and headed for the north of the town and Audi's technical centre or TI

(*Technische Entwicklung*). Leaving the cars in the main car park and passing through security, we turned left along the front of the engineering buildings. Inside the gates was a row of cars for senior management. I noticed a spotlessly clean black Audi 100 two-door sedan parked in front of a small name plate that read: Herr Piëch. Martin and I moved on.

We passed through the same glass doors at the front of the design building that I remembered from my first visit and I followed Martin immediately to the left. Entering through a set of steel doors, I found myself in a well-lit office room about 15 metres long, which reminded me of the automotive studio at the RCA. There was a row of four double facing desks to the right and up against the wall on the left a large draughting table. A space had been cleared for me at one of the desks. This, Martin explained, was the exterior studio. Unlike Vauxhall's, the designers worked separately from the modelling studios, which were situated in an adjoining workshop. Martin had said I should prepare myself for a smaller operation than where I had come from – he was right.

I noticed a few familiar faces from the Audi party during my first visit and slowly introductions were made. Apart from Martin, there were three other exterior designers. Gerd Pfefferle, Charly Witowski and Joseph Dienst, known to everyone as Jupp. In addition to these three, there was also a studio engineer named Hermann Bühler.

Pfefferle and Witowski had started their careers as apprentices at Audi and not studied art formally like Martin and myself. Pfefferle had come from the tool-making workshop, and Witowski had first worked in Ingolstadt as a shop window decorator before joining the design department as a trainee. Jupp Dienst was quite a bit older than the others and may have had some kind of art education. He was in any case a gifted artist, and apparently the author of the famous DKW Auto-Union SP coupé that borrowed its looks from the first Ford Thunderbird. By now, Dienst's creativity had gone off the boil, and he spent most of his time designing details such as wheel caps and badges.

The first formal design head of the Auto Union team in Ingolstadt was a man named Rupert Neuner, who had come from Mercedes Benz in 1963, bringing with him a designer named Claus Luthe. Luthe would go on to succeed Neuner as head of design, and incidentally created the NSU Ro 80. By the late '60s, Auto Union became Audi NSU and Luthe eventually moved to BMW in the mid-'70s, leaving the way free for Hartmut Warkuss, who had started his career at Ford, in Cologne, before moving to Audi to take over the department in 1976. Warkuss could also put some great designs to his name, including the elegant Audi 100 coupé of the early '70s.

Apart from its small in-house team, Audi also used the help of Giugiaro's Ital Design for design input – witness the first Audi 80, which borrowed parts from the Giugiaro-designed VW Passat. This meant that up until Martin's arrival at Audi, it had been up to Warkuss, Pfefferle and Witowski to handle the weight of Audi's exterior design. Under Luthe, Audi had also

*The Audi modelling hall 1978.*

*Poor image of the first Audi 100 wind tunnel model 1977.*

Czechoslovakian designer named Pavel Husek, who had influenced the Mk2 Audi 80 and its coupé derivative that were just about to enter production at the time. Husek had recently left the company and though some of his sketches were still pinned to the wall, he had left a big void in the department's pool of creativity.

Greetings exchanged, I moved to the desk I had been allocated. Just then, the office door opened and Herr Warkuss came bounding in. He greeted each staff member by hand, addressing them by surname and making firm eye contact, before delivering the customary '*guten Morgen*.' It then came to my turn.

"*Guten Morgen, Herr Birtwhistle, wie geht es Ihnen?*"

"*Danke Herr Warkuss, mir geht's gut.*"

I had been attending German lessons in Luton in the months before joining Audi Design and knew a few rudimentary sentences. What was very clear was that there would be little daily business carried out in English, so a fundamental understanding of the language was a priority.

As I would soon learn, Warkuss greeted each member of the team with a handshake every morning and was on surname terms with his entire staff. It would soon become very clear to me how wide the margin was between addressing a German by their first name in the '*Du*' familiar term, as opposed by surname and the formal '*Sie*' form. In professional life it was usually surname terms.

Right away, Pfefferle, Witowski and Dienst told me to call them by their first name, but for the time being I would remain on surname terms with Herr Bühler. This clear divide in German society is very difficult for English speakers to grasp, but it is a clearly defined rule that still applies to this day, and for many Germans, if you inadvertently use the '*Du*' form they may be quite offended. I had a lot to learn.

Martin suggested I take a tour round the department with him and we moved into the neighbouring modelling hall. As with the design office, the Audi fabrication area was a big contrast to Vauxhall, and once again reminded me in many ways of the Chrysler studio in Whitley.

The room was a long factory-style hall with high skylight windows and a central metal floor capable of taking three full-size clay models. On each side of the room were a collection of workshop benches, metal cabinets and desks for the modellers. A clay model stood on the central measuring plate, and, unlike most of the industry, was coloured green. Not only that, it had a kind of marble-like pattern to its surface, which the modellers would tone down by brushing talcum powder on the surface.

Martin introduced me to the chief modeller, Wolfgang Holzinger, who I remembered from the lake-side summer party.

We moved out of the modelling hall and through the adjoining presentation hall, which thankfully was similar in dimension to what I knew from Vauxhall.

Beyond the presentation hall was the second design studio, similar in size to the exterior studio, devoted to interior design, and where John Heffernan, who had also decided to join Audi, would commence work the following month.

The most senior interior designer was Horst Kretchmar, and he was assisted by Manfred Schäffler. Both spoke good English, but only Manfred suggested that I call him by his first name – wow, this was so complicated!

Also in this area was a small team devoted to colour and trim, where the interior materials and body colours were developed. This team was led by another Audi veteran named Erich Angerhöfer. Angerhöfer had worked originally for Mercedes and was rumoured to have worked in the team that designed the legendary gull-wing Mercedes 300 SL. It was amazing to think that under these roofs the creators of two of the most iconic car designs in history – the NSU Ro 80 and the gull-wing Mercedes – had at one time resided.

Angerhöfer was assisted by a rather matronly lady, whose name fails me. Since both were approaching retirement age, a new recruit had joined the team, also on that day. Her name was Sabine Zemelka. She represented the new young generation of Germans, and we were on first name terms straight away.

The first day concluded, Martin and I returned to his flat, with many impressions in my head, plus a list of tasks I would need to address.

My search for a flat seemed to have come to an end quickly. One of the young Audi modellers, Sepp Kolb, knew of an apartment that was available in a large block close to the Audi factory. I agreed to go over and take a look. Once again, Martin had to help with translating and we met the owner at the flat. It was on the eighth floor of one of two large apartment blocks literally opposite the main Audi factory. The two blocks were located in a park setting and were not as bad as they sound.

I had earlier asked Martin if he knew anyone who would like to share a flat with me, but apparently people didn't share flats in Germany. Another custom I learnt was the fact that all rented accommodation came without furniture and often no kitchen fittings were installed either.

The apartment was quite small – basically a bed-sit with adjoining kitchen and bathroom; it was, however, quite new. The owner said I could buy the kitchen fittings off him, so I agreed to take it. I had only been occupying Martin's spare room for a week, but I sensed that, understandably, he didn't want me there for too long. This would be a good starting base, and wouldn't require much investment to furnish. Martin said there was a new Swedish furniture warehouse just north of Munich – Ikea – and suggested we drive there with a rental van to buy some furniture.

The first evening at my new address, I lay on my mattress, the room illuminated by the large blue 'Audi NSU' sign from the Audi works across the road. It occurred to me that this was the first time in my life I had lived anywhere on my own. Reading, Luton and London, plus my friends and family, were miles away, and I could easily have drifted off into loneliness, but for me this was just another path to cross. I had always set new goals in my life and this new adventure in Germany was the next hurdle – it was exciting.

Despite having the famous four rings of Auto Union on its grille, in the '60s and '70s, the name Audi was not really considered a sporty car manufacturer; that domain was the clear property of BMW, Mercedes and Jaguar. In the UK, Audi was hardly known, and in Germany they were definitely the car of choice for pensioners. The saying went that if you couldn't afford a Mercedes or a BMW, then you bought an Audi. Things, however, changed for the better when, in 1976, it introduced the second version of its top of the range model, the Audi 100, featuring a new petrol engine with five cylinders. The key driver behind this car was Audi's technical development chief, Ferdinand Piëch. Piëch, who, through his mother, Louise, was related to the Porsche dynasty, had joined Audi in 1972, having left his previous position at Porsche when its board of directors declared that no family members could work for the company. Regarded as a brilliant engineer, Piëch had an ambition to make Audi on a level par with Germany's two major players in the prestige segment: Mercedes and BMW. The new Audi 100 was the first model created under Piëch's leadership as development chief and sent out a very clear message that Audi meant business.

In 1977, the latest Audi 100 had only been on the market for a little over one year. Plans, however, were already being developed in Ingolstadt for its replacement model, named in-house as the C3. The C3 was my first job as I opened my sketch pad in the first week of September.

Martin Smith and Gerd Pfefferle had been briefed by both Piëch and Warkuss as to the strategic thinking behind the new Audi 100 model, which was developed in two stages.

The German government had announced a competition – open to all the country's car manufacturers – to develop a car to showcase new technology. It was named the *Bundes Forschungsauto* – roughly translated: German future development car. The results of this competition would be displayed as prototypes at the 1981 Frankfurt Motor Show. The idea behind the project was to demonstrate how efficient future cars could be developed.

Piëch reasoned that this would be a good way of motivating his engineers and designers to new levels of creativity and at the same time give the automotive world a clue as to what Audi was planning for its next top model. Top of Herr Piëch's list of ideas in creating a vehicle as efficient as possible, was class-leading aerodynamics. A low coefficient of aerodynamic drag over the car's body (called CD in the industry) would mean that the car could be pushed through the air with less engine power, resulting in a drop in fuel consumption. Piëch wanted the lowest CD figure for a car in the Audi 100 segment, and reasoned that the only way this could be achieved was to develop the basic body form in the wind tunnel first, before the designers even contemplated how the car could look. In addition, it would necessitate a body surface as smooth as possible, and to reach this goal Piëch had come up with the idea of having a cabin where the window glass was as flush to the body surface as possible. On the conventional cars of the day, the side window glass was set in frames mounted to the top of the doors, which acted as tracks for the glass to run up and down in. Piëch proposed moving the side glass to the outside of the door window frames. Studs attached through the glass would run up and down tracks set into the door frame, which in effect enabled the glass to be secured to the outside of the door frames. This system would result in a car cabin both smooth and free of the protruding edges that would normally interrupt the airflow over the body. It was a simple but brilliant idea.

Warkuss got the exterior team together and proposed first developing a design for the *Bundes Forschungsauto*, commencing with a basic wind tunnel shape as Piëch had proposed, and learning from this, development of the design of the next production Audi 100 followed.

Martin Smith was given the task of developing the first aero model, and since my German was still minimal, I would work with him, while Gerd Pfefferle and Charly Witowski commenced ideas under Warkuss' eye for the production car. In addition to this, as was the case with almost all the future design programmes from the VW group, Giugiaro's Ital Design also got a contract to develop a proposal.

For aerodynamic work, Audi used the VW wind tunnel located at its HQ in the north German town of Wolfsburg. Without wasting any time, Martin and I set off to spend a week working in the wind tunnel and hopefully develop the optimal body form to be used as a base for further design work.

As a starting point for the wind tunnel work, Audi had built a full-size model armature, including the planned glass surfaces. It was also fitted with an engine compartment and full underbody, including exhaust system. In order to measure correct CD values, one had to take into account the drag that is achieved when air flows through the engine room and also under the car. Only in this way can you achieve the final production car's true aerodynamic CD figure. Piëch was aiming for a figure below CD 0.3.

For those who are unaware of the principles of automotive drag coefficients, here's a simple explanation: in a range of CD 0.5 to CD 0.2, the lower this figure is, the better the aerodynamic value of the car body. As an example, the CD value for a VW Beetle was around CD 0.48, and the figure for most cars built in the late '70s was probably about CD 0.36. If you could achieve a value under CD 0.3, say CD 0.28, this would be a real breakthrough, a first in the industry; this was Piëch's goal.

Martin and I arrived in Wolfsburg in the late afternoon. Joining us was Audi's chief modeller Wolfgang Holzinger, who had travelled up separately with a couple of colleagues; they would model the aerodynamic armature in the wind tunnel.

Audi had booked us into the Holiday Inn in the centre of the town. As we approached Wolfsburg it was possible to make out the four chimney stacks of the VW factory's own power station, which shadowed the town. We checked in, and having some time on our hands Martin suggested we take a look at the East German border. Wolfsburg is situated on the extreme east side of what was then West Germany, at about the same latitude as Berlin. Berlin itself was like an island, and only accessible by car via corridor autobahns, one of which ran out of Wolfsburg. We drove out of town and just before the border left the autobahn and headed into the country. It was eerily quiet, and from the small country road we were on it was possible to see the fences and watchtowers of the border. We stopped. On the other side of the fences you could make out a track, along which a military jeep was driving. Were they watching us? Who knows? It was in any case chillingly disturbing to think that the free society as we understood it ended on that line. We drove back to the comfort of the Holiday Inn. I will never forget the menacing image of that border fence.

The next morning, we drove to the other side of another kind of security fence and through the gates of the VW development centre, taking the road for the wind tunnel. The VW site, including the production factory, was the size of small town. The company had taken massive strides forward from the days where the only VW vehicles produced were the Beetle and 'Bulli' bus, of camper fame. The new models, in particular the Giugiaro-designed Golf, Scirocco and Passat were huge sales successes. Yes, VW was a company on the move, of that there was no doubt.

Making our way to the corner of the development centre where the wind tunnel was located, a matt black NSU Ro 80 passed us, making a noise like a jet. You could make out a heat haze following it. Martin commented that it was powered by a gas-turbine engine. I wondered if it would ever see the light of day.

The VW wind tunnel was not the first one I had visited, as while I was at Vauxhall I had been to the one at the independent research facility MIRA. Unlike MIRA, which used four ex-Lancaster bomber propellers to generate wind, the VW tunnel used a single fan.

Arriving at the tunnel, we parked and walked into the control room. Through a glass window you could see that work was already being conducted by the aerodynamic engineers on the armature. The model was secured to a large metal plate in the floor, which was essentially a large set of measuring scales. Once the air stream generated by the tunnel fan met the model, sensors in the plate would send information to the wind tunnel computers and a drag coefficient could be calculated. Also of importance was the calculation of lift or downforce. A good drag coefficient is fine as long as the car does not take off at the same time.

Holzinger and his team were securing bits of polystyrene foam and cardboard to the surfaces with tape, and the rough edges were filled in with clay; it wasn't the prettiest of objects. Once everything was suitably secure, the model was left, and everyone retreated to the control room. The technicians threw a few switches and slowly the sound of the massive tunnel fan could be heard gaining speed. The speed increased, and readings began to show on a screen next to the switchboard. Figures showed that already the basic form was good, with readings below CD 0.3. The chief aerodynamicist explained that we needed to reach a figure in the tunnel of around CD 0.25, which in production car terms would represent around the CD 0.3 mark that Piëch was expecting. The reason that we needed to achieve a much lower figure with the model, was that we would have to compensate for details such as door shut lines and similar obstacles, which would affect the final production model's CD values when it left the factory.

The next couple of days were spent experimenting. Modifications to the body at strategic points, located mainly around the back and front of the car, were modelled in and tested. A critical area was the angle of the rear screen in combination with the height and length of the rear trunk or boot. There were no typical wheelarch swellings and the body side was very smooth, pulling in abruptly in plan shape after the rear wheel opening. Another point of detail was to create as many sharp edges at the extreme end of the car, allowing the air to leave the body as cleanly as possible and not create a vortex or mass of swirling air directly behind the car. This meant that the lower part of the rear bumper had the look of a spoiler, a feature usually only seen at the lower front of cars. It was exciting and fascinating work. Ideas were modelled onto the armature and the team crowded excitedly around the control room monitor to see if any improvement had been achieved.

After three days we ended our work and returned to Ingolstadt, able to report back to Piëch and Warkuss that we had hit the sub CD 0.3 figure comfortably. Now we had to design something without disturbing the basic form we had ended up with.

John Heffernan, my colleague from Vauxhall, arrived with his girlfriend, Patricia Roberts, a few days before the beginning of October. He had purchased a black Willys Jeep, which they drove down from the UK. Pat was a very successful knitwear designer based in London and was far from happy that John had decided to move to Germany. She had a thriving business in London and there was no way she could give it up to join him. It would be a challenge for them both. Nevertheless, it was great to welcome them, and for me it was almost like a big brother turning up. I hadn't socialised much with John during my Vauxhall days, since he lived in London, but a close working relationship had developed between us. I admired his experience and wisdom, so I was naturally happy he was joining me on my adventure at Audi.

As with myself, John's first priority would be to find accommodation, which didn't take him too long; he acquired a large top floor flat located in a four-storey building right in the middle of Ingolstadt, just a stone's throw from where Martin was living. With both John and Martin living in the middle of the town, with easy foot access to all the bars and restaurants, I was beginning to think I had been too hasty getting a flat outside of the centre, but a move so soon was out of the question, and my key priority was getting a grasp on the German language and not worrying about whether I could walk to the nearest bar.

During the week, my life after work fell into a routine. I would cook myself a meal before sitting down to learn German for two hours. As a target, I hoped to be able to get a rudimentary German conversation together by Christmas. It was a somewhat monastic lifestyle, but I had to do it. Few people in Ingolstadt spoke English, and I needed to be able to converse in German in order to survive. Only when that was in place, could I start to think about the bars and hopefully get to know some of the locals, or more specifically, *Fräuleins.*

The design process at Audi was not dissimilar to Vauxhall, from sketch through to clay model; the Audi design centre just lacked the sense of scale that John and I had known in the Luton studios. There were very

*First Audi 100 sketch with NSU Ro 80 influence.*

few full-size boards for tapes or airbrush renderings – in fact, no one at Audi had done an airbrush, as far as I could make out. The exterior studio had one board that was used for full-size tape drawings, but doing an airbrush there was out of the question. I had noticed two large movable boards in the presentation hall, which would have to suffice once we got that far, but first we needed some sketch ideas.

Martin and I began to create illustrations based on what we had learnt in the wind tunnel, with Gerd Pfefferle and Charly Witowski joining in. It was, to me, immediately clear that both the German guys lacked polish in rendering skills, but then, who were they to learn it from? Jupp Dienst certainly knew how to draw but, after a few weeks of getting to know him, it was quite clear his heart was no longer in the job. He was a passionate fan of the German sci-fi comic *Perry Rhodan*, which had a similar format to the war comics we read as school boys. When Warkuss was out of sight, Jupp would pull out from under his desk the latest Perry Rhodan illustration he was working on. He dreamt of getting one of his pictures published in the magazine, but had yet to achieve this goal.

Back in the real world, we continued proposals in line with Warkuss' directions. He was looking for a theme that drew some flavour from the NSU Ro 80. It had to have a classic stance and proportion, featuring Audi's signature 'three-light' DLO, meaning that the side glass of the cabin had three separate windows with a thin rear C pillar. This was in contrast to the designs of Mercedes and BMW, that generally featured two main side windows on their cabins and a much thicker rear pillar, which, in BMW's case, always featured the so-called Hofmeister bend (in other words, a hockey stick-like shape): a feature seen on all BMWs to this day.

After a few weeks' work, Warkuss selected a couple of directions and I volunteered to attempt a full-size airbrush, which Martin was keen to witness, having not yet done one so far in his career. We decided to do our proposals in parallel, so that Martin could pick up on the technique as I moved along. Audi did not have the vellum paper I was used to, so I had to compromise by using a thin plastic drawing foil. We did manage to find a couple of airbrushes and some Flo-Master ink to get started.

After a week, we had two full-size renditions finished. Warkuss was satisfied and gave the okay to move forward into full-size clay models, one handled by Charly Witowski, who had completed his proposal as a tape drawing, and the other by Martin; I would contribute with further design input.

Audi used a method of building up its full-size models that was quite a bit quicker than the traditional Vauxhall process of taking individual points from the tape drawing and slowly building up the surface, with the modellers scraping away the clay to the points and smoothing the surface with large curved rulers.

At Audi, they would secure a large plywood board to cut to the shape of the curvature of the body in plan to the floor. Using an electric winch, a template of the body side, secured to a special carrier, was then dragged down the entire length of the model armature.

*Poor image of full-size airbrush for the Audi 100.*

This way, the basic form was filled in very quickly. Once the basic form was in place, tapes of feature lines on the side of the body were applied and the modellers would start to free model the forms.

The first thing I would learn when working with modellers from the south German state of Bavaria, was that they needed a constant replenishment of beer to keep them motivated. In south Germany drinking beer at work was no different than drinking a glass of milk in England. It was a tradition found only south of Frankfurt and was usually confined to companies where there was hard manual work. In the engineering offices of Audi, it was rare to see anyone drinking a beer, although some may have had a bottle under their desk on a Friday, but in the modelling studio it was the norm, and tolerated as long as no one got drunk. Beer, which was usually the Bavarian wheat beer or *Weizen*, was drunk from one-litre glass '*Krugs*,' which were shared by the team working on the model and placed in the aluminium measuring blocks or 'castles' used for measuring points on the model. On a summer day, it was not unusual for a team of modellers to get through a complete crate of beer during a hard day's work, and the designers were expected to pay for at least one litre or '*Mass*' during the day and as a rule. In addition to this, if a large modification to the model was to be made, it would also cost the designer responsible a *Mass* as a form of penalty. Today, with all sorts of health and safety rules in place, I assume the days of beer consumption in the work place is no longer tolerated. Then, however, it was quite normal.

*Full-size clay models of the Audi 100.*

The process from developing full-size clay models through to a final form capable of being signed off for production was still a long and laborious task, and the development of the C3 Audi 100 continued on through several stages for the next couple of years. Martin continued developing his model as the base for the *Bundes Forschungsauto*, while Witowski continued with the key in-house proposal. In Turin, Giugiaro's Ital Design was also pushing ahead with a model, with Piëch and Warkuss no doubt visiting him in Italy to brief and guide him. No information, however, was forthcoming as to how this design was progressing, and it was quite a while before all three designs were seen together.

In the interior studio, John Heffernan had already developed a strong direction, which would go through to the production development, his wide breadth of experience from Vauxhall clearly helping him.

1977 drew to a close, and with it came my first trip home for Christmas. I couldn't wait. The Christmas that year was very intense, and I soaked up all the things I missed about my family and home country, as it was a short break.

It seemed like I had only arrived the day before, when Martin met me in Reading and we set off for the trip back to Ingolstadt. Having sold the Manta I needed to hitch a ride back. This time, the departure didn't seem so heartbreaking for my mother. In 1978, there were things to look forward to: my sister was planning to marry her boyfriend, a doctor at the John Radcliffe Hospital in Oxford, and then later in the year, my parents wanted to visit me in Germany.

I had decided to sell the Manta over Christmas, since running a UK-registered car in Germany was only tolerated for a short period. To replace it I had ordered a new VW Beetle cabriolet on the company car scheme, but it wouldn't be delivered until March. Back in my small apartment, I was thinking about how I could survive the next three months without a car. Martin and John agreed they would pick me up each morning, but it wasn't a situation I was comfortable with. John realised I was struggling to come to terms with my living location out of town, and proposed that I move in with him. His flat in Josef-Ponschab-Straße in the centre of town was huge and he had a spare room, so why not? And with little love left for my first flat, I joined him in the centre of town.

The C3 project and next Audi 100 were planned for release in 1982, but before that, Audi's next new model was the Audi 80 sedan. It was scheduled to go on the

*My sketches for the Audi 100 as depicted in a specialist magazine of the time.*

market in 1978, with a coupé version to follow a year or so later. The design work on both these cars was complete. The exterior of both designs had been an in-house effort under Hartmut Warkuss, with the interior design having influence from Giugiaro. Although the basic design of both Audi 80 models was complete, Ferdinand Piëch had yet another idea up his sleeve, with the aim of further lifting Audi's image.

During winter tests of Audi 80 prototypes in Finland, an Iltis jeep, which VW built for the German army, and which used much of the running gear of the Passat and Audi 80, was being driven as a support vehicle. Piëch had noticed that in snowy road conditions the Iltis, with its AWD system (all-wheel drive), could run rings around the accompanying front-wheel drive models in terms of traction in these difficult conditions. Getting together with an Audi engineer named Walter Treser, Piëch realised that if a production car was fitted with a similar drivetrain, Audi would have a formidable weapon to compete in the Monte Carlo Rally, famed for its snowy conditions.

*Initial sketches for the Audi Quattro.*

On returning to Ingolstadt, Piëch instructed Treser to build a prototype based on the outgoing Audi 80 two-door sedan. Fitted with a turbocharged five-cylinder engine destined for a new Audi 100-based model (later named the Audi 200), and employing the four-wheel-drive system of the Iltis jeep, it signalled the birth of the yet to be named Audi Quattro.

With Audi 100 development proceeding well, Piëch instructed Warkuss to form a small team to develop a design for his latest brainchild. Warkuss immediately gave Martin Smith the task of heading up design work for a special version of the forthcoming Audi 80 Coupé as a base for a future rally-car.

Martin set to work, and since my role in the Audi 100 programme was now fairly negligible, he asked me to assist him with the first sketch ideas.

Piëch's brief for the Quattro was that it should keep the character of the Audi 80 Coupé it was based on, but, due to its larger wheels and wider track, employ wider wheelarch houses. Modifications could also be made to the front and rear bumpers in order to stretch the difference to the normal coupé. Martin and I started on the first ideas.

Warkuss had given Martin complete leadership responsibilities for the design development, and he would from then on work very closely with the programme chief engineer Walter Treser. Treser, in contrast to Piëch, who was noted for his somewhat icy demeanour, was a very friendly and approachable person. He had joined Audi from the Italian tyre manufacturer Pirelli, and in his spare time had done some motorsport. He was therefore very well qualified to develop a vehicle all about power and road traction. He would test the latest ideas for the future Quattro in the Audi 80 sedan that his team had cobbled together.

Painted dark yellow with a black vinyl roof and crudely widened wheelarch houses, this had to be the ultimate Q car. Treser would tell tales of his adventures on the autobahn, and the disbelief of BMW or Porsche drivers when he approached them at high speed and were unable to shake him off their tail.

The Quattro project was not dissimilar to the Vauxhall Chevette HSR, in that it was a heavily modified body of an existing model. The Audi Coupé had a strong feature line crease on its body side and Martin figured that if the fenders were widened in the now trendy teardrop form at that line, he could create quite a unique look for the car. It would also help to keep expensive tooling modifications to a minimum, since the upper part of the bodywork would remain the same as the normal coupé.

A metal prototype body of the forthcoming Audi Coupé had been secured to one of the measuring plates in the fabrication hall, and work commenced on the model, widening the body sides below the body side crease line on the front and rear fenders – it worked.

At the front and rear, the design of the car proved to be quite a challenging task in that the front and rear bumpers had to accommodate a separate insert to adhere to the different regulations for bumper protection in Europe and the US. Individual, more harmonious solutions for each market, that integrated nicely with the fenders, had to be reluctantly abandoned for cost reasons.

Martin finalised the design for sign off by mid-1979, and at the 1980 Geneva Motor Show, Audi revealed the Quattro – to the astonishment of the international motor press – resplendent in a pearlescent white paint finish, also a first.

The Quattro went on to conquer the world of rallying, and Piëch's theory that it could win the Monte Carlo Rally was proven right. The car paved the way for four-wheel-drive to become the norm in world rallying. More significantly, the Quattro name pulled Audi out of the quagmire of conservatism and eventually brought it to the same level as Mercedes and BMW, as a truly sporty and prestigious brand.

Audi remained faithful to the signature fenders of the Quattro, a design feature that became a legacy repeated in many future Audi models, right up to the present Audi A6.

Having completed my first full year in Ingolstadt, and thanks to diligent

*Me standing next to the Audi Quattro model armature. The clay applied to an Audi Coupé body.*

*Audi Quattro.*

*Faded memories of my VW cabrio in 1978.*

studying, I could now hold a reasonable conversation in German, albeit slightly tarnished with Bavarian colloquialisms associated with that dialect.

In the spring of 1978, my new VW Beetle cabriolet was delivered, finished in silver paintwork, which I drove proudly to the UK the summer of that year, to attend my sister, Helen's, wedding.

Later the same year, my parents would visit me, enjoying their first experience of my new life in Germany. Regretfully, it was somewhat spoiled by the extreme back pains my father had been troubled with during their stay. Returning to England, he visited his doctor and after a few tests, he learned that he was suffering from the very early signs of bone marrow cancer. The doctors were, however, optimistic they could halt its development, so for the time being we were all relieved.

With my ever-improving language skills, I was soon managing to engage in conversations with the locals in the many bars in Ingolstadt. A popular location was a small bistro that, despite its compact dimensions, could accommodate 50 guests at a squeeze. It was called M's Corner, named after the owner Hans Meyer. Located in the Friedhofstraße, on a street corner opposite one of Ingolstadt's old town gatehouses named the *Kreuztor*, it was probably the most popular bar in town. One key attraction of M's Corner was that you could park directly opposite, and it was therefore the location of choice should you own a flash car that you wished to show off. M's Corner would soon become the regular hang-out for Audi designers.

There were plenty of great looking girls in M's Corner, but getting past a somewhat rudimentary conversation was not so easy. I thought I was on to a winner one evening when during a conversation with a blonde German girl named Irmgard, who, eager to practice her English skills, said that she had 'hot legs.' If that wasn't a clear signal, I don't what was. As it turned out, in the cramped confines of the bar she was feeling physically hot and needed some fresh air. Oh well!

Apart from the town's bars, another meeting point was the newly opened squash club. Squash was new to Germany and quickly became the 'in' sport. The courts in Ingolstadt were also a draw for all the English ex-pats from miles around, which included many of the engineers developing the military Tornado fighter jet at the nearby MBB Company in Manching.

The club's coach was an English guy called Mike, and I enrolled for a few lessons. My talents in the squash court were of questionable potential. Nevertheless, it led to an encounter with a girl where quite a lot more than her personal charms would be revealed on a first date.

During the lunch break at Audi, I used to go over to the main engineering building where they had a very comprehensive library. Here, they had a very good selection of all the latest English car magazines. Not only that, the girl who ran the library was very pretty and, like Irmgard of M's Corner, was very keen to practise her English with me. One day, she asked me if I played squash; having just started the sport herself, she was looking for a partner. We booked a court and a few days later met for a game.

The game was of a level acceptable for two beginners, and she seemed satisfied with my performance. After the half hour in the court we made our way to the changing rooms. All fairly harmless until, quite out of the blue, she said, "I'll see you in the sauna."

I was used to the changing room facility at the squash court by now, and had seen the adjoining door with "Sauna" on it, but had not fancied squeezing myself into a wooden box with a load of sweaty guys. Now, it seemed, things were not quite as I had imagined, and I was curious as to what went on behind that door. I showered and, protecting my modesty with a towel, went on through.

Behind the door was a well-lit room with a selection of wooden benches and yellow plastic beach loungers. To the right, at the end of the room was a wood-panelled wall with a dark glass door. To its right, there was a large, open, tiled area with a selection of showers and hoses attached to the wall. There were a few guys walking around naked, all unfortunately supporting manhood of impressive proportions. I glanced down quickly to the shadows behind my towel – it would be difficult to compete. I didn't have time to expand on these thoughts, as another door at the other end of the room suddenly opened behind me and my squash partner came cheerfully into the room, a yellow towel wrapped around her to hide anything of interest.

"Oh, you've waited for me," she said cheerfully, and moved confidently to the glass door. I followed as she pulled it open and stepped inside.

The heat hit me immediately. Inside, it was dimly lit. Thank goodness I thought. There were about five or six people inside, all seated neatly in rows on towels. My friend dropped her towel and positioned it lengthwise on the second seating row where she climbed up and seated herself, pulling her knees up towards her. She patted on the wooden surface in front of her indicating that I should join her. So far so good and the warmth of the cabin felt surprisingly pleasant, I could see the attraction. I began to sweat, trying to avoid any erotic thoughts that I may regret when later standing to leave the cabin. I explained that this was my first sauna, which triggered an amused tittering.

"Ah, you English are so prude, nudity is very normal."

Apparently so.

One of our fellow sauna companions asked if he could do an *Aufguss*. This involved pouring water over the lava stones on the sauna oven that was located near the door. Nobody had any objections and he went ahead. I was unprepared, and everybody knew it. Suddenly, I was hit by a massive rush of heat that ran down my body like a wave of fire, and everyone burst out laughing; they were a friendly bunch and my inhibitions soon started to leave me. Boy, was I sweating now.

About fifteen minutes had passed and my friend told me we needed to get out and shower. I pulled open the door. What now? Discarding her towel on a nearby bench she walked over to the shower area and, standing completely naked, called to me.

"Peter, come, shower me down, please."

By this point, nothing more would surprise me, so I moved over to her. I grabbed one of the hose pipes attached to the wall, turned the tap, and proceeded to hose her down. It was freezing cold and she screamed with delight.

"Now it's your turn!"

I handed her the hose pipe and turned my back to her. There was a limit to my endurance, and there was no way I was going to have her spraying me with ice cold water down there.

Showering over, we pulled the towels around us and went over to the loungers to relax.

The same ritual would be repeated before we returned to the individual changing rooms and then met afterwards for a drink in the bar.

Yes, Germany provided a whole load of new experiences – I liked it!

With the next Audi 100 launch date planned for 1982, and the end of 1979 drawing near, it was time to select the final design to go forward for production development. The two in-house models of Witowski and Smith were undergoing final refinement before being presented to the VW board, many of whom would fly down to Ingolstadt for the event. The wild card was the proposal from Ital Design, of which nobody in Ingolstadt had seen anything in model form. Ferdinand Piëch was a friend of Giugiaro, so it can be assumed that he had visited him in Turin to check the progress, but all we had seen of his ideas were some initial blue illustrations that he had sent to Ingolstadt at the beginning of the sketch phase. These were the typical side and end elevations that the Italians always produced, and never did the designs any favours. They were basically coloured-in technical drawings, and you could put a ruler to them to check their accuracy. This style of illustration went down very well with Audi's engineers, in particular, Herr Dr Baylis, who was one of the original Auto Union engineers. Baylis had made his way up through the ranks only to have the rug pulled from under his feet when Ferdinand Piëch turned up on the scene. Nevertheless, he still wielded a lot of power in the halls of the engineering building, and made it his job to check the work of the *Stylisten* almost on a daily basis. He arrived with his junior engineers early in the morning, very much in the same way a doctor would do the ward rounds at a hospital. This situation was made even more believable due to the fact that the engineers all wore white coats with their title and name embroidered on them.

Baylis entered the studio, ruler in hand, and placed it on one of our illustrations to check the wheel dimensions. Unlike the work of the Italian's, all our sketches were 'cheated' and the designs usually sat on wheels of unrealistic proportion. Baylis called for one of the junior engineers to bring him the specification of the car.

"Jarwohl Herr Dr Baylis," they screamed, offering him the requested paper.

Memories of childhood war films came to mind.

Heated discussions followed, as to why our proposals were so inaccurate, leaving it to Herr Warkuss to calm the storm. Warkuss had no fear of any of the engineers, not even Herr Piëch. It was said that on one occasion, when Piëch told him to change a particular design, he declined, saying:

"If you want that done, then throw me out and get an actor to do the job."

For many years, there would be a cool relationship between Warkuss and Piëch, but over the years this would warm and they would become probably the most innovative designer/engineer duo the automotive world has ever seen.

***

In the '70s, the stability of industry and politics in Germany was being threatened by a terrorist organisation called the RAF – Red Armee Fraction or Baader Meinhof Group. They had assassinated several key German industrialists and politicians, as well as police officers, and the country was on high alert for whenever they would next hit.

When the senior board members of VW arrived in Ingolstadt, the town was ready. There was a massive police presence to be felt. VW's president was a man called Toni Schmücker and he was under 24-hour surveillance.

Walking into the design department for the Audi 100 design review was like something out of a film. There were both uniformed and non-uniformed police everywhere, most armed with some kind of automatic weapon.

Jupp Dienst had joked, "Don't make any quick action to comb your hair or you'll get shot."

But the look on the faces of the security men present clearly said that this was no joke.

The design review was taking place first in the presentation hall, and then out on the parking area Audi Design used as a viewing yard. Only senior staff members were invited, so all we could do was observe from the confines of the design studio.

The models were rolled out into the daylight; this was our first chance to see all three design proposals together. The guests from Wolfsburg walked out with Piëch and his team. Amongst the group, Warkuss and Giugiaro could be seen. There were machine-gun-toting police everywhere.

Giugiaro's design was, unlike the two painted clay models from Audi Design, made of a hard material, which was immediately obvious. Somehow, hard models reflect light differently and you could see it. It was a good design and couldn't have been from anyone but the Italian master. In some ways, the body stance had the feel of a BMW and it seemed quite heavy looking. What was definitely better than the two proposals from Witowski and Smith was the front-end treatment, which had a centre peak, unlike the Audi proposals which both had a single smooth curve for their front plan shape.

Giugiaro presented his design and was followed by Warkuss who, quite rightly, was pushing for one of his teams' designs to be chosen, but which one?

The review would take time. Piëch took Toni

Schmücker round the designs. Warkuss once more took the stage and his body language said it all: he was focusing very much on the Witowski proposal.

You could sense a look of frustration on Giugiaro's face.

More discussions followed until finally Schmücker and his immediate reports appeared to have made a decision. However, from our vantage point it was difficult to judge the outcome.

Handshakes were exchanged and Warkuss was grinning widely: they had obviously gone for one of the in-house proposals.

Aside from the crowd, Giugiaro was standing on his own looking somewhat forlorn. I think I saw him wipe away a small tear. He'd lost the battle and was finding it hard to take in. My memories of his proposal are very distant, but I believe it was ahead of its time. The very solid and heavy stance was, for many, too uncomfortable. Giugiaro was always one step ahead, but this time it was maybe one step too many. He was understandably disappointed.

The crowds dispersed, as did the armed guards. Once more peace returned to the studio. We would soon hear of the day's outcome.

It took some time before Warkuss got the team together. He explained the board's unanimous decision. As predicted, the Witowski proposal had been selected as a base to move forward. It would need more refinement, particularly at the front, where the proposal from Giugiaro was better, and they would use some of that flavour in the model. The Smith model would continue as Audi's proposal for the Bundes Forschungsauto, to be displayed at the 1981 Frankfurt Motor Show.

Martin, like Giugiaro, was disappointed, but reading between the lines he had as little chance as the Italian. During the months prior to the presentation, it had been clear that Warkuss had been keeping a close eye on the progress of Witowski's model. It was the combined effort of the established Audi team, who had worked together long before Martin or I had turned up on the scene. The outcome was understandable, and as the final proposal took form it would prove to be a good choice. The 1982 Audi 100, with its smooth body surface resulting in a CD value of 0.28, would be a ground-breaking design. Slowly, the automotive world would be taking more and more notice of this dynamic company from Ingolstadt.

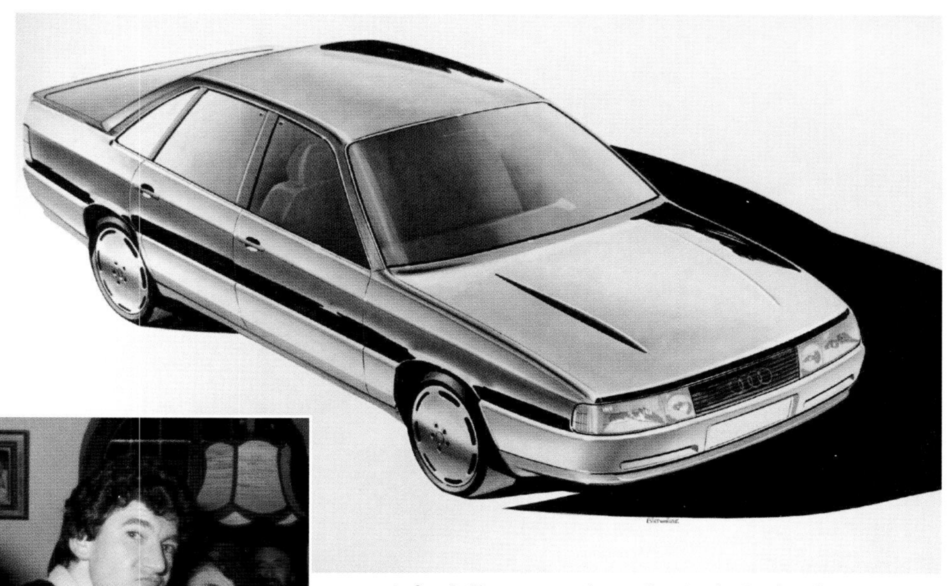

*A final illustration from the Audi C3 development.*

*Myself, Grahame Thorpe, and Jupp Dienst.*

# Chapter 15

## *Schreyer, Mays, and the rest*

**In 1978, as I had expected, John Heffernan returned to England and joined with Ken Greenley to form a design consultancy. As Heffernan and Greenley they went on to design several noteworthy cars for both Aston Martin and Bentley, and to this day we remain in touch.**

I was sad to see John go, as we had built up a strong relationship, and I look back with fond memories at those times sharing a flat with him. New faces, however, would soon arrive at Audi, and Warkuss' ambition to build a design team worthy of supporting his and Ferdinand Piëch's plans would gain momentum over the next two years.

In 1980, several significant new designers were to join the team in Ingolstadt. For the last two years, Audi had been sponsoring a student at the Royal College of Art: his name was Peter Schreyer. Schreyer had done an internship at Audi in the summer of 1978, and not only was he a lovely guy, he had so impressed Warkuss that he decided to send him to the RCA. In 1980, Peter Schreyer graduated and came to Ingolstadt, and he would take the empty desk left by John in interior design.

Peter Schreyer was, however, just the start of Warkuss' recruitment drive. In the summer of 1980, Martin and Warkuss attended the graduation show of key rival to the RCA: ArtCenter in Pasadena, California. Here, they met another young designer who also took up their offer to join the team.

*The Audi Design team, 1979. From the right: me, Peter Schreyer, J Mays, Martin Smith, Charly Witowski, Hartmutt Warkuss. (Audi AG)*

The designer in question had agreed to join the team in late summer of 1980, and on the day of his arrival from the States, Martin came to my desk and asked if I would drive to Munich airport and meet him. He gave me a piece of cardboard with the American designer's name on it: J Mays.

I drove to the airport in Munich Riem and waited at the arrivals gate, holding the name card. This must be him, I guessed, as a gangly boyish figure approached me. He was wearing a bright beach shirt hanging outside of his beige chino slacks, and had brown Timberland deck shoes on his feet with no socks. He could only come from California.

"Hi, my name's J."

We drove back to Ingolstadt, and it was soon clear that J was a nice easy-going guy. He came from the US state of Oklahoma, where apparently it was normal to name boys with just an initial. J began to tell me a bit about his upbringing and education. He had first studied journalism, but his passion for cars combined with his artistic talent had led to him to apply for a place at ArtCenter. We motored on as the conversation lapsed. My passenger was jet lagged and slightly nervous at the speeds allowed on the German autobahn.

*He'd need to get used to that*, I thought, and let him sleep.

In late 1980, the key project that J Mays, Charly Witowski and I would tackle would be the replacement for the present Audi 80. As with the Audi 100 programme, Ital Design would also be involved. Prior to J's arrival, both Charly and I had started preliminary sketches and completed some scale models. At the time, Martin Smith had been supervising the production introduction of the Audi Quattro and also the build of the Audi Forschungsauto. After completion of those projects he was promoted, together with Gerd Pfefferle, to the 'rank' of *Abteilungsleiter* or department head. With the promotion, Martin would also move to head up the interior design department.

The new Audi 80, which was known as the B3 model in-house, was launched in 1986. As with the Audi 100, Ferdinand Piëch had ambitious plans and at initial briefings with Herr Warkuss had indicated that he wanted to target an aerodynamic value of around CD 0.23.

Below CD 0.25 cars start to take on very dramatic forms, which also drives expensive development costs. At that time, aerodynamics was a hot subject widely recognised for pushing down fuel bills, and several companies had shown concept vehicles with CD figures targeted at the numbers Piëch was aiming for. Those cars, however, were only studies. Piëch wanted a production car with the same numbers, which would be very difficult to achieve.

Wary of the ungainly looks that could result from a design primarily formed in the wind tunnel, Piëch and Warkuss decided to take a twin approach: one model was developed at the wind tunnel of the Italian design company Pininfarina, while a second model, following the same path as the new Audi 100, was developed at the VW wind tunnel and modelled in-house. The Pininfarina model, under the eyes of Audi designers and engineers, would target Piëch's wish for CD

*First sketches for the Audi 80.*

*Me next to an Audi 80 scale model.*

*In a bar with Gerd Pfefferle.*

*Audi 80 aerodynamic development in the VW wind tunnel. That's me crouching at the back.*

0.23, while the second development would aim for a more attainable figure of CD 0.25.

Warkuss got the team together and gave J Mays the job of working on the aerodynamic proposal in Italy, while Charly Witowski and I continued to develop ideas for the more conventional approach, working together under Gerd Pfefferle.

Prior to this decision, the team in Ingolstadt had been working on various sketch proposals, and joining the team, J Mays had already started to make his mark on the studio presentation walls. It was soon obvious why Warkuss had been keen to hire J Mays; he was a very special talent. J was gifted with a very sensitive illustration style and his understanding of dramatic perspective was outstanding. Compared to the RCA, where the innovation and function of a design was often deemed more important than the level of visual presentation, ArtCenter is a complete contrast. Here, illustration is taken to an almost religious level of finish. Looking at J's first efforts on paper this was only too clear.

What was also apparent was J's admiration for the Ital Design study 'Medusa,' which had been shown earlier that year, and he was not concerned about borrowing some of that design's flavour for his own work.

J's first ideas for the B3 Audi 80 had a beautiful fluidity in their execution, not unlike that of the Medusa. Warkuss recognised this and reasoned that this look may suit the 'super-aero' model that would be developed in Italy, hence the decision to give J this task.

The development of the B3 Audi 80 ran through to 1982, also including a coupé derivative. The first stage would include a selection of proposals for the sedan, featuring aerodynamic targets between CD 0.23 and CD 0.25. In all, there were three proposals, including one from Ital Design. The individual teams set to work and J Mays left for Turin with a Spanish design engineer named 'Speedy' Gonzales. The outcome of their work would be shrouded in secrecy until the model review in Wolfsburg in the spring of 1981.

Meanwhile, the number of designers in Ingolstadt continued to expand. Martin had hired a new designer from England for the interior team, named Graham Thorpe, an RCA graduate who had been working at Ford. Also for the interior studio, he hired a German designer from Munich, Mike Ninich, and a further ArtCenter graduate named David Robb joined the team. In the exterior studio, an Austrian named Erwin Himmel started as an intern. Himmel had impressed Warkuss in a similar way to Peter Schreyer, and like him, would be sponsored by Audi at the RCA before joining the team after his graduation.

With the ever-increasing headcount, it was decided to expand the design building. A new interior design workshop was added at the far end of the building, with a second floor studio for interior design above it. In addition to this, room was found for an advanced design studio in which Dave Robb, Peter Schreyer and Mike Ninich would later reside.

In terms of the design process at Audi during the '80s, there were no real advances. The development in computers was still very much in its infancy and digital displays were just beginning to show themselves in various applications. Developing a design from sketch through to clay model and the final design sign-off was still a laborious manual process and very much the same as that I had first experienced working as an intern at Chrysler UK in 1972.

Back home in Reading, my father's cancer was getting worse, and, unknown to us all at the end of 1980, he only had months to live. He had decided to stop working, but despite the obvious pain he was suffering in his back, he kept going without complaints or admitting the fact he may not have long to live. He kept active and still drove regularly to the meetings of the local church council, with which he was very much involved.

I visited Reading for Easter of 1981 and celebrated with the whole family. It would sadly be the last time we would meet as a complete group. From then, my father's illness accelerated quickly. After the Easter weekend, and before leaving Reading to get the bus to Heathrow Airport, I had been down to the local shop. I had already said goodbye to my father but walking back I saw him driving towards me in his blue Vauxhall Cavalier, on his way to another church council meeting. He waved; it was the last time I saw him alive.

Ever since I went away to boarding school at Earnley I had written a weekly letter to my parents.

"It's what you do," they had always said. In return, they would take it in turns to reply. The weekly ritual of writing home returned upon my moving to Germany. As mid 1981 approached I had no idea my father's health was rapidly deteriorating as the weekly letters continued to arrive, always positive in their message.

In the August of 1981, I drove to what was then Yugoslavia with Dave Robb and an American VW designer named Ed Golden. We had stopped off at the Istrian harbour town of Rovinj. My parents had visited it a year or so before and during one of the last telephone conversations with my father, he had warned me to be careful not to tread on any sea urchins when swimming.

I was lying on the beach when suddenly I heard someone call my name:

"Peter."

I looked up but there was nobody to be seen. Then, for no apparent reason, I burst into tears. I later learnt it was the exact time that my father passed away.

It wasn't until the next day that, with the help of Peter Schreyer, the authorities were able to find me. I called home and we packed the car and set off back to Germany. Driving back through the Alps, it was a glorious sunny day and I had time to reflect. Father would have loved this, I thought.

All three models for the Audi 80 were reaching their completion. The design review would take place in VW's new presentation hall in Wolfsburg, as it was now deemed too much of a security risk to bring senior management to Ingolstadt.

A large entourage from Ingolstadt made its way north. J Mays arrived from Italy with his extreme aerodynamic direction, as did Giugiaro with his next proposal for Audi. Would history repeat itself?

The evening before the design review, the three models were positioned in VW's large 'Wahlhalle,' as it was named (translation: Voting Hall).

J pulled the covers of the super-aero proposal, which was very close to Piëch's desired CD 0.23.

We gasped in shock; the car was certainly very different. In order to achieve the low aerodynamic values J had had to create a form that was something close to the shape of a large light bulb. There were no hard edges, except at the tail end of the car, which was cut off sharply. It was packed with loads of technical innovations that could only be the results of Piëch and his engineering team's thinking. The cabin was teardrop formed and featured double curvature side windows. In order to reduce any unnecessary drag, the engine compartment was completely sealed off, and cooling air was directed through slits at the lower front edge of the car through channels to a fully enclosed radiator system, which was basically a tube with a fan at the end of it. Air was forced through the tube and exited the system via ducts behind the front wheelhousing. All this resulted in a design which, to be honest, looked a bit of a blob. Were Audi bold enough to turn this into a production car? It would be a challenge.

You could say what you liked about J's work, but it did make our proposal, which was basically a small version of the new Audi 100, look somewhat tame.

The car from Giugiaro had also relied on intensive wind tunnel work and lay somewhere between the in-house model and the proposal from J. Unfortunately the Giugiaro design had very little of the character of his beautiful 'Medusa' study.

On the day of the design review, as with the Audi 100 presentation, only senior managers were allowed to attend, and we watched the proceedings on monitors in a large control room. Once again, machine-gun-wielding security men were everywhere, and, as we found out later, the outcome of the day's review would be the same as the last. Giugiaro was thanked and received his cheque, and the Mays model was appreciated for its innovation, but would have cost too much and been very risky at this time as a production proposal. Once again the in-house direction got the thumbs up, and Warkuss and Pfefferle left for Ingolstadt with smiles on their faces.

This, however, was not the end of the story. Now that a key direction had been selected, two further proposals were developed in Ingolstadt, ending in the selection of a design that featured a combination of ideas. The final B3 Audi 80 that hit the showrooms in 1986 would benefit from the J Mays signature, and a small bit of Medusa – take a look.

In 1982, a shockwave rolled through the Audi Design department. Martin Smith had been approached by BMW's chief engineer, with an offer to join the Munich company as an eventual replacement for its design chief, Claus Luthe.

Martin accepted the job, and, in a second shock, took J Mays and Graham Thorpe with him. It was a big blow for Warkuss and the Audi Design team.

Thankfully, the recruitment drive over the last few years meant that Audi Design, despite the loss, still had a strong team to support Warkuss. Horst Kretchmar took over the provisional role of leading the interior studio and the young guns such as Peter Schreyer were moving up in the department. In addition, Erwin Himmel had graduated from the RCA and joined the team, and it was immediately clear that Warkuss saw great potential in the young Austrian.

Maybe in response to me staying with the team and not joining my long-time friend Martin (although Martin had kept his plans very much in the dark and had not asked me to join his team), Warkuss gave me a promotion to *Referent*. Being a German term, I had no idea what *Referent* really meant, but later learnt it was a team leader without a team. Although a promotion was good, I still felt that my career at Audi was levelling out somewhat.

As fortune had it, my next project would revive my enthusiasm.

# Chapter 16

## *Audi Sport Quattro*

**Hartmut Warkuss and Gerd Pfefferle came over to my desk. They had just visited the Audi Motorsport department, located in an old Edeka supermarket building near the main Audi production facility. I knew the location quite well since Peter Schreyer and I had designed the stripe motive for the first rally Quattros, driven by Hannu Mikkola and Michèle Mouton, and had to visit the facility quite often to prepare the cars for painting.**

Warkuss said that there was a monster (or did he say monstrosity?) being built in the motorsport department and, as he put it, they needed to send in the fire brigade before things got out of control; I was to be the fireman.

The primary reason for the development of the Audi Quattro was to win the world famous Monte Carlo Rally, and after that the World Rally Championship. In his days prior to Audi, as development chief for Porsche, Piëch knew the value of motorsport as a platform to promote a company's image, and to this end had developed the famous Porsche 917 to win the 24 Hours of Le Mans outright. Now at Audi he wanted to crush the world of rallying in the same manner, and the Quattro would be his tool.

Piëch's plan almost panned out the first time the Quattro participated in the 1981 Monte Carlo Rally. As soon as Hannu Mikkola hit the snow covered stages, he started to build up a massive lead, which left the competitors dumbfounded. Unfortunately, a combination of technical problems affected Mikkola's driving and he left the road, damaging his car so badly that he had to give up.

Things went better on the next Rally in Sweden, where Mikkola won and laid out the path leading Audi to win the 1982 World Championship.

As the saying goes, 'the competition never sleeps,' and soon Audi's dominance in world rallying looked threatened. Many manufacturers were developing special rally cars, which, unlike the Quattro, had very little in common with their production derivatives. Hannu Mikkola and Michel Mouton began to struggle, and a solution was needed to remain competitive. Mikkola had two key issues with the Quattro that he claimed were hindering his performance. The car's wheelbase was too long, making it hard to position on tight stages, and the front screen was so flatly angled, that he had problems with reflections in the screen on night stages. Put shortly, he needed a car with a shorter wheelbase and a more upright front screen. This was the birth of the Audi Sport Quattro.

By the middle of 1982, although Audi Motorsport was heading towards a win at the World Championship, they realised they needed to react quickly to Mikkola's wishes, and in a shroud of secrecy the decision was made to go ahead with the development of a homologation special based on the Quattro, to address all the issues. Key proviso: the car still had to have the character of the original car in order not to dilute its mission to promote Audi's production vehicles.

On the drawing board, the issue of shortening the wheelbase of a standard Audi Quattro was fairly simple; you just split the body behind the door, where there was space to remove some length in the car. But how do you solve the problem of the front screen angle? It then occurred to the development team that the two-door Audi 80 sedan had a taller roof than the Audi Coupé on which the Quattro was based, and with it a more upright front screen. This then was the answer. They took an Audi 80 two-door sedan, shortened it and then created a rear end that looked similar to the coupé.

Audi Motorsport began to develop a first prototype, which was the monster that Warkuss had been referring to.

The Motorsport department had always come to Audi Design when needing help on any styling matters, but this was usually confined to graphic treatments on the rally cars. When it came to building a complete new body for a car, they were out of their depths.

Realising this, they called Audi's chief modeller, Wolfgang Holzinger, and asked him if they could borrow his knowhow to solve some of the body modification problems on the prototype. Holzinger obliged and helped them model in wider wheelarch houses, plus the rear roof pillar treatment to take on the character of the original coupé. To this, they mounted moulded off copies of the front and rear bumpers used on the existing rally cars, which had by now been butchered to take in extra cooling air and were a mess of parting lines and crude openings.

Holzinger did his best, but the end result was a long way from perfect and a good solution was needed to clean up what was, in fact, a pig's ear.

Audi Motorsport had not considered asking Warkuss for help; they were happy with their work, and by the time he and Pfefferle had got their eyes on the prototype, they had already commissioned a company in Switzerland (called Seger and Hoffmann) to start making the tooling needed to build the production version. Warkuss was not happy.

With no time to spare, Warkuss told me to head off to Switzerland and sort things out. It was early November of 1982 and I had to get the job completed by Christmas. I would leave after the weekend.

On the Monday morning, I arrived at the studio with my suitcase and filled a large box with all the equipment I thought I would need. Warkuss had said he would organise a car for me. I waited, and soon Gerd Pfefferle entered the studio, car keys in hand, with a suspiciously big smile on his face.

"Congratulations, Birty, here are the keys to your VW Jetta."

My first big business trip and I had to drive a VW Jetta. Not only that, it was painted in Lind Green, the favourite choice of most German pensioners. Didn't they have anything better for me?

Pfefferle explained that secrecy was paramount, and driving a Jetta I would go unnoticed. How right could he be? I didn't believe a word, but there was no time to argue – I needed to get on the road.

The company Seger and Hoffmann was located in a small Swiss town called Steckborn, on the north western bank of the Bodensee (better known to English tourists as Lake Constance). Located at the foot of the Swiss/Austrian Alps, the lake is a favourite holiday venue for retired German couples. Pfefferle was right; my transport was not going to be noticed and I really did blend in nicely.

Originally, the company built boats and it was through its knowledge of moulding fibreglass that it had expanded into manufacturing racing car bodies, which by 1982 was its core business. Of particular interest in the motorsport world was the fact that it was one of the few companies in Europe using the relatively new material, carbon fibre. In the world of Formula 1, McLaren had been the first team to develop a carbon-fibre chassis or monocoque and they were soon followed by other teams, including the German team ATS. ATS were quite new to the F1 business and its cars were powered by a BMW four-cylinder turbo motor. It was Seger and Hoffmann that supplied the carbon monocoque for ATS, and it also made the bodies for the Sauber SHS C6 sports car.

The company's two founders, Herr Seger and Herr Hoffmann, had split responsibilities: Seger ran the manufacturing and Hoffmann was the business director.

I made my way to the Bodensee and crossed the lake by taking the car-ferry from Meersberg to the town of Constance, passing over the German border at the far side of the town and on into Switzerland. From there, it was a short drive north along the bank of the lake to Steckborn and Seger and Hoffmann's business site.

It was misty over the lake and, being November, a location that in summer is a bustling holiday location was at this time of the year practically abandoned. Driving on, I passed through several small villages on the lake, all of which had the character of ghost towns. It was all very spooky.

At the far end of Steckbach, I pulled off the road to the right and into the car park in front of a collection of buildings. I parked the Jetta next to a number of sailing boats on trailers, walked past the boats and into the reception area. I was expected, and a good-natured gentleman with a welcoming smile came out to meet me; it was Herr Hoffmann.

"*Gruetzi*."

*What's he saying?* I thought. Greetings – that was my first introduction to the Swiss German dialect. Fortunately, Herr Hoffmann also spoke good English, but I was quite sure the fabricators I would be working with probably didn't, and the Swiss dialect was, for me, completely incomprehensible, sounding not unlike somebody about to vomit.

We exchanged a few pleasantries then, looking at the impressive Rolex on his wrist, Herr Hoffmann suggested we go for a business lunch, something he obviously did on a regular basis. It was a good opportunity to go over my job for the next few weeks. I waited while he picked up his car and was pretty excited when a white BMW M1 coupé came into sight and drew up to a halt. Herr Hoffmann reached across and opened the door. *That's more like it*, I thought, *no chance of being mistaken as a pensioner in that.* As Herr Hoffmann blasted out of the company yard and on to the road, all I could think was that Seger and Hoffmann were obviously doing good business.

We drove a short distance to a restaurant, which was the only one open at this time of the year.

We ordered, with Herr Hoffmann recommending the local Bodensee specialty Bodensee Fenchel, which was a fresh water fish similar to herring.

Getting to business, Herr Hoffmann's jovial manner changed a bit. His company had been given strict instructions to complete work on the master model it was building for Audi Motorsport by the end of the year. Up until now everything was going to plan but my intervention had come as a bit of a surprise to them. Of course, I was welcome to make adjustments to the model but there was very little time left for major surgery. I could start right away.

We drove back to the factory, a description that flatters somewhat, since it was more like a group of large out-buildings. Herr Hoffmann dropped me off at the reception and drove the M1 behind one of the nearby boat sheds; it was Switzerland, and flaunting any kind of wealth was not a good idea if you ran a business.

I walked inside.

"Hey, how ye doin'?"

A young guy, called Jim, walked over holding out his hand. An American, he was working as a contract engineer for Seger and Hoffmann on some of its racing car programmes. He had long, blonde, shoulder-length hair, and was the last person I would have judged to be an engineer, looking as he did, as if he'd just arrived from Woodstock. We exchanged a few words. He was working in a different area to the one I'd be working in, and he was intrigued about what we were doing. I decided it best to avoid an answer.

"This place is dead. How about coming over to my place for a meal sometime?" It sounded a nice idea.

By now, Herr Hoffmann was back from hiding his car, so we walked through a security door and into the main workshop area. The room was not dissimilar to the fabrication hall at Audi, being a large hall with several metal floor-mounted measuring plates and a scattering of workshop machines at its periphery. To my right was a large metal cabin, which I assumed was a paint-booth, but I was wrong. This, explained Herr Hoffmann, was the latest toy – a computer-controlled water-cutter; the very latest in model finishing hardware and certainly not something I had heard of before. In the booth stood a mechanical robot, with what looked like a spray nozzle attached to its outstretched arm. In front of the robot was a metal rig with a carbon fibre moulding, looking like the inner structure of a door, attached to it. Hoffmann explained that carbon fibre was very difficult to cut cleanly in the conventional manner with saws. This machine used a jet of water under very high pressure, which guaranteed a clean cut. He explained that the idea of using water to cut cleanly through objects was inspired by unfortunate circumstance occuring in submarine warfare in the last war. Apparently, if submarines in deep water, with a high outside pressure, had suffered damage to their hull and were leaking water, then the spray of water leaking into the vessel was strong enough to cut a sailor clean in two. I looked, with some alarm, at the robot standing menacingly in the middle of the booth, with warning lamps all around. Later, I learnt that when in use, warning buzzers would sound, with large safety doors simultaneously sliding into place. No wonder.

We moved on. Looking around, of more interest to me were the racing cars, or parts of them, sitting on various benches or measuring plates. At the far end of the room I could make out the body of the Sauber SHS C6 in its distinctive BASF red pinstriped livery, and of even greater interest, standing on some metal feet, was the monocoque of the 1983 ATS Formula 1 racing car. Later in the week I would see ATS's technical director, Gustav Brunner, inspecting the black carbon tub of its next year contender.

Finally, we arrived at the area of the workshop where the model of the modified Quattro was standing. The car was finished in what looked like (and was) grey body filler, and was surrounded by modellers who were busy sanding the surfaces; there was dust everywhere.

One of the fabricators put down his sand paper and walked over.

"*Gruetzi,*" he said. He introduced himself as Herr Seger. In comparison to Herr Hoffmann, Herr Seger appeared to be a shy and unassuming person and looking at his dust covered overall, was obviously a hands-on man, leaving the business side of the company to Herr Hoffman. That, as it turned out, was very much the case.

I walked over to the model with the two company owners. As it stood, the car looked like a mirror copy of the model that Audi Motorsport had developed, which Warkuss had described as a monster. The basis was there but I realised almost every surface and detail of the car needed to be either refined or, in the worst case, redesigned. I needed to make a plan, and then develop a few quick sketches of what I wanted to do. What worried me the most was the fact that the car was made of a material as hard as concrete and I wondered how on earth we could model in that?

I explained to Herr Seger that Herr Warkuss had sent me to direct the model development at short notice and that I needed at least 24 hours to get some sketches together. In the meantime, they could keep sanding the key surface areas of the body that looked OK to me. I decided the best thing would be to go to the hotel Seger and Hoffmann had arranged for me and start sketching.

I left the premises and drove towards the centre of Steckborn, pulling the Jetta into a car park in front of the hotel. Not the first time a Jetta had graced the car park, I guessed. The hotel was a large villa at the side of the lake, which looked eerily empty and dark; I walked in.

As I had suspected, there was nobody to be seen. I called and eventually a man, who turned out be the hotel owner, came into view.

"*Gruetzi.*"

I was lucky, he told me, as this was the last week of business before closing for the winter. As it turned out, they might as well have already been closed, since the restaurant was not open and I would have to get my breakfast the next morning from the local bakery in the town. The owner showed me to my room and explained that as I was the only guest in the hotel it would be a good idea to take a key should I decide to go out later. Apparently, there would be no one there after 8 o'clock to let me in. I took the key, thankfully, as there was no way I was staying there for the rest of the evening.

He showed me to my room, leaving me wondering why I was on the third floor if the place was empty. No matter, I unpacked my case. Taking out a sketch book I had brought with me, I began to think about how I could fix the monster.

Before I left Seger and Hoffman, I had taken a few Polaroids of the car for reference. Warkuss was right: the car was far from pretty, but it had a robust and powerful stance and I was sure I could turn it from a pumpkin into a carriage. I began sketching.

The engineers at Audi Motorsport had told me that the critical issue at the front of the car was creating an area for as much cooling air as possible. This meant using smaller headlamps than the normal Audi Quattro in order to create a wider top grille opening. Also, the front bumper and lower valance needed as many cooling openings as possible. Putting it mildly, the front needed to be like a cheese grater. Another area where there were cooling issues was over the ducting for the turbocharger on the upper right side of the engine; here, some kind of vent in the bonnet was needed.

Starting with the front of the car, the bumper and valance area of the car as fabricated by Audi Motorsport had 24 air holes in it, and the grille it was proposing was a very crude trapezoid block that it had taken over from the existing rally cars. I decided the front bumper needed a much less cluttered form that integrated better with the front fenders and minimise the number of inlets to six main holes fitted with a smaller black mesh. The main grille just needed a more refined solution, with recesses to take the new smaller headlamps, and for the bonnet I decided it would be best to create some ideas in black tape on the actual model, which I could do at a later date. My first priority was to get the basic form resolved.

At the rear of the car, as with the front bumper, it was just a case of calming things down and getting a more integrated feel. I started to feel better about it and closing my sketch pad, pondered over the next issue as to where I could get something to eat.

I walked down to the reception where I found the hotel owner about to close for the evening. For a meal, he suggested I drive back into Constance, since everywhere local would be closed – I had no choice. Constance, however, seemed a good idea since I could also use the time to look for a new hotel. Driving back towards the German border it was clear to me that there was no way I could stay much longer on the misty banks of the Bodensee.

Constance, similar to Steckborn, was also pretty empty, but at least a few restaurants were open. I found a bistro, went in, and made myself comfortable at the bar, where I began to peruse the menu. This part of Germany is called Baden-Württemberg and like the

Swiss they have a strong dialect very different to the Bavarian accent I was used to. The culinary specialities in the menu were also special to the region. I'd tried the *Bodensee Fenchel* at lunch; what else?

"*Was sind Maultaschen*?" I asked.

The barman tried to explain but immediately I got lost.

'*Maul*' is the mouth of some kind of animal, usually a dog, and '*Taschen*' are bags. Dogs' mouth-bags. I thought I'd give them a miss and ordered *Wiener schnitzel* – I couldn't go wrong there.

There's nothing worse on a business trip than sitting alone in a restaurant. I tried making conversation with a couple of girls at the bar, but they looked at me as if I was about to molest them. I wished I was back home in a local pub.

No point hanging around here, I thought, and paying my bill, I made my way back to the Jetta. Somehow, it looked at home on the lonely streets of Constance.

I drove back through the Swiss border control, the guard looking briefly at my passport.

"*Haben sie Waren dabei*?"

Smugglers probably used Lind Green Jettas, I thought, reflecting on Pfefferle's point of not getting noticed. I turned right after the border control and on towards Steckborn. I noticed a couple of army trucks at the side of the road. There were some military police standing with lit batons, but I thought nothing of it and motored on.

I got back to find the hotel bathed in darkness. Why hadn't they at least left one light on? I locked the car and walked up to the door. After an age trying to find the door lock, I eventually made my way into the reception area. Thoughts of the *The Shining* came to mind. The reception area was even darker than outside, and I realised that up until then, never in my life had I needed to find the light switch in a hotel lobby. I rubbed my hands up and down the walls near the entrance – nothing. I could just make out the reception desk: telephone, note pads, but no switches. It was no good – I needed light and the only source of that was the Jetta.

I went back to the car, climbed in, started it, and drove carefully up to the hotel entrance. I nudged the front wheels up on to first step and put the headlights on full-beam. To my satisfaction the reception was now bathed in light. Back in the lobby, I soon noticed a brass switch panel on a column; I breathed a sigh of relief.

Back in my room, I finally collapsed on my bed. What a day. Little did I know, things weren't over yet.

I lay back, my head racing. Visions of Jack Nicholson and Herr Hoffmann patrolling the corridors with axes troubled my thoughts. Had I locked the door?

Finally, sleep drifted over me.

It must have been around three in the morning when my room suddenly filled with the beams of flashing blue lights, followed by an explosion of noise that made me sit up in bed, expecting a plane to suddenly crash into the room, any minute.

At least it wasn't Jack Nicholson, but what was going on? I went over to the window, opened it and stretched my head out to find the source of the commotion. The noise had subsided to a low rumble, and looking down, a young soldier, protruding out of a hatch on the top of a tank's gun turret, greeted me.

"*Gruetzi*."

My mind flashed back to the border crossing and the military police at the road side. I was sleeping in the middle of a military exercise; and I thought Switzerland was neutral. Later, I would learn that the Swiss Army practice war one week a year. That week.

The second time I woke, the tanks had moved on and Steckborn was back at peace with the world.

I left the hotel and found the bakers for breakfast. It was full of young soldiers who all looked as if they hadn't got much sleep either. I must have looked terrible as I slowly sipped my coffee and reflected. It had been a night to remember.

I put my selection of sketches on the bonnet of the model, and Seger and Hoffmann looked at them with a degree of mistrust. It was clear they weren't happy.

"You want to redesign the whole car?" Herr Hoffmann exploded. All traces of my jovial lunch host from the previous day vanished in the dust on the floor around the model.

"We have to," I explained, my description of a 'pig's ear' lost on them.

Herr Seger was more accommodating: there was no time to argue, time was running out, they'd do their best.

Herr Hoffmann left the meeting and I walked round the model with Herr Seger, pointing out what needed to be modified. One thing was clear, I had to be very sure of what I instructed them to do. This wasn't clay, so there was no room for mistakes. I suggested we start at the rear of the car and move forward, as I needed time to consider how to handle the air vents on the bonnet. There was a technical drawing of the body on the wall, so I pinned some tracing paper on it and gave them a new bumper section to model in. Two guys grabbed a couple of pneumatic drills and started to chisel off the old bumper profile like it was stone.

Modelling in a hard material was new to me, though not unusual in the car business. Clay modelling had roots in North America and had been imported to the studios of England and Germany, but in Italy, arguably the home of automotive styling, they still used a plaster material called gesso. The material Seger and Hoffmann used was not dissimilar to gesso, and the more I watched these craftsmen work with it, the more admiration I had for them.

At the side of the body, Herr Seger had returned to what he had been doing before the meeting and was laying a thin coat of the body filler material they used and refining the top of the wheel fender flairs. I looked on, and to my relief, realised he was a true artist. He modelled the surface using just a steel ruler; it looked perfect.

On the plus side, applying black design tapes to the model surface was much easier than clay, since once applied they stuck and stayed there.

I readjusted some lines on the rear pillar and taped a line over the rear section of the roof. The Audi Coupé had a small step at the rear of the roof which was missing on this model; they would need to raise the surface and model it in.

Back at the drawing, I defined a new section for the front bumper and valance and realised that, similar to the rear of the car, some major surgery was required. The team now had plenty to get on with, so I thought it best to leave them in peace, allowing myself some time with my sketch pad to work out some of the details.

Outside in the reception area, I bought a cup of coffee from the vending machine; to this day I can still taste it.

"Hey buddy, how's it goin'?"

It was my American friend Jim, from the first day.

"Fine," I told him. "A few starting problems but things have calmed down."

We got into a conversation. He was from Florida and had been working for a Trans-Am racing team. Trans-Am was a touring car championship in America. Through the 'Racing Mafia,' as he called it, he had met Herr Hoffmann and been persuaded to do some work for them. He said he'd have gone back to the States long ago but he'd met a Swiss girl and decided to stick around.

"Why don't you come over for dinner on Thursday?"

I planned to leave for Ingolstadt on Friday afternoon, so this would work for me – better than sitting alone in a bar in Constance.

"We can have fondue."

We left Seger and Hoffman right after work on the Thursday and I followed Jim to his house. Hopefully I would remember the way back.

Jim lived in a town south of Steckborn called St Gallen, in an unremarkable block of flats at the edge of the town. I parked the Jetta and we went in.

Jim's girlfriend greeted me. "*Gruetzi.*"

She shook my hand and beckoned me into the kitchen where, on the table, there was a heavy saucepan on top of some kind of burner. This, explained Jim, was Swiss fondue. In the saucepan was a mixture of melted cheese which was bubbling away. It smelt like old socks, I thought to myself. We sat down. Jim poured some wine while his girlfriend explained, in a broad Swiss accent, how to eat the cheese mixture. It was very simple. You just put a piece of bread on a long fork and dipped it in. The fondue was great and the atmosphere was very relaxed. There was no difficulty in finding things to talk about, with both me and Jim being nuts about motorsport.

Jim's girlfriend smiled as we talked, but didn't have much to say. I guessed she didn't speak much English. I sipped on a glass of white wine, well aware that I needed to be careful since I was driving.

We finished the fondue and Jim beckoned me to follow him. We went through to the entrance hall where he opened what looked like a broom cupboard. There was shelf with a collection of crumpled, rolled up newspapers, which were covered in some kind of Asian writing. It smelt dusty and sweet. He selected one of the newspaper rolls and took it back into the kitchen.

Jim unrolled the newspaper and revealed what looked like a pile of small twigs. By now, it had clicked and I was pretty sure it was some kind of intoxicating plant. It didn't faze me. If that was what they were into, fine, but it wasn't really for me.

Jim took out a small pipe and pressing some of the twigs into its bowl, lit up. He drew a couple of times and passed it on to his girlfriend, who did the same. The room began to fill with the sweet smell of whatever was burning in the pipe. She offered it to me.

I'd tried inhaling a joint years ago, but as I didn't

smoke, it always ended in a coughing fit and I was in no doubt that this would be the same. OK, I thought, just a small suck.

I took the pipe and drew in. As expected, I almost burst out in a coughing fit. Whatever was in that pipe was very strong and my head started to feel light almost immediately. I passed the pipe back to Jim, declining the next time. Unknown to me, I was, however, still inhaling this jungle weed. It was a small room and no windows were open. I got up from my seat, my head spinning. I needed some air and fast.

I went into the neighbouring sitting room and opened the balcony door, stepping into the cold November air. There was no doubt that I was not in a good way. Struggling to get my thoughts together, I knew I had to make my way out, but first I needed to clear my head. I started taking in deep gulps of air, but it didn't seem to make much difference. I staggered back into the kitchen, but Jim and his grinning girlfriend had obviously decided to finish off the pipe in the bedroom since the kitchen was now empty. I thought it best to return to the balcony.

Slowly, my head began to function, and though I wasn't one hundred per cent, it was time to leave.

I wobbled my way back to the Jetta. Concentrate, I told myself, and I turned out of the car park and back towards Steckborn. It must have been about one in the morning and fortunately the roads were quiet. I needed air and wound down the window, sticking my head out into the rush of air passing the car.

Thankfully, the route back to the hotel was very straightforward, or at least it would have been, if not, once more, for the Swiss Army. Looking ahead, I made out a jeep with the large sign that I knew all too well from the German autobahns: "*Militair Konvoi.*"

I drew up to the jeep, my head still out of the window. Did military police stop civilians? I hoped not.

I pulled out and passed the jeep, still taking in gulps of air, and immediately realised that my next obstacle would not be quite so easy to pass, as the rear end of a large tank weaving its way up the road came into view.

Unfortunately, the tank's exhaust was belching thick black smoke right in the path of where I wanted to overtake. I decided it was best to hold back.

Then came my chance, as the tank began to pull up. I moved out, my head still partly out of the window; on top of the slowly diminishing influence of pot, I now had the diesel fumes to contend with. Drawing level with the tank, I realised that the convoy was probably the entire Swiss Army on the move. The opposite lane was clear and slowly I began to make progress, until at last a military policeman with an illuminated baton came into view, signalling the end of the convoy. He waved me on.

Still in one piece, and now thankfully with a clear head, I arrived back in Steckborn and pulled into the hotel car park. It was early Friday morning. Thankfully, I would later be returning to Ingolstadt for the weekend. What a week!

The weekend flashed past. It was great to catch up in M's Corner and tell my friends about my adventures down at the Bodensee.

Monday arrived soon enough, though, and after a stop at the Audi studio to say hello, I was on my way south again. I had given Herr Warkuss and Gerd Pfefferle a quick bit of feedback and we agreed that they should visit Steckborn by the end of November. I figured that by then things would be pretty advanced.

I arrived back at the Bodensee in the late afternoon and checked into my hotel, which, thank goodness, was now located in Constance. This meant I would need to cross the border twice a day, and with that, as I would learn at the end of the week, came the increasing inquisitiveness of the border guards.

Since my departure the previous Friday, Seger and Hoffmann's team had made real progress on the model. The front and rear bumper profiles were now modelled in and the large teardrop fenders had benefitted from Herr Seger's magic touch. Everyone, including Herr Hoffmann, could see that there was light at the end of the tunnel.

My next job was now to define the six large air intakes on the front bumper and start designing the various air ducts on the bonnet.

The centre profile of the bonnet surface had to be raised because of the large intercooler placed in front of the engine and here, as with most of the front of the car, cool air needed to pass through.

I taped on three rectangular slots at the front of the bonnet bulge. Now all that remained was the intake above the turbocharger on the right side of the bonnet.

Since the previous Friday I had been sketching ideas of how this area could be handled. I didn't want just a hole with a piece of black mesh inserted. Thinking about turbochargers, many of the Formula 1 cars of the day had turbocharged engines; maybe I could draw on their designs to influence the bonnet detail. Soon, I had an idea.

The 1982 Ferrari Formula 1 cars featured a vent on their side pods with angled vanes. I had seen the cars that year at Hockenheim, where unfortunately Didier Pironi had had a very bad crash in practice. Putting that fact aside, I thought I could steal the Ferrari solution for the Quattro bonnet.

I marked out a shape on the grey surface of the Quattro model and taped in a row of angular vanes. The result looked good, and I instructed the fabricators to cut out the area so we could model in the angular blades. To call this area an intake is not strictly true, the idea of the vent being, in fact, to let hot air from the turbocharger out. Previously, Audi Motorsport had just cut a crude hole in the bonnet and filled it in with a piece of wire mesh that could have come from a chicken coop.

Towards the end of the week, Seger and Hoffman's craftsmen had modelled in what I wanted, and looking at the front of the car, it was now really taking on a strong character. I hoped Herr Warkuss, and of course Herr Piëch, would approve.

The week had passed by quickly and we agreed that my presence was not be required for a couple of weeks, so to give the team time to double over the model and slowly get it ready for presenting to Audi's senior management in mid-December, I packed the Jetta and drove back to Ingolstadt.

Returning to Steckborn a couple of weeks later, I was pleased to see that the car was symmetrical and almost finished. I would spend a few days correcting some of the lines, before greeting Herr Warkuss and Gerd Pfefferle to show them the final model. Warkuss expressed himself satisfied and agreed for me to end my time in Steckborn and return to the Audi studio.

I am pretty sure Ferdinand Piëch and his reports would visit Seger and Hoffmann, but that piece of information I never found out, only learning that in early 1983, work had started on production tooling and that a prototype of the car, to be built by the Bauer company in Stuttgart and named the Audi Sport Quattro, would be shown at the September Frankfurt Motor Show.

My involvement in the development of the Sport Quattro was now concluded, or at least I thought so. However, one day in the spring of 1983, I looked up from my desk to see Herr Piëch standing there with a large brown envelope in his hand; Warkuss was with him. I immediately thought the worst – the Swiss authorities had traced the Jetta's number and I was wanted for smuggling.

"Herr Birtwhistle, I need you to do a job for me," said Herr Piëch in his usual, slightly intimidating style – it was almost as if someone had opened a fridge door nearby.

As it turned out there was no

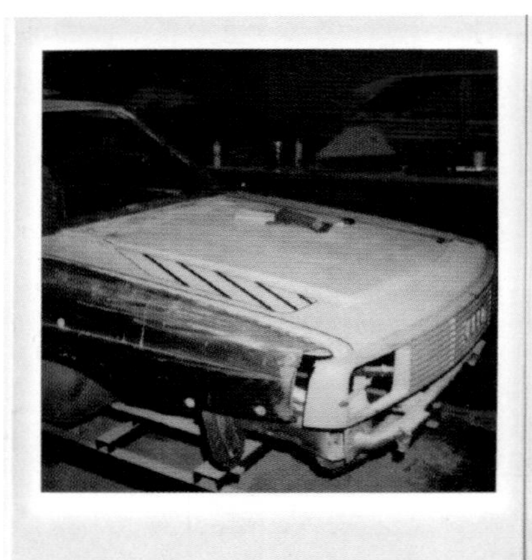

*Polaroids of the Sport Quattro hard model. The only images of its development.*

need for alarm. Audi wanted to start drumming the propaganda machine by releasing limited information in the motoring press about the new Quattro 'rally weapon.' Piëch's idea was for a photo-realistic illustration of the car to appear on the cover of the famous German motoring magazine *Auto Motor und Sport*. With the image on the cover, inside would be an article all about the car.

*Auto Motor und Sport* was Germany's key voice of the car industry. It was thought in many circles that the key powers in the boardrooms of the leading German manufacturers would instruct the influential magazine on exactly what should be written about their products. Nothing could be proved, of course.

*AMS*, as it was known, used a graphic artist called Mark Stehrenberger to create illustrations of upcoming products, and their likeness to the final product was so close that it was pretty clear he was being guided by photos supplied to him from industry sources.

Piëch explained that he didn't want to give Stehrenberger the job of illustrating the new Sport Quattro, but did want an illustration done to look as if it had come from him. He asked if I could do it.

Though he said 'could,' he actually meant, "You will do it."

"Of course. I would be pleased to do this task," I replied.

Piëch continued. He told me that when I had finished the illustration, which Warkuss would check, I was to put it in the envelope that he then gave me, and post it with no cover letter. I looked at the address, which was in Los Angeles. Thoughts of *Mission Impossible* popped into my head.

The discussion was over, and Herr Piëch left in a manner not dissimilar

*Top: Auto Motor und Sport cover, June 1983, with my illustration. (Motor Presse Verlag)*

*Left: Illustration close-up with my hidden signature. (Motor Presse Verlag)*

to that of Darth Vader, with Warkuss winking at me as they walked towards the door.

I was slightly irritated, not about imitating Mark Stehrenberger, but realising that I wouldn't be able to sign the illustration. Here was a chance to get one of my drawings on the cover of *AMS*. I needed to find a way of secretly signing it, and I knew how.

As I wrote earlier, in the 1960s, American designers had hidden messages, sometimes pretty crude ones, in their drawings by disguising them as patterns. Sometimes they would be formed in some crazy paving or as a wavy reflection in the car's body. I decided in an instant to apply the same kind of feature in my Sport Quattro illustration and set to work.

After two days, I had completed the job, and in the reflection on the door of the red car I had weaved in letters to spell out 'Birty.' I had naturally not said a word to anyone, and looking over my shoulders as I worked, nobody had spotted my hidden message. I took the finished artwork to Herr Warkuss who, with a casual perusal, approved the picture and instructed me to post it. I sealed the envelope and took it over to Audi's in-house post office. Soon, it would be on its way to Los Angeles.

Later, as planned, the issue of *AMS* with the Sport Quattro article appeared on the news stands, and there on the cover was my illustration. Inside, there was also a double spread of the illustration. The magazine did the rounds of the studio and congratulations were made for my good job in mimicking Stehrenberger. Apparently, no one had spotted my hidden message, and since no summons to Piëch's office came forth, the great man had seemingly not discovered it either. Within two weeks, the magazine was old news and the source of the illustration still only known to myself and a few colleagues.

It was only a few weeks later when I bumped into my old friend Peter Stevens.

"Hey, what have you been up to?" he asked. To this day he is the only designer who had seen my name on the side of that car, adding, "I saw that illustration and thought immediately, oh dear, Stehrenberger's had a bad day today." – Thanks, Pete.

After the adventures in Steckborn and the motivation of working on my own project, I now had to get back into the everyday work in Ingolstadt.

The exodus of Martin and the team he took with him to BMW was forgotten. With the new Audi 80 design completed, final touches were now being made to the coupé version of the car. I had started on this project with J Mays before moving down to Switzerland.

J had put together quite a radical proposal, which featured a wraparound glass cabin where the body pillars were hidden under the glass. It looked great but Warkuss felt it was too progressive, and when J left, he asked the new young Austrian designer Erwin Himmel to take over the project. On my return to Ingolstadt, Erwin was doing a good job but Pfefferle suggested I pick up on the design and come up with some ideas for the opposite side of the model to that which Erwin was working on. I completed some sketches and did a full-size airbrush that Warkuss approved of.

I remember from the time back at Vauxhall, when I once had to share a model with Wayne Cherry, that differences of opinion could soon come to the surface, and it wasn't too long before I realised there was a lack of harmony working this way with Erwin. Warkuss' body language also sent out the message that he was pleased with how Erwin was handling the design. I began to wonder why I was bothering.

*Audi Sport Quattro.*

In the end, differences were smoothed out but for the first time during my time at Audi I began to lose some enthusiasm at work. Like Martin and the other guys, maybe it was time to start looking around.

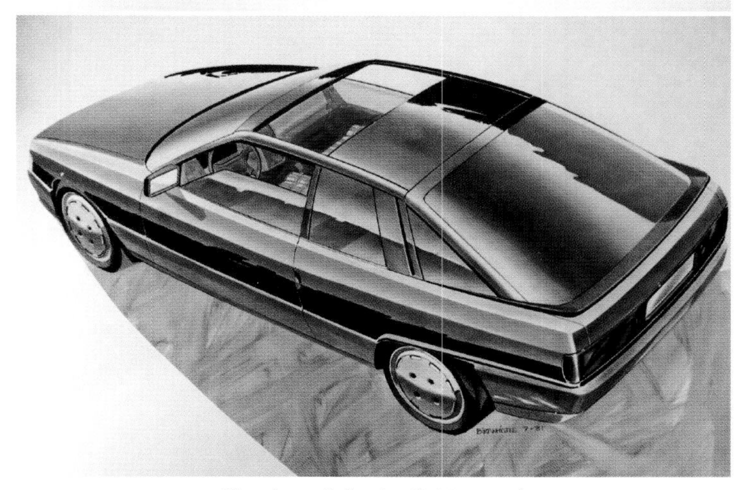

*Sketches of the Audi 80 coupé.*

As chance would have it, I got a phone call from an old friend who worked at Porsche. His name was Dick Söderberg. I had met Dick for the first time when I had visited Martin in Stuttgart when he was working for Porsche. Dick, who was a studio chief, asked me to keep it a secret, but told me he would be visiting Audi the following week to be briefed on a new project that Piëch wanted Style Porsche – the Porsche design department – to handle. He'd be staying in Ingolstadt and he asked if I fancied meeting him for a drink. He had a few ideas that he wanted to pass my way.

I met Dick at the Rapensberger Hotel in the centre of town and after a meal we walked down to M's Corner.

Dick was American and, like most Americans, he called me 'Birdy' rather than 'Birty.' He had family roots in Sweden and had started his career working at GM before moving to Europe to avoid being drafted into the army and ending up in Vietnam. When Tony Lapine, who Dick knew from his GM days, set up the new Porsche studio, Dick applied to join Porsche and got the job.

M's Corner was packed, as usual, and Dick and I made ourselves as comfortable as we could, perching our beers on top of the cigarette vending machine by the door. All evening the conversation had been dominated by small talk but eventually, he got to the point of why he was keen to meet up with me.

During the last few years, Porsche R&D had established itself at the new research centre located west of Stuttgart, near a small town called Weissach. Since its inauguration, the design department had expanded parallel to the engineering department, and apart from working on its own automotive projects, was also functioning as a design consultancy. Dick himself had been working for the past couple of years on the cockpit design for the Airbus passenger plane, which was a good example of the variety of work that Style Porsche handled.

With its growing number of projects, it now wanted to increase its headcount and was looking for three assistant studio heads.

*Full-size airbrush of the Audi 80 coupé.*

*My half of the Audi 80 coupé model.*

Dick asked me, "Would that interest you?"

The question had come at just the right time, and with no hesitation I said yes.

Dick went on to say that I should write to Porsche Weissach's personal chief, Herr Fischer, and describe our meeting and my interest in joining Porsche. Assuming that went OK, he would send me an invitation to visit Weissach for an interview.

The brief business side of the evening was concluded, and we continued with more casual conversation before calling it a day.

I said goodbye to Dick at his hotel and walked the few hundred metres to my flat in Josef-Ponschab-Straße.

Reflecting on our conversation, I felt that it was an opportunity I needed consider carefully. I wasn't desperate to leave Audi, but I did feel that I may have achieved all I could in terms of career development. Perhaps now *was* the right time to move on. I got back to the flat and started to draft a letter of application to Herr Fischer.

***

Dick had mentioned during our meeting that it wouldn't be long before we learnt of what he and Porsche were involved with at Audi.

Herr Warkuss called me into his office with Gerd Pfefferle. He explained that he wanted Erwin Himmel to finish off the design work on the Audi 80 coupé and asked me to start developing ideas for a new derivative of the present Audi 100 model. The project in question was yet another idea from Ferdinand Piëch.

Ferdinand Piëch was not only a powerful man at Audi, he was also a member of the Porsche dynasty and therefore fluent in the product and business strategy of Porsche. Always on the lookout for a way of moving Audi closer to the market territory of Mercedes and BMW, he realised that the then flagship model of Audi, the 200, with its turbocharged five-cylinder engine, was still lower in the pecking order when it came to engine specification and more importantly market prestige. Piëch reasoned that the best way to catch up to the two rivals from Stuttgart and Munich was to offer a product not only with the Quattro AWD system but also an engine bigger than the Audi five-cylinder, preferably a V8. This was where his relationship to Porsche came into effect, and he figured that the V8 motor in the Porsche 928 would fit well into the Audi 200 engine bay.

Warkuss explained that development work on the Porsche-engined Audi was already well advanced. A modified Audi 200 had already been fitted with the V8 from the Porsche 928 and mated to a strengthened Quattro AWD system. The car in question was sitting in the engineering workshop and we made our way over to look at it.

In the workshop, we headed for a black Audi 200 crouched in the corner of the large hall. Warkuss explained that the car would need some cosmetic work without losing the character of Audi's family of models. The prototype's modifications were pretty crude, but it already looked very menacing. The nose had been lengthened to accommodate the Porsche motor, which, although shorter than Audi's five-cylinder, needed more space for radiators and other ancillaries. At the side of the car the wheelhouses had been widened and it was sitting on the familiar disc-like, Porsche 928, forged wheels. Although still recognisable as an Audi 200, the prototype said one clear thing: "I mean business."

We returned to Warkuss' office where he went on to explain the second part of the development, which in turn explained Dick Söderberg's visit to Audi the previous day. Apparently, Piëch had reasoned that if the mechanics of the car were sourced from Porsche then why not have a body design handled by Porsche to round off the package. He felt that a Porsche flavour to the exterior design would give more prestige to the car when it reached the showroom.

It was easy to understand Piëch's strategy, especially as the car was built at Audi's factory in Neckarsulm, where they also assembled the Porsche 924 and 944 models.

Warkuss, ever the diplomat, had naturally gone along with Piëch's plan, and welcomed Dick Söderberg into his office for a briefing and a run through of the project. We all knew, however, that Warkuss was burning to give Piëch a counter proposal from the Audi in-house team.

Sure enough, Warkuss asked me to start sketches and said Erwin Himmel would join the development as soon as he had completed the Audi 80 coupé. I set to work, thinking how ironic it was to be competing with Style Porsche – a team I was now keen to join.

About a week later a letter arrived with the famous Porsche typeface on its envelope. I opened it. Herr Fischer had reacted positively and arranged interviews at Weissach for the following week. The next day, I booked a day off for my visit.

The village of Weissach is situated in a hilly and partly forested landscape, about 20km west of Stuttgart, at the northern border of the famous Black Forest. It was here, in 1962, that Porsche had built a test track, and in 1971 it moved its development centre to the same site, having run out of space in the Zuffenhausen factory in Stuttgart.

In 1983, the centre in Weissach comprised four key areas of activity: engineering, design, motorsport, and a unit that worked for the German military, where such projects as the Leopard Tank were developed.

I drove up the winding lane out of the village of Mönsheim, close to the Stuttgart/Karlsruhe autobahn exit for Weissach; there was no name on the exit signs, mind you. It was only when I was almost on top of the centre that it first came into view. Porsche had done a good job of hiding the place.

I parked in the visitor's car park and made my way to the security gatehouse. It wasn't long before Herr Fischer greeted me and we made our way to his office. Looking around, it was obvious Porsche was expanding, as there was building construction evident everywhere.

Herr Fischer gave me a small introduction to the setup at Weissach and moved on to the terms of the position I was applying for. An unexpected highlight of the position was that I would be able to lease a Porsche.

It wasn't a long discussion, and Herr Fischer said he'd take me over to Herr Lapine's office. I duly followed him.

We moved to the main foyer of the engineering building, which was dominated by a McLaren Formula 1 car; Porsche had recently become engine supplier to McLaren. Impressed, I continued to follow Herr Fischer. The Porsche design department was located at the far left of this building. The entrance to the department was made up of two unassuming wooden doors; it reminded me of Vauxhall. There was a metal name plate that read Style Porsche. Herr Fischer pressed a button and the doors opened with a buzz; we went through. On the other side of the door, we were greeted by a tall, striking woman called Margot; she was the department assistant.

"Ah, Herr Lapine is expecting you."

Herr Fischer shook my hand and left.

Margot – she didn't offer me her surname – asked me to follow her through a door on the left and into Herr Lapine's office.

Anatole Lapine, to use his correct name, was Latvian by birth and had left Europe for America via Germany, where he had worked briefly as an apprentice at Mercedes. In America, where he was living as a refugee, he had got a position at Fischer Bodies in Michigan. Bill Mitchell found out that an ex-Mercedes employee was working for a GM supplier and gave him a job in his famous X-Studio, where Mitchell created all his special cars such as the famous Monza Coupé. It was working here that Mitchell decided Lapine had the temperament to shake things up, so he sent him to Germany to work at Opel.

It was after a tour of the Opel design studio that Lapine approached Ferdinand Porsche and offered to set up a new design department for Porsche; he joined the company in 1969.

Lapine's big claim to fame at Porsche was the design of the 928 coupé, which was created under his eye by Wolfgang Möbius, a designer he had taken with him from Opel. Porsche aficionados will strongly protest, but the Porsche 928 is in fact a typical GM design. Take a look at the style of the cars that GM was creating at this time and it is easy to understand my thesis. In particular, the design of the 1973 XP-897GT 2-rotor Corvette comes to mind. The stance and basic graphic of that car are uncannily similar to those of the Porsche 928.

By training, Lapine was an engineer, but he knew what good design was and made sure he had good designers working for him. His great passion was motorsport and was quite a competent hobby racer. When he joined Porsche, he apparently had a clause in his contract drawn up allowing him to use the Porsche design facility to work on his own collection of sports and racing cars.

I had been warned by Martin Smith that Lapine had quite a strong temperament, but at the time of our first meeting, he was still recuperating from a mild heart attack and was being careful not to exert himself. I walked into his office.

"Ah, Mr Birtwhistle come in, call me Tony."

Although born in Latvia, Tony Lapine was very much an American and the loud checked trousers he was wearing that day just emphasised the fact.

He beckoned me to take a seat, then began quite a long dialogue, the topic of which was mainly himself and why, as a designer, there was no alternative but to work for Porsche. He finally got to the point of my visit and said that if Dick and the 'other guys' were happy, then he was happy. He picked up his phone and asked Margot to call Dick.

After a few minutes Dick Söderberg – followed by the 'other guys,' Wolfgang Möbius of 928 fame and Porsche's interior design chief, Ginger Ostle – entered the room. I had previously met all three on a few occasions, and the atmosphere was warm and friendly. Going through my portfolio seemed more like a procedure. Slowly, I began to feel at home.

A few questions were asked, and it occurred to me that Möbius didn't appear as if he wanted to participate. Maybe he was just a quiet guy. The few times in the past I had met him, I always found him a nice and mild kind of character. In the end, I brushed it aside. Lapine looked at their faces and smiled.

"Well, Birdy, do want the job?"

I did.

"Fine, welcome to Style Porsche."

The meeting was over, and Tony Lapine said Herr Fischer would send me an official offer.

I made my way back to Ingolstadt, wondering if Porsche had filled the other two assistant studio chief positions. I thought the meeting had gone well and liked the fact that Lapine, Dick and Ginger spoke English as a first language, though still slightly concerned that Wolfgang Möbius had been so non-committal. Möbius was head of exterior design, and Lapine had indicated I would be working with him.

It would be a while before I could inform Herr Warkuss of my intention to leave Audi. My meeting with Porsche had been in July of 1983 but due to my contractual arrangements with Audi, I could not hand my notice in to Herr Warkuss until September. In my mind I had left Audi, but I could not afford to show any signs of easing up the pace, and returned to the Audi studio determined to do a good job until the end.

I had already started sketching for what would be my final project at Audi – the V8 Porsche engined saloon – before taking my trip to Weissach. It was interesting, I thought, that no mention had been made during my visit concerning this project. Warkuss was determined not to let Style Porsche win this competition and was very enthusiastic about the first sketches I completed. On this car, there was freedom to treat the front and rear fenders in a different way to the standard Audi 100/200 body, so my immediate thought was to apply a teardrop treatment similar to the Sport Quattro I had been working on. Warkuss, however, wanted to keep that look as a Quattro only design signature and requested something more classic. I continued sketching and it occurred to me that the project was not dissimilar to the V8 Ventora I had worked on at Vauxhall. Hopefully, this one would see the light of day.

The Porsche proposal for the V8 saloon arrived in Ingolstadt much sooner than we had anticipated, and was pushed into the Audi viewing hall. We went as a group and had a look. It was a shock for us all. Probably the most controversial part of the design was the front end, in the area that was usually taken up by a grille. Porsche had decided to give the car a complete 'soft nose,' and rather than a normal black grille featuring the four chrome Audi rings, they had instead used the four circles of the symbol as an air intake. The soft treatment continued throughout the design, which I guess was understandable, knowing the form language of most Porsches, in particular the 928. Interestingly, they had chosen to handle the fenders with the soft teardrop treatment that I had initially proposed, which I felt worked quite well, though it was an opinion not shared by the rest of the team. Overall, however, the car looked fat and was out of place in the Audi landscape.

I didn't attend the reviews with Warkuss, Piëch and the Porsche designers, but the message got through that Porsche had been instructed to revise the design. How would it end up? I wouldn't find out; the time of declaring my intention to leave Audi was getting nearer.

One last note before closing the Audi chapter of my life. The V8-engined version of the Audi 100 went ahead, but not with the Porsche motor, which Audi gave up on, preferring to develop its own design, and the car eventually came to market as the Audi V8 in 1988. Interestingly, it was the first Audi to move away from the clean horizontal front grille, featuring as it did a separate 'mouth' framing the Audi rings. It was, in a way, the start of Audi's move to give its cars more character to the front design, ending many years later in the single frame grille that graces its cars today. When I left Audi, the V8 front end treatment was yet to evolve, but I remember the last time I witnessed Herr Piëch at his merciless best was during a design review of the model, concerning its front design. The Porsche proposal may have been controversial, but it made

*Sketch idea for the Audi V8, featuring Quattro-style wheelarches.*

*A more conventional solution for the Audi V8.*

*Rear view of Porsche proposal for Audi V8. Note the Porsche 928 alloy wheels.*

*Good times at Audi!*

Well, at least there would be overtime to earn that weekend!

My final meeting with Herr Warkuss didn't last long as I told him that I had decided to take a new job in order to further my career.

He understood, shook my hand, and thanked me. He excused himself to go to a meeting, and left me alone in his office. I wouldn't see him again.

Piëch realise there needed to be a stronger design message at the front of all future Audis.

It was a Friday afternoon and the mood was good. On Fridays, especially in summer, the Audi modellers usually consumed a fair amount of 'weizen' beer in anticipation of the weekend. People were laughing and joking.

The latest iteration of the model was prepared for Herr Piëch's final meeting of the day before he left for the weekend, maybe to travel to the Porsche/Piëch family seat in Zell am See, Austria.

We waited, some nervously, with the merriment subsiding as Piëch entered the modelling hall with Warkuss. Pfefferle went around the model explaining the various modifications, ending up at the front of the car. Piëch was satisfied, apart, however, from the treatment of the front grille.

"A good job, gentlemen, but Herr Pfefferle, I am not happy with the front grille. We need another solution."

Pfefferle grinned nervously.

"Of course, Herr Piëch. We'll get working on a new solution."

Piëch smiled, which was not always an indication that he was completely satisfied.

"Good. I look forward to seeing the results first thing on Monday. I wish you all a pleasant weekend."

I thought Pfefferle's face was about to turn inside out.

I returned to my desk, and after a little while Gerd Pfefferle came to me, having been giving an update by Warkuss. I was to clear my desk and leave within a couple of hours. It was what I had anticipated and hoped for.

Gerd was genuinely sad about my decision, and said that he had been hoping for me to eventually take over Herr Bühler's position in the studio. Hermann Bühler was the studio engineer, and mainly dealt with production feasibility issues on the models: he was definitely not a designer. It would have been a good position but certainly not creative, and I knew that it was definitely the right time to move on.

When I'd started at Audi it was recognised as a rather conservative car manufacture, the model of choice for pensioners, but when I left the foundations were in place that would allow the company to grow into the one we know today. The overseers of that great revolution at Audi, Ferdinand Piëch and Hartmut Warkuss, went on to greater things and would build VW into a group of companies, many of which are the design leaders of today.

I'm often asked who I think was the greatest car designer of my generation, and my answer is very clear: not one person, but two – the tandem of Ferdinand Piëch and Hartmut Warkuss.

*A selection of my Audi artwork.*

# Chapter 17

## *Let them sit on bricks*

**I was to commence my new job at Porsche on January 2, so I reasoned it best to return from Reading (where I had celebrated Christmas) to spend New Year in Germany. I had given up the flat in Ingolstadt, and already had the keys for the new one I had found in Leonberg. I drove over to Stuttgart on January 1.**

J Mays invited me to see in the first day of 1984 at his flat in Munich – George Orwell's predictions for the new year were a bit wide of the mark. As it struck midnight, I wondered what this significant year would bring me.

Leonberg is a small town directly west of Stuttgart, with an attractive marketplace, but, as I discovered walking around on the evening of January 1, it was completely dead. Well, it was the day after New Year celebrations. I would later realise that I had been spoiled in Ingolstadt, with its catalogue of bars at virtually each street corner. Giving up on finding anywhere open, I walked back to my new address, a modern two-room apartment with a large terrace looking to the west. I didn't want to live in Stuttgart due to the drive it would involve each morning, so Leonberg seemed a good compromise.

My furniture from Ingolstadt would arrive in a day or two, so I had set up a temporary arrangement, comprising a camping mattress and sleeping bag, plus a folding chair and table. I also had an electric cooking ring plus eating utensils, my hi-fi and TV. I'd survive.

I can't say that I slept well in this new place. I'd been comfortable in Ingolstadt, with friends and an established social life, but I'd been in this situation before. If I kept moving forward, things would turn out for the best.

The drive to Weissach the next morning took about 20 minutes, and I arrived at the front reception desk in good time. Once again, Herr Fischer met me and we went to his office to go through all the administrative formalities before going over to the design department.

Herr Fischer pressed the button at the entrance door to design and, instead of the department assistant opening it, it was Tony Lapine whose face appeared at the door. Apparently, most of the team had taken the week off after New Year. I followed him to his office.

Tony Lapine (I could call him Tony) showed me to a seat. He explained that all the 'chiefs,' as he called his direct reports, were on vacation that week. Before leaving for the Christmas break, they had got together and decided it would be best for me to start work with Ginger Ostle, who ran the interior studio. I was surprised. I had stated at my interview that my strengths were in exterior design and, at the time, Lapine had said I would start in the exterior studio.

Lapine reasoned that I would, naturally, get my opportunity in exterior design, but that this was the best solution to commence working at Style Porsche, going on to explain that he expected all assistant studio chiefs to be capable in all disciplines within the department.

I was irritated, but this was no time to get into an argument so I accepted the new situation, wondering if Wolfgang Möbius' distance towards me during my interview had been a signal that he didn't want me working with him.

Introduction complete, Lapine motioned for me to follow him into the studio area. We walked a few steps down an adjoining corridor, turned left and through a door into a well-lit hall: it was the interior studio and my new workplace. The studio was long enough to take two cars nose to tail, and had central floor-mounted measuring plates. To the left was a row of desks, and on the right some movable boards, similar to the type I remembered from Vauxhall. As Lapine had mentioned, there were no staff in sight, except for one. A slightly ruddy-faced guy with unkempt curly brown hair got up from his desk and walked over to us. I guessed he was about my age.

"Rob, I'd like you to meet our new assistant studio chief, Birdy," Lapine announced. "Get him settled in and explain what's going on."

Then he left us. I'd need to get used to Lapine calling me 'Birdy.'

Rob Powell was English, had worked at Porsche for a few years, and was an RCA graduate too, although we had never crossed paths before. He took me over to a desk that had been cleared for me. We exchanged introductions and it immediately struck me that he wasn't the liveliest of characters, opening the conversation with a line along the likes of:

"I hope you realise what you've let yourself in for." Perhaps he had been hoping for my job.

We sat down and he began explaining what was being worked on in the studio. They had just finished the interior for the Porsche 944 face-lift, plus its new Turbo version, and the next key job in the studio was the interior for the production version of the 'Gruppe B' study that Porsche had shown at the Frankfurt Motor Show the previous September. The Gruppe B was based on the 911 body, but apart from the cabin, doors and bonnet, was all new. The design, which had by now been given the Porsche name 959, was basically intended as a run of 200 cars, to enable Porsche to compete in a new sports car segment. Ferrari, it was rumoured, was also planning a similar vehicle (later to be revealed as the F40), and since sports car racing was Porsche's primary source of media exposure, it had no alternative but to take this route. Whatever the resulting car would be, it would be as exclusive as it was expensive.

I had seen the study at the Frankfurt show and remember thinking that I needed to adjust myself to Porsche's very fluid style of design, which was certainly a contrast to Audi.

The exterior of the 959 was being handled by Dick Söderberg in the neighbouring studio. Rob Powell continued, expanding more on what Style Porsche was working on apart from the 959.

In many ways, Style Porsche was like a consultancy, working not only on design proposals for its own car programmes but also on a wide range of products labelled loosely under the heading of transport design. With a comparatively small model range and the long model life of its cars, there was capacity to handle outside projects.

For Porsche, this was a good opportunity to keep the in-house design department fully occupied and, probably of more importance, earn money to fund the Weissach centre.

At the time I joined Porsche, Lapine's team had just finished designing the cockpit for the Airbus A320 passenger airliner, which, up until then, Rob Powell had been working on with Dick Söderberg. The Airbus project complete, they had a new project commencing, namely a complete range of fork-lift trucks for the company Linde, which Dick would handle in his studio.

Rob Powell then went on to describe a further project that had started as an independent programme but had since been taken over by Porsche. My ears immediately pricked up.

In the late '70s, Porsche Engineering had developed a four-cylinder engine for the Spanish car company Seat, which was to be used in the first car developed by them as an independent company. Up until then, Seat had produced cars that were basically re-badged Fiats, the Italian company having partly owned them for many years. Fiat and Seat had parted company, hence this new model, named Ibiza.

Once the engine development for Seat had been completed, Porsche was anxious to continue working for the Spanish company, and had proposed developing a small mid-engined sports car for Seat, using the engine they had developed for the Ibiza.

As a company, Porsche had a history of similar such developments, having worked with VW on the 914 model and the 924, which was originally planned as a VW before Porsche took it over.

As is the case of the 924 for VW, Seat had deemed the idea of a sports car to be too ambitious to market and declined the offer. Therefore, Porsche had decided to continue the car themselves.

By the time I arrived at Porsche the project had been given the Porsche project name 984 and instead of using the Seat engine it had decided to develop a new boxer or flat four-cylinder engine for it, similar in layout to the engine used in the 911.

Having given me a quick run-through of Style Porsche's present workload, Rob took me on a studio tour.

Basically, the design centre comprised of four studios. Next to the interior studio, and practically part of it, was the new colour and trim studio, which was run by a lady named Katherine Mueller Goodwyn, usually called 'Katyl.' Moving out of the interior studio and into the main connecting corridor, we walked along to the next door on the left and entered the advanced studio of Dick Söderberg. We went in. Similar in layout to the interior studio, it was also devoid of staff. Placed on the central measuring plate was the clay model of the Gruppe B-derived Porsche 959 that Rob had described to me. Very much a work in progress, its basic shape resembled the study shown in Frankfurt, but had developed quite a few extra air intakes.

We left the advanced studio, returning to the central corridor, moving along and through a final door at the end and into what was the exterior design studio. Also similar to the previous studios but wider, enabling it to house parallel measuring plates. On one of the modelling plates was a wire frame buck of what I assumed, and Rob immediately confirmed to be, the new small sports car – the 984.

A wire frame buck is basically a model constructed from metal rods that form a frame approximately the same shape as the car's body. It can be best compared to a boat or aeroplane frame or skeleton before the outer skin is applied. The main purpose of this is to check the package layout and size before moving on to full-size exterior and interior clay models.

Looking at the buck from some distance you could assess the design reasonably well and it immediately struck me that this was quite a small car. Also of interest was the fact that rather than a traditional soft top it apparently had a hardtop that slid back under the rear engine cover.

Apart from the buck there were no models to be seen and no illustrations on the boards. I assumed they had just started on the design. Rob turned around and we headed out of the room. I felt myself hoping that I would get the opportunity to work on that new project.

Before moving back into the interior studio, Rob suggested we take a quick look in the fabrication workshop, which was across from the main corridor we had come along.

Walking in, it was very clear why Rob wanted to show me the workshop. At the end of the room, which was surprisingly small for a design studio workshop, was an open-wheeled racing car standing on a measuring plate. It looked like something from the '50s and was most probably American, maybe an old Indianapolis 500 race car, I thought.

This, Rob explained, was Tony Lapine's latest hobby car. It was an old American dirt track racer called a Mukowski, which Lapine was in the process of modifying to compete in classic car races.

Having started his career as a design engineer in Bill Mitchell's X Studio (where Mitchell and Japanese designer Larry Shinoda created the Corvette designs and their racing siblings), Lapine was a confirmed motorsport fanatic. On joining Opel, he had worked on a two-door Opel Record that had comprehensively trounced the BMWs and similar racers of the day. The Record was painted black and yellow, and in typical Lapine style, adorned ironically with a taxi sign. This ironic streak in him, to make fun of the established racing fraternity, was also very obvious when he designed a graphic scheme for the Porsche Long-Tail 917 LH Le Mans car. The car was notably ugly, and when Lapine compared it to a pig, it motivated Porsche Motorsport to have it painted in pink and adorned with markings to look like a butcher's cuts of meat.

Upon seeing the car at Le Mans, the world famous racing journalist Denis Jenkinson confronted Lapine angrily, saying, "This isn't motorsport."

In reply, Lapine looked at Jenkinson up and down and commented, "Judging by the clothes you are wearing, you are the last person to talk to me about style."

The car in the fabrication workshop at Porsche was just one of several vehicles that Lapine owned, including a Porsche 356 Carrera, a heavily modified MG TC and an MV Augusta motorbike, the last of which he had sitting in his office.

Apart from the 356, Lapine had tampered with the mechanics of all the cars in his collection and the ones he raced were certainly non-standard, to the extent that you could maybe call him a bit of a cheat.

Lapine figured it was crazy to restore an old car to its original specification, arguing that in the days it was developed the mechanics were not the most efficient. To this end, he took on some light modifications in order to improve the design.

His MG was fitted with a modern Fiat twin-cam engine and had Alfa-Romeo brakes and axles, to which extra wide wire wheels were fitted. In America, this kind of modification was fondly known as 'hot-rodding,' and in motorsport it was halfway accepted. In Europe, however, as Lapine discovered to his cost, this kind of modification was not tolerated in competition circles.

Once he had completed the MG, he had spoken to his inner circle of colleagues at Porsche who had helped him create it.

"Let's go and show those limeys how to win a race."

The race he was referring to was the official MG owner's club hillclimb event in Prescott, England.

Lapine had put the MG on a trailer and driven all the way from Stuttgart to Prescott with his support team, only to be informed by the scrutineers that his car was ineligible. He had no choice but to pack up and return to Germany.

This unfortunate misinterpretation of the rules summed up the man, and looking at his latest project now, it sounded as if he was going to do the same thing with this old American dirt track car. This time, however, it looked as if he may well succeed, since there was no record of the original car's specification.

Rob explained that Lapine had thrown away the old Offenhauser four-cylinder engine fitted to the car, and replaced it with a new fuel-injected small-block Chevrolet racing engine. In addition to this, he had sent Style Porsche's chief modeller, Peter Reisinger, down to the Motorsport department and told him to raid the parts bins for any unwanted components. Reissinger had returned with various titanium bolts and other parts that were leftovers from the Porsche 917 Le Mans programme. Lapine would no doubt find a way to include these parts in the build, in an attempt to out-do the competition.

The inspection of Lapine's latest project concluded the studio tour and we returned to the interior studio.

I spent the rest of the day sorting out my desk, and when I left in late afternoon, I asked Rob if he fancied meeting for a drink later on. Looking at me with a deadpan face, he declined the offer. Are they all like this at Porsche, I wondered?

It had been an eye-opening first day for me at Porsche, and driving back to Leonberg I felt excited at what I had seen, although at the same time perhaps a little disappointed that I was having to start work in the interior studio.

The remainder of that first week was very quiet, with the full studio manpower not returning from the New Year break until the following Monday.

Although I should have spent the weekend

*The Style Porsche viewing yard. Always an interesting collection of cars. From L-R: 928, 911, 911, 944, 959, and Tony Lapine's 356 Carrera.*

*Tony Lapine's US dirt track racer, sitting in the Style Porsche viewing yard.*

organising my flat, I couldn't face the thought of spending it on my own. Rob Powell had shown no interest in introducing me to the nightlife of Stuttgart so I had called a friend of mine who was working as a designer at the Ford design studio in Cologne.

Ed Golden was American and had started his career working for VW North America in Detroit before being sent to the VW studio in Wolfsburg, where our paths had first crossed. Ed and I hit it off from the start, as he possessed a very similar sense of humour to mine. When he moved to Ford he had often asked me to visit him, but the journey from Ingolstadt was at least 500km and quite a distance just for the weekend. Stuttgart, however, was quite a bit closer, so I called him and arrange to visit for the weekend.

The visit turned out to be just the right tonic after a pretty lonely first week in Stuttgart. Although at the time it would have been unfair to criticise Stuttgart's potential for building a social life, Cologne was a city I immediately fell in love with.

Ed had developed a great circle of friends, and the city, with its old town dominated by the magnificent Cathedral, was buzzing.

Returning to Leonberg on the Sunday evening, I had a feeling I would be visiting Ed more often in the future. Little did I know, I would build a lasting bond with this wonderful metropolis on the Rhine.

After a quiet first week at my new job, it was with much anticipation I walked through the doors into Style Porsche on the Monday morning.

The studio was already filling up with staff and I could see Porsche's head of interior design, Ginger Ostle, standing by his desk.

I knew Ginger pretty well and he greeted me enthusiastically. I'm sure he realised that I had not been expecting to be working with him but now was not the time to go over the ifs and whens, and he immediately suggested meeting the other members of his team.

Rob Powell needed no introduction. He smiled, but I was sure his weekend had not been as good as mine. There were two other designers I had yet to meet: Freeman Thomas was a young American, and a graduate from ArtCenter. He greeted me with an almost stammering "Hey hey." The other designer was a Frenchman named Olivier Boulay, who was an Architecture graduate. Both he and Freeman had not been at Porsche long.

As well as the designers, there were also two studio engineers to meet, Reinhold and Reiner, who, from their appearance, immediately struck me as being the modern equivalent of Laurel and Hardy.

The colour and trim team, who resided in the adjoining studio, had by now come over to say hello: the head of the department, Katherine Mueller-Goodwyn, and her colleague Vlasta Rubyr. In addition to the two C&T designers, there was also an American graphic designer who worked with them, called Cheryl Dimpson. It was a friendly, welcoming group.

Introductions over, Ginger said we may as well visit the other studios to meet the rest of the design team, so we made our way to the adjoining studio where we found Dick Söderberg standing by the clay model of the 959 with a few guys. He smiled warmly and introduced me to his three designers, two of whom were ex-RCA, and all of whom I had met before at one of the numerous designer gatherings that took place around Europe. Steve Murkett was English and quite new to Porsche, and the other RCA guy was Roland Sternmann, a German who one could easily have taken for an Englishman, so good was his command of the language. In addition to these two was Benjamin 'Bengy' Dimpson. He was from the Philippines, an ArtCenter graduate and the husband of Cheryl Dimpson, who I had just met. We exchanged some small talk and then left for the exterior studio. Walking in, I saw Tony Lapine standing with Wolfgang Möbius and Peter Reisinger.

Lapine called out. "Hey Ginge, Birdy, come over." We walked over to them.

As before, the welcome from all three was warm; maybe I was wrong in my assessment of Möbius.

There were two designers reporting to Möbius in the exterior studio, Ranjit Bhambra, an Indian, and another German, Roland Heiler, both also RCA graduates. As in Dick's studio, pleasantries were exchanged and I was also introduced to an English engineer name Rob Owen, who was taking measurements from the 984 wireframe buck.

By the beginning of the '80s, model measuring equipment was linked to an electrical calculator that had a digital display for the XYZ coordinates, making life easier for the clay modellers and engineers. However, there were still no computers to be seen in the design area, a situation that would slowly change in the coming years.

Returning to the interior studio it occurred to

me that apart from Lapine and his four chiefs none of Porsche's designers had much more than three years' experience under their belts. Why was that, I wondered?

Back at my desk, Ginger Ostle explained the project I would lead, which, as Rob Powell had indicated the previous week, turned out to be the interior for the Porsche 959. I would work with Rob Powell and Freeman Thomas.

Mounted on the studio's measuring platform, there was a metal floorpan with front and rear bulkhead, which represented the layout of the 959; this was basically modified from a Porsche 911. The 959 had an AWD system and therefore a larger centre tunnel to accommodate the propshaft taking the drive to the front wheels and with it the resulting higher gearshift position to that of the normal 911. The design work basically involved creating a central console to take the suspension control switches, which then stretched back between the front seats to house the gearshift and handbrake before widening out to accommodate two foldaway emergency seats, in a similar fashion to a 911. These seats, in fact, were only big enough for children.

Having explained the project, I asked Ginger if that was all the studio had to do – it was. There were things in the pipeline, such as the 984, but they weren't at a stage where we could start design work yet.

Three designers to handle what an interior middle console seemed excessive but that was the way it was. Getting together with Rob and Freeman, we decided to divide the design work into three parts. I handled the centre console, Rob the rear seat area and Freeman the graphic and colour break-up.

The idea of three designers working on what was a relatively small part on an interior seemed pretty crazy, but for me it was to be the first taste of a working process that was a part of the company's philosophy. Everything Style Porsche handled would be laboured over to an extent unique to the industry. On the one hand it made the company what it was, in its efforts for perfection, but on the other, it was a process that was way longer than it needed to be, or at least that's how I saw it.

For the next few months we three musketeers would battle together on the interior of the 959, and if you included clay modellers and engineers, there could be as many as six individuals trying to find a place to stand around the model.

Working in such close proximity, I naturally began to get to know the designers a bit better. Rob Powell's character remained the same – a person who seemed to have accepted his fate at Porsche, with no apparent drive.

Freeman Thomas, on the other hand, was bursting with energy, and it was clear from the start that he lived for cars, and in particular Porsche. In fact, there seemed little else on his mind. It was clear that Freeman was slightly hampered working in the interior studio, however, it didn't stop him continually drawing his dream, which was a modern interpretation of the legendary Porsche 356.

The 356 was not a model of the famous brand that got to me in the way it got to Freeman, but nevertheless, like him, I had joined the company to fulfil an ambition of designing cars for Porsche. So, baring this fact in mind, when only after a few weeks at Porsche, Tony Lapine called me to his office one day and asked me what I wanted to

*Full-size rendering of the 959 instrument panel: basically, based on the 911.*

achieve at Style Porsche? My answer was probably the same as all designers joining the company and I replied without hesitation that I wanted to design the next 911.

The dream of designing a new 911 was in the minds of most of the Weissach designers, but for Porsche AG, and in particular the head of the company, Ferdinand 'Ferry' Porsche, son of the company's founder Ferdinand Porsche, and father of the 911, there was still a lot of life in the existing car. Neither he nor the company were about to make the decision to develop a completely new model in a rush.

Plans for the next evolution of the existing 911 were, at the time, well advanced under the project name of 964. The 964 was, in effect, a major facelift rather than an entirely new model. Having said that, it was rumoured that the car was 80 per cent new. Porsche saw no reason for changing the shape of his beloved 911, but after many years of production the tooling for the existing car was getting old. Many of the press tools for the body parts needed to be remade. In addition to this, the car required a new platform to accommodate 4WD, and a new coil sprung suspension.

In the mid-'80s, Porsche's key market, the USA, had declined, and sales of both the 928 and the 924/944 were not meeting their targets. It had been expected that, with the introduction of the 928, the 911 would eventually be dropped from production, but it continued to soldier on, and was, for many, still *the* Porsche. No, the original 911 was not going to die, and Ferry Porsche was happy with that, a situation that nevertheless caused some frustration within the design team who were all keen to develop a new car.

Dick Söderberg was responsible for the 964 design programme. Before Dick and his team could commence work on the 964, the Porsche 959 design development had to be finished.

It was at the 1983 Frankfurt Motor Show that Porsche unveiled its proposal for a car to compete in the newly created Group B sports car category. This, incidentally, was the same time and venue that Audi chose to show the world its new rally weapon: the Audi Sport Quattro.

By September of 1983, I had left Audi, and it was therefore interesting for me, when visiting the show, to see the final production version of an Audi project I had been involved with sitting under the same roof as a new product from my new employer-to-be.

The Porsche Group B study was about as far removed from the original 911 as you could get. Apart from the roof, screens and doors, it was pretty much all new. The bonnet was basically the same, but this had also been modified at the front, where it curved down at its leading edge, similar to the original 911; this part may in fact have been an original part from the first model.

In line with the Porsche design philosophy, the Group B was almost devoid of any hard lines or creases, with soft organic forms swelling out to cover the widened track over the wheels. Probably the most controversial part of the design was an integrated rear spoiler, which was a kind of hoop in its form.

Looking at the car, and in particular the rear spoiler, it struck me that this feature was very similar to the Greenwood Corvettes of the late '70s, some models of which also feature a hoop-like rear spoiler. The rear spoiler of these heavily modified Corvettes didn't integrate flush with the body, as was the case in the Group B, but there was nevertheless an uncanny resemblance.

I had only been at Porsche for a week, when I asked Ranjit Bhambra, one of the designers who had worked on the Group B, what the influence had been when tackling the rear design of the car. Immediately, he took out a crumpled photo from his wallet. It was a picture he had taken of a car parked in a New York street; it was a Greenwood Corvette. As they say: there are no new designs, just old ones revisited.

Once all the excitement of the Frankfurt Group B concept had subdued, it was up to the company to get on with the job of making it feasible for manufacture.

Dick Söderberg had led the Group B development, alongside the previously mentioned Ranjit Bhambra and Roland Sternmann. By the time I joined Porsche, unlike the interior, the exterior development of the project, now named 959, was well on its way. Once again, Dick Söderberg was responsible for the design work, but his team now consisted only of him and Roland Sternmann, with Ranjit Bhambra having moved on to help Wolfgang Möbius with the small sports car project I had seen on my first day at the company.

The task of changing the Group B concept into a vehicle that was feasible for production, albeit of limited numbers, was not inconsiderable. In fact, the design needed be completely reworked. Most of this was due to aerodynamic adjustments, cooling requirements and manufacturability.

Upon my arrival at Porsche, Dick and Roland had

*The 959. The only time I've driven one!*

completed most of the main bodywork and were occupied with smaller details such as the air-intake on the rear fender, directly behind the door. This feature was a good example of the Porsche design process in its purest form. The intake was about 20cm in diameter and was very organic in its shape. Indeed it could have been influenced by some orifice from an animal or fish. This similarity to nature even prompted Tony Lapine, at a studio review, to open the meeting with a quote from Shakespeare's *Julius Caesar:* "Lend me your ears".

Dick and Roland spent days adjusting the clay model in this area until they were satisfied it had the correct flow of light reflection. For me, it was a good example of why Porsche took the time to refine its designs to such a high degree. The design of soft organic forms should have the tension and strength of an athlete's muscles. If not, there is the danger it can become ungainly and fat. This was the *raison d'etre* of Porsche's design language.

The reason for building the 959 was for it to compete in a specially designated class of sports car racing. Like Porsche, Ferrari was also developing its own car for this class of vehicle known as Group B. The Ferrari in question was named the F40, and although very different in style to the 959, it also featured a rear spoiler that was an integral part of its rear bodywork. Regretfully, neither cars would see true competition together on the race track. The powers that be in the FIA decided at a very late stage that the Group B class of car would be better suited to rallying and so the sports car racing series was abandoned. So as not to lose face, Porsche went on to develop the 959 for 'rally raid' events such as the Paris-Dakar. Here, in 1986, on its second attempt and in the colours of the famous Rothmans cigarettes, it won the desert rally while in the hands of René Metge, with Jacky Ickx finishing second in a similar car.

The final swansong for the 959 in competition, did, however, see the car finally compete on a race track, and in addition to my small contribution to the interior design, involved my personal hand on the exterior design of the car, albeit not in a particularly memorable way.

Since the 959 was originally intended for circuit racing, it was decided to enter a modified version at Le Mans in 1986. For this purpose the car was renamed 961 and had its bodywork specially developed for the famous race.

Tony Lapine came to me one day and told me to go to the Stuttgart University wind tunnel where a scale model of the race car was being developed.

If the 959 was a caricature of the Porsche Group B concept, then the model I was confronted with upon my arrival at the wind tunnel was a similar exercise on the 959.

Memories of my work on the Audi Sport Quattro came back to me, since this development was also a bastardisation of the vehicle it was based on.

The key aim of the wind tunnel work was to create a shape that handled well at the high speeds attained at Le Mans and with as much downforce as possible. Porsche race department engineers had already developed a final form they were pleased with and asked me to spend some time 'cleaning it up,' as they put it.

If I'm truthful, I was never a big fan of the 959 – I found it a bit fat and bloated. The 961, however, made the 959 look positively graceful. The car was heavy and distorted, and with its rear spoiler having grown in size, was very much the ugly sister of the 959.

Getting to work, it was made clear to me that I couldn't change much. Dutifully, I spent just a few days adjusting some lines and surfaces and that was that.

Porsche only built one 961, and at its first appearance at Le Mans in 1986, painted resplendent in white, it went on to achieve seventh place. Spurred on by this

*The racing version of the Porsche 959, the 961. (Porsche AG)*

*Porsche 961.*

success, the company continued the programme; the following year, this time painted similar to the Paris-Dakar cars in the Rothmans colours, the car was entered once more. Here, it was not so successful and crashed out to retire. The sole 961 can be seen today at the Porsche museum in Stuttgart.

A couple of years passed by at Style Porsche and I began to realise that much of the work handled was what I would describe as piecemeal projects, many confined to colour and trim projects. Top of the list of these quick projects was developing special one-off cars for customers ranging from celebrities (such as the tennis star Martina Navratilova, a confirmed car nut), to business moguls. One such customer was the head of the hamburger chain Burger King, who wanted a 911 Turbo finished in the same colours as its corporate identity; the car was painted in a dark brown metallic, with gold wheels and a red and gold metallic leather interior. TV stars like Don Johnson of *Miami Vice* and members of European royal families would all get their Porsches made to order, visiting the studio for personal consultation. It was interesting work, but was not the real reason I had joined the company and the opportunity for working in the exterior studio was still out of reach.

Creating one-off vehicles for the rich and famous meant that individual leathers and cloths needed to be developed, where cost was of no account. For an industrial weaver, supplying cloth to Porsche for such contracts meant producing massive lengths of material for just one car. Different colour nuances and patterns would be produced until a final direction could be agreed upon. This was a very expensive process, which of course didn't concern these wealthy customers, and was a point that Tony Lapine was well aware of when he came to my desk one day.

"Birdy, what do people sit on?" I was slightly confused by his question. In fact, I had no idea what he was going on about.

"They sit on walls!" he continued. Again, I was not sure where this was leading. He went on. "And what are walls made of?"

"Bricks," I replied.

*Football. This time the Style Porsche team. I'm bottom row, second from the right.*

Tony paused and then replied. "Let them sit on bricks!"

I had given him the correct answer – bricks!

Still, even having answered his question, I was very much in the dark. Lapine explained his idea, which was to contact Porsche's cloth manufacturer and get them to weave some samples with a pattern resembling a brick wall, remarking that a seat cloth with a brick pattern could be a unique idea.

Lapine left the studio and I went over to Katyl Mueller Goodwyn's team in the colour and trim studio to explain Lapine's request.

They reacted with just as much confusion as me. "What's this all about?"

There was no programme that required seat cloth with a brick pattern. However, they would send some design samples to the supplier and see what resulted. Tony Lapine was always coming up with special requests and he expected them to be followed up – that was that.

After two or three weeks, several one and a half metre wide rolls of cloth arrived in the studio; they were Lapine's samples.

We unpacked the rolls and laid them out on the floor of the colour and trim studio. In all, there were three or four design patterns to look at. Lapine was called into the studio by Ginger to see the results.

After a few deliberations it was agreed that the designs were interesting, but it may be difficult to use the idea at the present time. The samples were stored for future reference. Lapine was satisfied and that appeared to be the conclusion of the exercise.

Nobody thought any further about the brick-patterned cloth, assuming that Lapine would let the idea pass by. Little could have been further from the truth, as he knew exactly what he wanted done with the cloth.

Months later, after the dust had long settled, I was in the fabrication workshop, walking by Lapine's single seat racer, when I noticed a familiar cloth pinned to the seat of the car –it was that cloth! A light went on in my head: home to all single-seater racing cars in North America is The Indianapolis Raceway, better known in racing circles as The Brickyard. Tony'd had no intention of the cloth being used in a Porsche. He'd only wanted it for his classic racer, and what better than a brick motif for this American racing car?

Tony Lapine would always find a way of achieving his goals. Never one to go by the rule book, you had to admire the innovative way he went about handling things.

I have related to that story many times and can provide no deep philosophical reason for liking it so much. Maybe it just sums up the whole mantra behind creating something unique.

Good design should always have an element of surprise and keep you guessing. When you realise its intention, it should leave you with a smile on your face, just as I had when I saw that cloth in Tony's racer.

# Chapter 18

## *A second category of designer*

**There is nothing more exciting than picking up a brand new Porsche from the factory in Zuffenhausen, especially if it's a 911.**

When joining Porsche, I had decided it prudent to start with a 944 coupé as my company car, and indeed the first two Porsches I ordered on the company leasing scheme were that model. In the end, however, you have to take the plunge and try the legendary vehicle with the famous Carrera badge on the rear motor cover.

My first 911, which I was able to pick up just before the Christmas of 1986, was a wine-red cabriolet with a beige leather interior. Like most of the designers, I could choose a car that was painted in a colour developed for evaluation purposes within the design department, which meant it was completely unique.

When picking up your car, you first had to report to Herr Freund, who delegated the company cars from an office in the old factory building in Zuffenhausen. Not only did Herr Freund have this responsibility, he was also Ferry Porsche's chauffeur, and with such close links to the company owner, he regarded every company car as his own personal property.

As was always the case at that time of the year, I spent Christmas with my mother, who was now living in the pretty Suffolk town of Southwold, situated on the east coast of England. The town had been the location of many of our family summer holidays over the years, since my mother's elder sister, Jean, had lived nearby, where her husband Geoff managed a farm. Although Uncle Geoff and Auntie Jean had moved around the country, they eventually retired to Southwold, where Auntie Jean ran a wholefood shop called Nutters, located almost opposite the Nelson pub. After the death of my father, my mother found herself visiting the town more and more and finally decided to move to Southwold with my grandparents. She bought a cottage on South Green, which she named Dolphin Cottage, after the Birtwhistle family crest: an ancient springing dolphin.

In the new year, 1987, I returned to Leonberg. I was excited because at last Tony Lapine had kept to his word and persuaded Wolfgang Möbius to let me join his team in the exterior design studio.

Since moving to the Stuttgart area my social life was still nowhere near as busy as it had been during my days in Ingolstadt. My

*Picking up my first Porsche 911 from the factory in Zuffenhausen – you never forget it.*

fellow designers at Porsche seemed to keep their private lives to themselves and I found myself visiting my friend Ed Golden in Cologne almost every weekend. Through this friendship I met a girl named Claudia. As chance would have it Claudia moved to Stuttgart to study fashion design, and we would meet and start a relationship to last many years.

My move into the Porsche exterior design studio was just one of a few personnel changes in the department. Ranjit Bhambra joined Dick Söderberg to commence work on a range of fork-lift trucks for the German company Linde. In addition to Ranjit's move, a new assistant chief had been recruited. Tony Hatter, another RCA graduate, whom I knew quite well, joined Dick's team. I had first met Tony at one of the many car designer events. After graduation, he had joined Opel, and now that he was moving to Porsche there was yet another GM man joining the team.

At the beginning of 1987, the 984 small sports car programme had advanced to a full-size clay model but its progress was quite slow. Porsche kept putting the programme on ice and months went by when no work was done on it at all. One of the key bones of contention amongst the development engineers was the retractable hardtop system. The original idea from the design department was to have the hardtop sliding back on rails then disappear under the rear engine cover. On paper this looked a nice simple solution, but it proved difficult to realise technically, with little or no space for the tracks to guide the hardtop into position above the engine. The idea was eventually abandoned, and the engineers went back to the drawing board to come up with a simpler solution. Their new proposal was indeed simple, but resulted in a big compromise for the exterior design. Instead of the hardtop sliding back parallel to the body form, the new system involved the hardtop hinging back in an arch from a point behind the two passenger seats, similar to opening a door. This seemed a good idea, but the problem was that as the hardtop swung back it became naturally inverted or simply put, it was upside down over the engine. This resulted in two key design issues: the hardtop needed a lot of space under the engine cover due to its inverted curved shape, the result of which was a very high rear end to the car in order to fit it in; and secondly, the hinge point for the roof needed to be positioned quite high behind the seat backs, which resulted in an unsightly kind of 'collar' around the rear of the passenger compartment.

Neither Tony Lapine or Wolfgang Möbius were happy with this solution and a debate between design and engineering went on for months.

Hardtop aside, the car had some interesting features, in particular the headlamp design that incorporated the latest ellipsoid lens lamps. As in the case of the 911 and 928 they were round in shape, and like the 928, lay flat at the nose of the car; however, *unlike* the 928, where the lamps popped up when switched on, the lamps on the 984 revolved in a screw-like manner to achieve a vertical position. It was a very simple and clever solution.

In terms of body styling, the car was very much in the style of Wolfgang Möbius – soft in its general shape with hints of the 928 front-end treatments.

Wolfgang Möbius was undoubtedly a very significant designer; his Porsche 928 was a masterpiece.

His true talent was in the way he, alongside Porsche's chief clay modeller, Peter Reisinger, could develop an interplay of body forms resulting in an elegance, tension and beauty comparable to the muscles of a well-trained athlete.

The front fender wing of the 928 is a great example of their work. In good light conditions, the shadows and reflections in the surface of the metal form a line that appears to be a stretched 'S.' It was referred to in Porsche design language as the 'S-highlight,' and is the key foundation of all body designs coming from the company, even to this day.

This sensitive construction of forms separates Porsche from all other companies, and modelling it is a very time-consuming process.

Body feature lines would first build up to a theoretical point in space, which, only when running correctly, would then be finally rounded off.

Only in this way could you achieve a sophistication of line and reflection in the metal that sets a Porsche apart from the rest of the automotive world.

For Wolfgang Möbius, the complex method of creating the surfaces of Porsches and their resulting S-highlights became a discipline that, for him, almost took on a religious significance.

For hours he sat at his desk with an A4 pad, ball-point pen in hand and draw the same 'S' graphic on the paper. He reasoned this was the purist form of design, a sheet of paper and a pen, and he almost shunned the typical coloured renderings the designers put upon the wall.

As I started to design my first Porsche exterior, I seriously worried if I could meet Wolfgang Möbius' high expectations.

*Sketches for the 984.*

The basis for the 984 were very much in place as I put pen to paper on the project, my first contribution to a Porsche exterior.

There was still quite a bit of design work to do, with engineers still figuring out how to solve the retractable hardtop issue, as well as where to take in air to cool the rear-mounted engine. After I had completed a selection of sketches, it was agreed I could commence a scale model to incorporate the latest modifications, before applying them to the full-size model.

As it turned out, the programme again ground to a halt while the powers that be deliberated if the 984 made sense in the Porsche family. My ideas were put on ice.

With both the 928 and 944 needing a replacement, the decision was made to combine both models into a front engine coupé fitted with a 'civil' version of the V8 engine Porsche had developed at the time for the American 'Indy' car series. In size, the car would be right between the 928 and 944. It made sense.

With the 984 still on hold, the exterior team started the first sketches for the proposed front engine coupé. I can't recall Möbius giving us a clear brief, but Roland Heiler had a good working relationship with him and advised us to take some of the flavour of the 984 and create a coupé back with a typical 911 DLO (side window) treatment.

After a few weeks of sketching, the boards in the studio were soon getting covered with images, all of which looked the same.

I approached Möbius about a design review, but he just replied, "There is nothing to review."

Slowly I began to realise that the man may have been a great designer but as a manager he lacked all traces of the so-called 'soft skills,' preferring to smoulder at his desk, cigarette in mouth, drawing his beloved S-highlights. The situation was, to put it mildly, very difficult. A wall began to build up between me and Wolfgang Möbius and I just couldn't understand why.

I approached some of the other designers, but their comments and advice were of little help. Over drinks one night, Ranjit Bhambra was quite candid, saying Möbius had no respect for me because I couldn't draw, while Roland Sternmann simply said, "The guy's an arsehole."

Despite all this, things had to move forward, and three scale models were started, though each designer, afraid to upset the applecart, created a model that,

*Prototype of the 984, now in the Porsche museum. Note the 'collar' at the base of the hardtop.*

*Side view illustrations for a front engine coupé to replace the 928 and 944.*

*Wolfgang Möbius (left) and Tony Lapine.*

similar to all the sketches, was pretty much identical to the other two.

Working with a Swedish modeller called Sven – what else? – I set about the model the same I always had, with a tape drawing in all elevations, while Sven started creating the forms guided by my sketches. Peter Reisinger would pass and in his usual jovial manner say: "Birty, Birty."

Similar to Möbius, he gave few comments.

A first design review was planned, where the head of R&D, Helmuth Bott, and his adjutants would check the progress.

We lined the three painted models up in a row and Herr Bott and his team entered the studio. One of the younger engineers asked me why the designs were so similar – I'm not sure if he got an answer.

Möbius had obviously already made his mind up and concentrated his presentation on the model that he believed should be taken forward; it wasn't mine.

As it turned out, it wouldn't have made much difference. This programme, like the 984, was still under review and we were asked to stop any further work for the time being.

The delegation left the studio and the models were pushed to the side.

I was frustrated: Möbius' body language and lack of communication with regards to my work was starting to drain my enthusiasm.

I had never had any problems with criticism and there was no doubt that the bar at Porsche was set very high. My design work was not up to Wolfgang Möbius' expectations, but a designer can only improve if you question them about their work, give them constructive feedback and if there are issues, agree on how to improve things to move forward.

I returned to my desk and pondered over my future. It was without doubt the lowest I'd ever felt in my career. Wolfgang Möbius was not motivating me and I needed to talk to him about that. I approached him the next day.

As it happened, he came over to me first and poured out his own frustration at what I had achieved (or rather *not* achieved) during my tenure in 'his' studio. To him, it was very clear that I didn't understand the Porsche design process and things had to improve should I wish to remain in his team. He concluded with a comment that I will never forget:

"There are two kinds of designer in the world, the good ones and the bad ones. You, without doubt, belong in the second category."

That was enough. I had to talk to Tony Lapine and I went immediately to his office.

Tony sat me down and pulled up a chair in front of me while I poured out my feelings of frustration and

*Side view proposals for a new 911 and a small mid-engined coupé.*

anger. He listened patiently but wasn't prepared to take sides. After a pause he said, "Well, Birdy, Wolfgang is a tough bastard, but I rely on him and trust his judgement. You need to go back and show him he's wrong. Don't get mean, get even."

I told him that I didn't deserve this kind of treatment, but I would do my best, and I left his office to return to the battlefield.

No matter how dark the storm, there is always light behind the clouds, and for me that light was my engagement to my girlfriend, Claudia, over Christmas of 1987. We had been together for almost three years and the time was right. Looking back, I wasn't very romantic about proposing – in fact it definitely wasn't a case of me going down on one knee. Nevertheless, we announced it to Claudia's parents on Christmas Day, who reacted with shock, expecting us to have arranged a big party for the announcement. After Christmas we drove to Southwold to break the news to my mother, who was delighted. There was no hurry to marry but it felt good to have a strong relationship with Claudia.

***

Wolfgang Möbius' vent of frustration in my direction seemed to have freed him emotionally, something that maybe he had needed to do for years. During the days following, he seemed a different person. I don't believe Tony Lapine had talked to him, but he was smiling and taking time to talk to people in the studio, a strong contrast to the usual sullen figure seated at his desk in the corner.

I had confided in Dick Söderberg. He also listened with understanding and tried to downplay the situation a bit.

"Yeah, Wolfgang is not easy, but you'll get over it." His recommendation, however, was very much the same as Tony's.

With most projects, except for the 964 (911 facelift), being put on ice, a period of tranquility came over the studio and the designers were given time to create freely. Tony and Dick's advice to me was right: I had to prove to them all that I could earn my place in that studio. I decided to develop ideas for a range of what I thought would be a good model line for Porsche and set to work.

Wolfgang Möbius was not a fan of fancy illustrations, preferring a much more purist approach. I decided to concentrate my efforts on clean well-proportioned side views of three models. A small mid-engined coupé, similar in dimensions to the 984, a classic front engine coupé with a convertible option, and my vision for the next 911. It was a pretty logical line-up, I agree, but I had seen no work in the department where anybody had considered a range of cars that had a close family look, so it was at least worth a try and I had nothing to lose.

Working in quarter-scale I drew up some technical packages – passenger space, engine layout, etc – and started to work on tape drawings of all three directions. It was important that the stance and proportion of the designs were perfect. One trick that many designers use to judge their work, which I employed, involved a mirror. When holding a mirror in front of a drawing, the flipped image you see in it will generally give you an idea if your work is well balanced or not. It is almost as if you are looking at somebody else's work.

Over the next few weeks, I began to get back the confidence I had lost after the bashing Möbius had given me. Tony Lapine's words were firmly in my head – "Don't get mean, get even."

Satisfied with my three tape drawings, I decided to do airbrush renderings of my proposals but stick to a more monochromatic approach, with plain black backgrounds.

Completing the first rendition, it was Dick Söderberg who first popped his head around my drawing board, and like the famous quote from *2001: a Space Odyssey* he said, "That's a great rendering, Birdy."

I felt a weight falling off my back. At last, a positive piece of feedback.

Over the next few weeks I completed my series of proposals, mounted them on white card and attached them in a row to one of the large studio boards. Looking back, I really believe those designs were a highlight of my career at Porsche. People started to take notice and visiting engineers to the studio made positive comments. Tony Lapine came by and expressed his approval.

"Well done, Birdy, now you're getting there," he told me – or something similar.

Despite all this, Wolfgang Möbius kept quiet, and when questioned about my work he just couldn't bring himself to comment. Things were all too clear, and with a heavy heart I admitted to myself that I was wasting my time at Porsche.

# Chapter 19

## *The highlight at the end of the tunnel*

**Some things in life are just meant to happen. I had been looking through a copy of the weekly German illustrated magazine *Stern* and a small article leaped right off the page and hit me between the eyes. It read:**

"The Japanese car company Mazda is to open an R&D centre near Frankfurt. It will include a large testing laboratory and a design studio."

I knew straight away that this was my chance. I had to share the information, but with who?

By the next day I knew who to talk to, and when nobody was near his desk, I walked over to Roland Sternmann and placed the open copy of *Stern* in front of him.

"Let's talk later," I said, and left him with the magazine.

Roland Sternmann's talent had been spotted by Tony Lapine. He had been working as a student doing a holiday job at Porsche and went to Lapine's office to ask if he could give him some feedback on his car drawings. Apparently, Lapine was so impressed that he agreed to sponsor Roland at the RCA.

Roland was, in a way, a bit of a loner, but he was a great designer and I knew he was also not so happy at Porsche. If Mazda was looking to build up a design team, Roly, as we all called him, would be good to have along.

I called Roly later and he said he was interested but asked how we move forward. One thing we agreed on was that neither of us felt ready to take on a managerial role in whatever design organisation Mazda was planning, and I suggested we approach Ginger Ostle for such a position. Roly agreed. In the meantime, he said he would contact a head hunter he knew and ask him what he thought or knew about Mazda's intentions.

We decided to meet Ginger at his apartment and explained about the Mazda R&D centre, telling him that we felt he was the right guy to lead such a venture. There was little hesitation – he was interested.

Roly had made contact with his man in the recruitment business, employed by a large company called Russell Reynolds. The person in question was keen to meet us and a meeting was arranged, once again, at Ginger's apartment.

As a group, we were well aware that there was much frustration among the design team at Porsche, and carefully spreading the word around, many soon expressed further interest in our plan.

By the time Roly and I met again at Ginger's flat we had a list, including ourselves, of around ten Porsche design staff interested in leaving the company for something new. For the man from Russell Reynolds, it was like winning the lottery. He said he would contact his partner in Tokyo about Mazda, but felt that such a 'package' may interest other companies. He would get back within a week.

I'm often asked why a designer working for Porsche would even consider working for a company like Mazda, which I guess on the surface is not an unreasonable question. But you have to consider the situation as it was in those days at Porsche. Sure, I had my personal issues with Wolfgang Möbius, but the overall situation was much bigger than some personal differences. I, and most of the designers working in Weissach, joined the company to design exciting sports cars, but, regretfully, at that time the company had no concrete plans to expand its range. There didn't appear to be any funds to finance new projects and as designers we had to satisfy our creative ambitions by working off derivatives of the company's existing models.

The 964 was a good case in point: apparently 80 per cent all-new parts, but still based on the original 911 body shape. A new 911 looked years away, and as it turned out, that was the case. Working for Porsche was a great privilege, above all having the opportunity to drive its amazing products. But even the thrill of driving a Porsche wore off, and as I often said, the girls of Stuttgart knew if you were driving a works car and that you probably couldn't afford it.

No, I became a car designer to design cars and that wasn't happening at that point in time for me and most of the design staff at Porsche.

At the time, I knew little about Mazda, other than that they were fairly new to the European market and made good looking cars. If they were setting up a design office in Germany, they meant business, and you could be sure there was a lot of design work to be done. For me, that sounded exciting.

We met the man from Russell Reynolds at his office in Frankfurt. Through his contact in Japan, Mazda had replied and were very interested to meet us. Alfa-Romeo had also expressed interest but was concerned that it couldn't match German salaries. We decided to meet with Mazda.

We had to wait another week before learning that the German personnel manager responsible for the recruitment at the new Mazda R&D centre wished to meet each candidate at a location near Weissach. Once again, Ginger's flat was to be put to use.

In the end, nine candidates gathered at Ginger's flat for the first meeting with Mazda. Ginger, Roly, myself, a senior modeller named Jim Howell, plus two of his younger colleagues, Daniel and Yvonne, a studio engineer called Dave Samways, and a further engineer and designer.

The name of Mazda's personnel manager was Bernd Lesny. Lesny was responsible for hiring all the new staff for the R&D centre, which, as it turned out, was to be located in the town of Oberursel, situated below the Taunus hills just north of the city of Frankfurt.

We gathered in the front room of Ginger's apartment; it was a tight fit. Each of us had been asked to bring along a resume and one by one we had individual meetings with Lesny, with the man from Russell Reynolds also in attendance.

Once all the interviews were complete, all candidates except Ginger, Roly and me were thanked, being told that they'd be contacted soon.

The small group remaining got together and Lesny explained that while he was very pleased, he was also acutely aware of how sensitive the issue would be if nine Porsche design staff all quit at once. It was a situation that needed to be handled carefully. To move forward, he proposed that the more senior members of the 'package' should attend a meeting at the site where the new Mazda centre was to be built, and it was agreed that the meeting should take place within the month.

Mazda was based in the Japanese town of Hiroshima, a name which needs no introduction. When the company was founded by a Mr Juriro Matsuda, it was originally named the Toyo Cork Kogyo Company, and was a manufacturer of, as the name implies, cork. In 1931, the company had developed a small three-wheeler truck called the Mazda-Go, which it used in its factories and was the beginning of what would eventually become its main business. During the war, it had made rifles, and due to its location in the shelter of a small mountain, the factory buildings had escaped major damage when the atomic bomb had been detonated over the city centre. For this reason, at the end of the hostilities, the company was able to quite rapidly resume production of cars on a large scale under the name of Mazda.

Mazda expanded as a company, but soon suffered financial problems and, in the '60s, needing to secure financial backing, went into partnership with the Ford Motor Company. With the support of Ford, Mazda developed rapidly as a global car producer, being particularly successful in North America, a trend that, from the '80s, also spread to Western Europe. With its products successful on both sides of the Atlantic, Mazda realised that if it was to be a real automotive force, it needed to develop and meet the engineering and design levels expected in these markets. Where better, then, than to gather that knowhow on the home soil of those continents. With this in mind, at the beginning of the '80s, the decision was made to build two research and development centres: one in Irvine, California and the other in Oberursel, Germany.

In Europe, Oberursel was the best solution for Mazda to put an R&D centre. Germany was after all the home of the world's most prolific car industry, so a good location for hiring industry talents such as our group from Porsche. Not only that, but with the closely accessible autobahn network and nearby Frankfurt with its big airport, the town was logistically well suited. Also important for the Japanese was the relatively large population of their own countrymen in the Frankfurt area ,with Japanese schools and, of almost equal importance, Japanese shops and restaurants.

*The Georg Schutz factory site in Oberursel, north of Frankfurt, where Mazda planned to build its European R&D centre.*

***

It was a Saturday, a sunny June morning in 1988 when four Porsche 911s converged on the address of Mazda's new European R&D centre in Oberursel. The town of Oberursel was very proud that Mazda had selected the town for this new venture. In exchange for the mini-bus that the company had donated to the town to transport old age pensioners, it had agreed to name the road outside the centre Hiroshima Straße.

The location in Oberursel where Mazda planned to build its European R&D centre was the site of a disused factory – named Georg Schutz after the founder of the family that had owned it. The red brick building with its two landmark towers had produced industrial wax.

As we arrived, we could see that most of the factory building was still intact, although that day's *Taunus Zeitung*, Oberursel's local paper, was full of images of the demolition of a tall chimney that had been pulled down at the site the previous day. After passing through the entrance gate, we were hurriedly beckoned into an empty factory hall, our Japanese hosts very keen to keep the four Porsches out of view from any inquisitive eyes.

What was once the entrance gate-house of the factory had been converted into offices and Bernd Lesny was there to welcome us and introduce the delegation from Mazda.

In order to emphasise the importance of this venture, Mazda had appointed one of its board members to supervise the project in Oberursel. His name was Seiji Tanaka. The Japanese are similar to the Germans in that first introductions are usually on surname terms. We would therefore address Mr Tanaka as Tanaka-san – I needed to get used to that.

With Tanaka-san was his right-hand man Shizimatzu-san, and from Mazda Design, none other than the corporate head of design, Shigenori Fukuda, plus two of his lieutenants: Yoshitomi-san (his first name eludes me) and Yutaka Shimazu. It was an impressive delegation.

Introductions were made, all of the Mazda team speaking good English.

Our group consisted of Ginger, Roly, me, Jim Howell, and Dave Samways. It was agreed that Ginger would meet with the Japanese delegation first, the rest of us joining later on. The rest of us made our way to the nearby spa town of Bad Homburg for a coffee.

After an hour or so enjoying the sun at a café in the centre of Bad Homburg, we returned to Oberursel, where upon arrival the mood seemed very good. It was now our turn to go through our CVs individually and listen to what Mazda was planning.

When my turn arrived, I went through my prepared portfolio, with Fukuda-san leading the meeting. They appeared impressed and I was amazed at their almost childlike enthusiasm.

Eventually, all meetings were concluded, and in a small speech addressed to all of us, a very happy Fukuda-san thanked us for coming and hoped we could agree to the terms of employment. Moving our cars out of the factory building, we set off back to Stuttgart, Bernd Lesny agreeing to travel to Stuttgart once he had drawn up contracts.

I could feel the ball and chain that Porsche had become being cut free. I couldn't have been more relieved in my life.

In the end, seven of the nine Porsche design staff signed the contracts that Bernd Lesny brought to Ginger's flat the following week, and after a celebratory glass of German sparkling wine we split up, Ginger agreeing to be the first to hand in his notice to Tony Lapine.

Ginger's announcement was a massive shock for the whole Porsche company, not just the design department. He had a great reputation within the company and was much liked by everyone, right up to the highest management. Apart from the fact he was joining Mazda, he said he had not been too specific about the nature of the job and urged Roly and myself to let the dust settle before handing in our notices. For me, however, the wish to close this chapter in my life as soon as possible couldn't come soon enough, and about two weeks after Ginger's departure, I walked into Tony Lapine's office.

I made it simple.

"Tony, I've decided to join Ginger. I never thought I would go to work for a Japanese company, but my mind is made up."

Lapine leant back in his chair (never one to show any signs of disappointment), smiled and said in his typical ironic way: "Jesus Christ, Birdy, is it any different than working for the fucking Krauts?"

Lapine knew the last year had not been easy for me, so he wasn't going to get down on his knees. He said I should go over to Herr Fischer in personnel and give him my letter of resignation. Before I left, he warned me in the friendliest of tones: "Birdy, be careful what you say to Herr Fischer. Just remember, never kick a piece of shit."

Lapine knew his boat was running into a storm and he didn't want me causing any trouble.

Herr Fischer took the news worse than Tony Lapine, and as Tony had suspected, he tried to get some kind of detrimental information out of me; it didn't happen.

I had no bone of discontent with Tony Lapine, in fact, I liked him a lot. When I joined Porsche, he had just survived a heart attack and the energy and fire he once had was gone. He was cruising, and knew his time there was also coming to an end.

He retired the following year, as did Dick Söderberg. In 1989, Dutchman Harm Lagaay took the helm at Style Porsche, and under his tenure the design team at last realised the opportunity to develop a completely new range of cars for the Stuttgart company.

When I started my new job, the Mazda R&D centre in Oberursel was beginning to take shape, but, in September of 1988, it was still quite a way from completion. It was therefore agreed that we would all travel to Japan to work in the studios there until Christmas.

I closed the door of the flat in Leonberg for the last time, asking the taxi driver to stop at a post box on our way to Stuttgart airport. I would not see Claudia for three months, so I posted her a card and the new Phil Collins hit *A Groovy Kind of Love.* Maybe there was a bit of romance in me after all.

Ginger, Roly and Jim Howell were waiting at the airport for me and together we boarded a flight to Amsterdam. During the move out of my flat, I had pulled a nerve in my back, so the guys had to carry my luggage for me. "Typical Birtwhistle," I heard, and not for the first time!

In Amsterdam, we met Tanaka-san, who'd been at our first interview in Oberursel, and we all boarded a KLM flight bound for Tokyo.

*The Style Porsche design team, 1988. A slightly confusing photo since everybody is standing behind a large mirror that's placed along the centre of the 944 cabriolet model.*

Tanaka-san was proud of his big coup in bringing a group of Porsche designers to Mazda and wanted to enjoy the glory of delivering them to Mazda's HQ in Hiroshima.

Taking our seats, the captain announced that there was a baggage handler strike in Amsterdam and that not all the luggage was on the plane yet; we would have to wait. Drinks were served and the waiting continued. After two hours, the captain informed us that some luggage was still not in the hold and offered us a choice: we could go on to Tokyo and risk arriving without luggage or leave the plane with the chance that our luggage had made it on board. For us though, there was no choice to make. We were committed. We were off to Japan with or without luggage.

The KLM 747 left Amsterdam, heading west. It stopped at Fairbanks, Alaska to refuel, before continuing on to Tokyo.

Landing in Alaska, it was already winter. Walking out onto the viewing platform while the 747 was being fuelled, it was weird to see snow on the ground having left the warmth of a late summer day in Stuttgart only hours before.

The KLM Jumbo resumed its course to Tokyo and somewhere over the Pacific we crossed the international date-line into October 1, 1988. Tanaka-san had ordered champagne and we lifted our glasses as he proposed a toast.

"*Kampai*. Welcome to Mazda!"

"*Kampai*," we replied in unison. It was the first of many more *kampai*s.

Having drunk his toast, Tanaka-san went on to present us each with a small package, which contained our Mazda business cards. For the Japanese, the business card is much more than just a small piece of card with your name and title on it. Japan is a fiercely status-orientated land and the ceremony of presenting your business card to a new colleague, whether from your own company or any other business contact, is a procedure of great importance in their society. When meeting someone for the first time, the card is presented, held forward in both hands, while bowing. The angle of the bow depends on the business or personal status of the person you are engaging with. The higher up the ladder or older the person is, the deeper the arch of your bow. Having exchanged cards or '*meishi*' as it is called in Japan, you don't just then stick it in your wallet, but instead take time to study and learn the name on the card before, in the case of a business meeting, placing it on the table in front of you. The Japanese don't expect foreigners or '*gaijins*' to copy their mannerisms, but some form of mimicking this protocol is not a bad idea.

For Tanaka-san, the presentation of our first *meishi* as Mazda employees was of great significance, but it was years before all the intricacies of Japanese etiquette were half-way familiar to me. I therefore unceremoniously ripped open the package:

**Peter Birtwhistle**
**Senior Designer**
**Mazda Motor R&D, Representational Office Europe**

Roland Sternmann and I had the same title, with Ginger Ostle, our future boss, named design manager and Jim Howell senior modeller.

It looked good, and flipping the card I noticed my credentials were repeated on the other side in Japanese '*kanji*' script. I had to trust that what was written was correct. The finer points of Japanese ceremony still lost on me, and with little respect for what Tanaka-san had passed over, I stuffed the small package in my hand luggage and signalled the flight attendant for a top-up.

Three months in Japan was, for me, not a hardship. Yes, I had to leave my fiancée, Claudia, back in Germany, but compared to Roland and Jim, who were both married with small children, I easily came to terms with the new situation, and what was three months really?

Slowly, the time difference and champagne took control and sleep beckoned.

Returning to my seat, I closed my eyes. It had been a long journey from Sundon Park, Luton to where I now found myself; a new journey was about to start.

What lay ahead in Japan?

*Farewell card from the Porsche team, painted by the great Dick Söderberg. RIP.*

*The legendary head of Style Porsche, Tony Lapine. RIP.*

# Chapter 20

## *Exploding underpants and wasabi*

**For me, Ginger Ostle, Roland Sternmann and Jim Howell, it was our first trip to Japan. In a way, we were the pioneers, with three further Porsche employees joining our team when we returned to Germany at the end of the year.**

Ginger Ostle had been studio head for interior design at Porsche, and it had been unanimously decided that he should lead us and the new venture. Although I had the title assistant studio head at Porsche, and was senior to Roland Sternmann, at Mazda we would be of equal rank. Jim Howell had been a senior modeller at Porsche and, once back in Oberursel, was given the task of building up the modelling and fabrication department. It was a great opportunity for all of us. That so many design staff should leave Porsche to join a Japanese company had caused quite a stir in the industry. Were we crazy? Time would tell.

Mazda was anxious for us to start designing straight away, hence our three-month trip to Japan. We would commence work at its advanced studio in Yokohama, close to the Japanese capital, Tokyo. First, however, we would be visiting the main production studio at the company's HQ in Hiroshima, in the south of Japan.

It was early evening in Japan as the KLM 747 approached Tokyo Narita airport. After touchdown, and having passed through immigration, we had two problems confronting us.

After a flight that had taken almost twenty hours, including the delay and stopover in Alaska, we waited at the luggage carousel for our luggage, which we gradually realised was still in Amsterdam.

In addition, the delayed arrival meant we had no time to connect with our onward flight to Hiroshima that was scheduled to leave from Tokyo's domestic airport in Haneda. Haneda was a good hour's bus ride from Narita airport, and the last transit bus had already left. We had no choice: still in the clothes we'd been wearing for over a day, and with no luggage, we had to stay the night at Narita.

While Tanaka-san was busy organising a hotel and a new flight out to Hiroshima in the morning, I wasn't concerned. In fact, I felt pretty smug. Ever the first person to worry about disaster, I'd had the good sense to put a couple of fresh shirts and underwear in my hand luggage, to get me by in the unlikely event of my cases being lost. My fellow travellers were not so smart, and were already wondering where they could get a fresh change of clothes at nearly seven in the evening.

When Tanaka-san returned from his phone calls, he had news of mixed fortunes. The flight to Hiroshima for the next morning was sorted, but he could only locate two rooms at one of the airport hotels. One room had a double bed and the other had two singles with arrangements for a mattress on the floor. Since he was the boss, it was agreed that Ginger would share the double room with Tanaka-san. Roly, Jim and I would take the other room. With my back still in a bad way, sleeping on the mattress on the floor was a good solution for me.

Feeling tired and sticky, we made our way over to the hotel. After checking in, we agreed to meet in an hour for a meal. I, with no worries about a change of clothes, made my way to the room and the shower, while Ginger, Roly and Jim headed for the airport shopping mall in search of some emergency clothing.

The shower brought me back to life, and it wasn't long before Roly and Jim returned from their shopping trip. They didn't look too happy, and, throwing their bags on the bed, said that Tanaka-san was tired and wished to eat as soon as possible.

With little time to spare, we met in a bistro near the hotel lobby, Tanaka-san judging it best not to confront us yet with the delights of Japanese cuisine.

Sitting down, it was clear that Ginger was slightly apprehensive of the night ahead of him and, with

Tanaka-san out of earshot, revealed that a Japanese double bed was little bigger than a European single.

Airport hotel bistros are the same the world over, and I'm quite sure the meal was nothing to write home about. We agreed on a time to meet for breakfast and left Ginger and Tanaka-san to discuss their sleeping arrangements – maybe Ginger would be calling for a mattress.

Back in our room, an interesting series of events unfolded. I needed to do some stretching exercises for my back and, reduced to a t-shirt and underpants, began to arch myself, stretching out my legs and arms on top of my mattress. There was little floor space, and once my mattress was down at the bottom of Jim's and Roly's beds, the floor was covered.

Meanwhile, Jim had been trying to call his wife, but to no avail. Homesickness had already cut in and he was not in a good way. With the realisation that the hotel phone system didn't handle overseas calls, I could see tears welling up.

On top of that, the shopping trip for clothing had not gone well. Both Jim and Roly were around six-foot tall and Roly in particular was quite sturdily built. Walking into the airport supermarket they had appeared like two giants whose spaceship had been diverted to Narita. It soon became clear to both of them that there was no clothing anywhere near their size. Ginger had apparently had more luck, being almost of Japanese stature. He was able to find some t-shirts, socks and underwear. Just as well, since I'm sure he didn't wish to share the double bed with Tanaka-san without some body coverage.

One small solution to Jim and Roly's problem, however, had been found. A dispensing machine located in the hotel lobby had various articles catering to air passengers who had lost their luggage. It offered toothbrushes with toothpaste, plastic razors with shaving cream, plastic hair brushes and other toiletry articles. Amongst the offerings was also a selection of paper underpants, which, as a last-ditch solution to their problem, both Roly and Jim had purchased.

With our room dimly lit, me straining on my mattress and Jim whimpering at the thought of his family on the other side of the world, Roly decided now as a good time to test his recent purchase, and he opened the vacuum-packed underpants he had got from the machine.

I watched from the corner of my eye in disbelief. Roly pulled back a small sash and with a hiss the paper bundle inside the bag gasped for air, the underpants expanded and took shape. I imagine the scene in that room must have resembled that of three astronauts in a space capsule. Roly stretched the paper underpants out to their full width and climbed into them. It was a tight fit but he was determined and with a bit of persuasion he slowly eased them up into position.

"They'll do!" he said triumphantly and turned around to sit on his bed.

This was too much for Japanese paper underwear technology, and with a loud crack that made both me and Jim come back to our senses, Roly's underpants literally exploded under him like a sausage being thrown into a hot frying pan.

It was painfully clear; Jim and Roly had to remain grubby until we reached Hiroshima and hopefully the reunion with our lost luggage.

Somehow, we got through the night, but as we met Ginger and Tanaka-san the next morning, we agreed that we had all experienced more comfortable nights in our lives. Ginger had survived the double bed with Tanaka-san, although also not without its peculiarities. We had all discovered before retiring, that the rooms were kitted out not only with the usual selection of towels but also night gowns to sleep in. Like us, Ginger had assumed they were dressing gowns and discarded his when climbing into bed. Tanaka-san had immediately corrected this misunderstanding and asked Ginger to attire himself in the garment while in bed. Apparently, Ginger had woken from his restless night with the waist cord to secure the gown still in place, but the rest of the garment twisted around his neck.

Checking out, we were all grateful to be on our way and followed Tanaka-san to the airport limousine that would take us to the domestic airport in Haneda in Tokyo Bay. 'Airport limousine' was somewhat misleading, since it turned out to be a normal coach.

The journey to Haneda from Narita is not a great distance, but as we pulled on to the expressway towards Tokyo, the traffic was almost at a standstill, and this on a Sunday.

Everything was new. The first thing that hits you when arriving in Japan is the mixture of sounds you are bombarded with. A cocktail of the Japanese language emanates from various loudspeakers, which for most visitors is completely incomprehensible. There are chimes and flute-like melodies signalling anything from a door opening to the start of some loudspeaker announcement. In the end, it all washes over you.

I gazed out of the airport limousine window as we crawled towards Haneda. Japan's entire population appeared to be on the road. Apart from small delivery trucks, everybody was driving either a Toyota Corolla or some kind of boxy van, most of which were painted white. Taxis passed by and I noticed that the drivers all wore peaked caps and dark suits, with white gloved hands on the steering wheel.

The journey to Haneda took a little over an hour and with no luggage to check in, we made our way quickly through to the departure gate and to our ANA flight to Hiroshima.

The flight south to Hiroshima takes about an hour, and we were lucky to get a good view of Mount Fuji as the plane moved up and out of the Tokyo airspace.

Today, Hiroshima has an international airport located to the east of the city on top of the mountain plateau that rises behind the city. Before the completion of this airport in 1993, all air traffic landed at a small airport located near the city harbour, which is where our plane touched down.

Walking out of the plane, the heat hit us immediately. Hiroshima is approximately the same latitude as Morocco, and at the beginning of October it was still very warm.

Tanaka-san arranged two taxis and we headed off towards the city centre and our hotel for the next week.

Despite the city being the home of Mazda, all the taxis seemed to be exclusively from either Toyota or Nissan. All looked the same, being rather nondescript upright vehicles with a side profile similar to the kind of shape a child would create when asked to draw a car.

As we approached the taxi, the driver used some kind of lever to opened the rear door for us, which, once we were seated, he also used to close and lock the door. Just as I had observed from the airport bus, our driver was also wearing a black suit, a cap, and white gloves. The seats' head and back rests had white linen covers on them with frilly lace borders – I wasn't sure how that would have gone down in London.

We set off, and glancing forward at the centre of the dash board, I saw a TV screen was broadcasting a baseball match, which the taxi driver was avidly following.

It had been over 40 years since the dropping of the atom bomb, and Hiroshima showed no signs of the terrible destruction that had comprehensively flattened the city centre. Like many American towns, the roads were laid out in a grid system, the city presenting itself as a typical modern business and shopping metropole. Positioned on a large river delta, we passed over several of the town's many bridges as we headed to the centre.

The city's main shopping streets were wide alleys with a tram system taking up the central part of the road. Of interest is the fact that the Germans helped Hiroshima build their tram system during the post-war re-construction and even to this day some of the old trams have German writing on parts of the rolling stock.

The taxis pulled up outside our hotel, which was located next to one of the river deltas and close to the main railway station at the edge of the main shopping centre. A fairly nondescript block, a large gold sign above the entrance read, "Hiroshima City Hotel."

We checked in, still with no luggage. Tanaka-san had asked Mazda to contact the KLM agents in Tokyo, and, assuming all was well, we would get our cases

*Hiroshima SOGO department store.*

on Tuesday at the latest. Nevertheless, with business meetings on Monday and Roly and Jim still in their attire from Friday's departure, a quick shopping trip was top of the list. In Japan, the shops were thankfully all open on a Sunday.

Tanaka-san, who had rented out his house in Hiroshima for the duration of his tenure in Frankfurt, was also residing in the hotel, and suggested that we make our way to the large SOGO department store. There was a large clothing department there and afterwards we could meet for our first Japanese meal. It sounded good.

The walk along the wide boulevard to the SOGO department store took about 15 minutes. Immediately, some of the many differences between Europe and Japan became clear. Cyclists were not permitted to use the road and had to share the pavement (sidewalk). The outer top surface of the pavement had a foot braille pattern for blind pedestrians. At traffic light crossings, the green man still signalled that it was clear for pedestrians to cross, but was accompanied by the tones of the Scottish melody from Robert Burns, *Comin' Thro' the Rye*. How out of place was that?

The SOGO store was located at a main traffic junction in the centre of the city. It could have been anywhere in the world, given that the floor layout was the same as all department stores, with perfume and leather goods on the ground floor.

Entering, we were confronted by two young girls dressed in uniforms, whose job was to bow and greet each customer individually.

"*Konnichiwa*," they repeated, while lowering their heads respectfully.

We took the lift to the men's department – signs fortunately had an English translation. As the lift doors closed a small gong rang out. It's a sound that you only hear in Japanese lifts, whether in shops or hotels, and to this day if you blindfolded me and played that sound, it would transport me right back to Japan.

We wandered around the men's floor, but I decided I could manage with the hotel laundry service to wash my two spare sets of clothes, so I left the guys to make their purchases. For Roly, it was still a struggle, but he found something in the end and we all made our way to the restaurant floor and Tanaka-san.

Eating out is a big part of the Japanese culture and all department stores have at least two top floors dedicated entirely to restaurants. A unique feature of Japanese restaurants is that the windows are full of life-like wax models of the dishes they serve, which makes ordering food very easy for foreigners. In addition to department stores, big cities will usually have a region in the town consisting of a maze of eating houses. In Hiroshima, it was called Hondōri and was famous for its buzzing restaurants and bars.

Tanaka-san guided us through the door of a typical sushi restaurant. Seats had been reserved for us at the counter where we could watch the dishes being prepared in front of us.

For all of us, Japanese cuisine was new, as in Europe it was still very exclusive. I recalled visiting a restaurant in Munich in my Audi days called Daitokai. There, you sat around a large table that doubled as a cooking surface, where a chef created the dishes in front of you, frying the meat, fish and vegetables on the hot metal surface. That style of eating is called *teppanyaki* and is not to be confused with sushi or sashimi. Sushi is usually a roll of boiled rice with some kind of fish or vegetable in its centre, held together with a thin wafer of dried seaweed. Sashimi is a cold slice of raw fish, tuna, salmon and squid being the most popular examples. Both dishes are served with a small bowl of soy sauce and a hot green paste called wasabi. The wasabi paste, made from a plant related to that of horseradish, is added to the soy sauce to taste. In addition to this, there will also be a small bowl of thinly sliced pickled ginger which can be added for extra flavour.

The restaurant that Tanaka-san had chosen specialised in both sushi and sashimi.

We took our places and a girl in traditional Japanese dress gave us a small hot towel to wipe our hands on. In front of us were some chop sticks and a selection of small ceramic bowls.

Tanaka-san looked happy. He had been away from Japan for several months and was ready to indulge in his favourite style of eating. As a company board member, he had an expense account to entertain guests, and therefore was not scrimping.

The first selection of sushi arrived and taking our chop sticks in hand, we all fumbled to place a roll on our plate. So far, so good. Tanaka-san raised his chop sticks and announced: "*Itadakimasu*," or in other words, "*bon appétit*."

I had some experience using chop sticks at Chinese restaurants, but Jim was already struggling, deciding to take one chop stick in each hand and use them like a knife and fork. Luckily, the restaurant staff remained courteous!

Having got the hang of manoeuvring the sushi, I dipped a roll in the soy sauce and then, forgetting what Tanaka-san had advised us, pushed it into the wasabi paste as if it were ketchup. Too late to realise my error, the wasabi immediately detonated in my mouth. Choking, I gasped for some water.

The portions kept arriving and Tanaka-san, oblivious to our various mishaps, was in sushi heaven. He gestured to one of the cooks behind the counter, it was time for some sashimi or raw fish.

By now, we had each consumed at least six to ten sushi rolls, but it was clear from the smile on Tanaka-san's face that things were by no means over.

I was coping quite well, but looking at Jim, the expression on his face said it all, though, not wishing to offend, he ploughed on.

The first plate of sashimi, a selection of raw salmon, tuna and squid, was very edible. Only the squid felt a bit like a piece of tyre inner tube. With each operation of the chop sticks, I got better, by now having also accustomed to the amount of wasabi paste that suited my palate – I could get used to this, I thought.

Tanaka-san pushed on, but the next selection of dishes he ordered certainly brought things to a close, drawing the blood out of Jim's face altogether.

Tanaka-san explained that the banquet would end with a selection of seafood that was the most exclusive of this style of eating, and presumably the most expensive.

A selection of shellfish was placed in front of us. There were various kinds of sea-urchins alongside a selection of sea snails that looked as if they had been pulled out of some beast's nose. My grandmother used to eat whelks and cockles, but these were of a wholly different calibre.

*Jim Howell's first taste of Japan.*

Jim bravely prised out the body of an orange sea urchin. Once it had entered his digestive system, though, his face turned a pale shade of green. It was time to go.

We made our way out into the fresh air, pleased to have enjoyed Tanaka-san's hospitality but thankful it had reached a conclusion.

By now, it was late afternoon and dusk was closing in over Hiroshima. We were all feeling the jet lag and decided to return to our hotel rooms. It had been an eventful 48 hours. The air was still and warm, and the sky was filled with a unique pink light, the likes of which I had never experienced anywhere in Europe. Yes, Japan was something completely different, and that felt good.

# Chapter 21

## *Mukainada*

**Travelling halfway around the globe, to the east or west, there's no doubt you are going to suffer from jet lag. Many seasoned travellers have their own methods of combating the effects, but as a rule of thumb the best strategy is to try to continue through the day of arrival and retire to bed at the right time in the evening. Travelling east is more difficult, since you effectively miss a complete night's sleep. Leaving Europe around midday, you will arrive in Japan in the early morning, around midnight in Europe. Therefore, the temptation to go straight to bed upon arrival at your hotel is understandable but not to be advised. If you do this, you will sleep for six hours, only to wake in the late afternoon local time; a sleepless night to follow is predestined.**

That first day in Hiroshima, we had made it through most of the day, and as a result of Tanaka-san's generous lunch invitation, had managed to stay on our feet until around six in the evening. However, returning to the hotel, we all decided it was time to 'hit the sack' and parted company to the peace of our rooms.

That first night, I was already asleep before my head hit the pillow, but at around two in the morning, I was sitting up in bed wide awake. I got up and made my way to the window. Looking down into the street in front of the hotel, all was quiet. Hiroshima was asleep, resting before the beginning of a new week.

The hotel room could not be described as spacious but it was adequate, not the plastic cube that friends had joked I'd be sleeping in upon my arrival in Japan. It had a single bed, a desk and chair, and by the window a small coffee table and armchair. There was a wardrobe that was enough for a couple of nights, but I began to wonder how I would manage in a similar-sized room for the three months planned in Yokohama, when my two cases eventually arrived.

I had brought a few paperbacks with me, but I was bad at reading books even at the best of times, and right now there was no way I could concentrate. There were too many things buzzing around my head.

There was a small TV on the desk with some kind of coin machine integrated into its base. Were you expected to pay for the TV?

After flicking through the channels, it soon became clear what the coin machine was for: adult movies. 100 yen for five minutes.

I zapped through the channels and eventually looked to see if I had any 100 yen coins. The menu gave a choice of a Japanese or a Western title. I selected the Japanese.

Japan has very strict rules about what can be shown on 'adult' films, and images of both male and female sexual organs, as well as the act itself, are strictly forbidden. The films shown on hotel pay TVs are therefore tailored to compensate for this. As soon as things start to get 'naughty' the offending body areas are covered by a kind of digital fog or pixels on the screen. Japanese erotic filmmakers know this and therefore spend a lot of time avoiding shots that could offend, preferring to build up a story first.

I put in my 100 yen. What soon became clear was that Japanese men had a very particular sense of sexual fantasy, the actors being mainly young women dressed in school uniforms. I'd seen enough, and changed to the Western film, which, predictably, was a German production. Well, at least I'd understand the plot, I thought. In contrast to the Japanese production, the story got straight to the point. The country's sensors had obviously seen quite enough of this German romance and suitably adjusted the film for Japanese viewing. To the panting and cries of delight from some presumably good looking *fräulein*, the screen was completely covered in digital fog for all the time that remained from my 100 yen coin.

I turned off the TV, and sleep took over again.

My alarm rang. Seven o'clock – about midnight in

Europe. I pulled myself out of bed, my bed gown around my neck, and moved to the bathroom.

My visit to the bathroom the previous day had been fleeting and in half sleep my level of concentration had been low. Only now did it strike me that the toilet was, like the TV in the room, also equipped with an extra console at the side of the lid. 100 yen for a pee? Now was not the time to figure out what the switch board was all about, so, after relieving myself, I flushed the toilet in the normal manner. The control panel had to wait.

I put on my remaining clean shirt and underwear, leaving my used garments in a laundry bag on the bed. They would be clean by the time I returned to my room at the end of the day. I wondered if our cases would also be at the hotel that evening.

Ginger and the guys were already having breakfast and we soon got talking about our efforts to sleep through the night. Yes, they'd also discovered the delights of Japanese erotica!

The breakfast room was mainly full of Japanese businessmen, all dressed in dark suits. Looking across to another table I noticed one of them indulging in a dish that, at that time of the day, appeared pretty unappetizing to me: a raw egg cracked over some steaming rice, with a bowl of salad, sardines and pickles to accompany it.

Tanaka-san met us in the hotel lobby and we made our way out to some waiting taxis that would take us to the Mazda HQ building.

Although we all had fresh shirts on, we were still attired in jeans and sports shoes. I hoped our new colleagues would have some understanding for the predicament we were in.

The taxis moved off and over the bridge next to the hotel, passing the main railway station and onwards towards the southern part of the city and the small suburb of Mukainada, where Mazda's sprawling factory complex was located.

The taxi made its way along a series of streets under the shadow of the elevated concrete structure that threaded the famous bullet train (*Shinkansen*) out of Hiroshima and on towards Fukuyama, the first stop on the way to Tokyo. Slowly, the shops and hotels disappeared, and the road became lined with industrial yards and car dealerships. Any street harmony was ruined because of the entanglement of service cables running above ground, a precaution apparently because of the risk of earthquakes.

The route followed one of the rivers that led down to the Japanese inland sea, although for us it was out of sight due to the factory buildings that lined its banks. On the other side of the road houses were packed together, spreading out and up onto the hills that circled the town. Japan, with only 17 per cent of its land surface habitable, makes good use of every part of that small area.

The taxis turned off the main road, and drove through a tunnel before pulling up in front of Mazda's main HQ building. The building was a white office block with the Mazda name written proudly in blue at the top. Facing it, as we got out of the taxis, you could see the factory buildings to the left, stretching out into the distance.

We followed Tanaka-san through the doors and walked up to a reception desk where three girls in blue uniforms welcomed us with a simultaneous bow.

*Mazda HQ building.*

The reception area was not unlike a car showroom, with a selection of Mazda's latest models on show, many of which were unfamiliar.

We signed in and followed Tanaka-san through a door at the far end of the entrance hall.

Back outside, we found ourselves standing in front of a brown brick building that Tanaka-san explained was the original company HQ and where the board members had their offices. Next to the entrance door was a large bust of the company founder Matsuda-san, who looked down sternly at us, the expression on his face not particularly happy.

The incumbent company president was called Furuta-san, and it was planned to meet him later in the week, hopefully attired in suits and not polo shirts and jeans.

In front of the building, protected from the sun by an awning, were a row of black 'Luce' limousines, sold as the Mazda 929 in Europe. Uniformed chauffeurs in black suits and white gloves, similar to the town's taxi drivers, were polishing the cars. The 929 was quite a sizable vehicle, but in Europe, where this class of car was dominated by the well-known prestige makers, Mazda had little chance in selling such a model in real numbers.

Tanaka-san led us down what looked like a normal street. To our left was the main engine assembly building, to the right, stretching down to where the road ended at the banks of the river delta we had followed in the taxi, was a seven-story building. This, Tanaka-san explained, was the design and engineering centre. Pipes and cables ran overhead, and the air smelled of sulphur. The surroundings couldn't have been more different to what we had left at Porsche's development centre in Weissach, at the edge of the Black Forest. Now, however, was not the time for second thoughts.

Passing through some more glass doors, we entered a small lobby painted in several shades of brown. Ahead of us was a small counter, and a metal plaque on the wall read: "Design." On the wall next to the sign was a framed automotive illustration, which must have faded over the years since it lacked almost any colour. Tanaka-san picked up a phone at the counter to announce our arrival.

It wasn't long before the door opened behind the counter and a gentleman in a black cord jacket came out smiling. His name was Norihiko Kawaoka. Tanaka-san said he would leave us with Kawaoka-san and bade us farewell, thanking us for the companionship over the last few days and wishing us good luck. He left through the glass doors, no doubt already planning the next sushi meal with his old colleagues that evening.

Kawaoka-san, who had a thick head of wavy black

*Mazda R&D Centre Mukainada*

*Peter Birt-o-whistle.*

hair and thick black-rimmed glasses, introduced himself. He spoke very good English, a skill he had picked up while working for a short time at the Opel design studio in Germany. First, we needed to pick up our identity badges, and with a burst of laughter he motioned us to follow him. He seemed a friendly guy. We followed him through the doors behind the counter.

This was our first experience of a typical Japanese office. The room was around 20 metres long and crammed full of desks. Seated at each desk was an office worker surrounded by mountains of paper, with enough space left to spread out a couple of sheets in front of him. Even at that time of the morning the air smelt stale. I noticed some large TV screens on the work surfaces with typewriter keyboards. They were personal computers, which had only just started to appear in the work environment.

Kawaoka-san called to one of the office staff, who came over with some small yellow badges in his hand. Reading from the *kanji* script on the badge, the office worker pronounced my name "Birt-o-whistle." Badges in place, we followed Kawaoka-san through a door at the end of the room.

The next office was pretty much the same as the one we had left, only not quite as many desks. At the end of the room we immediately recognised the face of Mazda's design head Fukuda-san, who was already up on his feet, a big smile on his face, walking towards us.

"Welcome to Mazda, welcome, welcome." He was very happy.

We exchanged greetings and explained about the lost luggage situation, which produced screams of laughter from all within earshot. The atmosphere was good and if our lack of suitable dress eased the mood, then it was fine with us.

Fukuda-san shared the office with all the chief designers, his status only noticeable by the bigger desk and larger office chair.

Fukuda-san, like Kawaoka-san, spoke excellent English. Before being elevated to the position of head of design, he had worked at Mazda's California development centre in Irvine, where a sister design studio to the one being built in Frankfurt had just been opened. He beckoned us to sit down with him and Kawaoka-san at a meeting table in the middle of the room. It was clear that Kawaoka was his right-hand man.

Fukuda-san explained that the week would be used not only to introduce us to Mazda Design and what the company planned for the future, but also to brief us on what projects they wished us to handle when we moved to the Yokohama studio the following week.

The schedule he laid out was pretty full. One day of the week would be devoted to a Mazda factory tour and a small bit of sightseeing in Hiroshima, including a visit to the famous Hiroshima Peace Memorial Park and the Hiroshima Peace Memorial Museum, which was dedicated to the atom bomb.

In the latter half of the week there would be a global design summit, where all the company's chief designers would get together to go over all the programmes running at the Hiroshima centre.

The week in Hiroshima would end on the Friday, when we would depart for Kyoto with a travel guide to visit the historic Japanese emperor's castle and the nearby shrines of Nara, before arriving in Yokohama on the Sunday evening. It all sounded good.

As a company, Mazda was certainly riding on a wave of success, but in terms of scale was still one of the babies in Japan as a producer of cars. Toyota was the true giant, followed by Nissan and Honda.

In the Japanese domestic market, Toyota was usually the make of choice if you didn't have too much knowledge about cars, and the company dominated the sales chart by a huge margin. Even in Hiroshima, which with Mazda boasted its own car manufacturer, it was still the models Toyota and Nissan that filled the streets of the city.

At the end of the '80s, Mazda realised that if it was to catch up with Toyota and Nissan, it needed to expand its range of vehicles and sales outlets, and to this end had created a range of new brands, which, in effect, was a similar strategy to that of the British Motor Corporation in the '60s, with its Morris, Austin, Riley, Wolseley and MG names.

Throughout Japan, Mazda had created five different sales outlets and, as we soon learnt, wanted to expand further. Apart from the Mazda name there was Cronos, Efini, Eunos, and Autozam, which basically sold the same cars but with different badges. In addition to this slightly complicated structure, the models that sold in Europe as 121, 323, 626 and 929, all had different names in Japan. For example, the popular 626 was named Capella in Japan.

Kawaoka-san brushed over this information as a prelude to our first meeting, which he took us to that afternoon. Here, Mazda's product planners outlined the next stage of the companies drive to catch up with Toyota and Nissan.

The room was already full of staff members, all wearing grey jackets with the Mazda name embroidered on the top left-hand pocket. We sat down.

In front of us on the table were (easily recognisable) technical package drawings, and pinned to the wall was a chart breaking down Mazda's model line-up into a kind of family tree.

Introductions were made, but it was soon very clear that, apart from Kawaoka-san, nobody spoke English. He said he would translate.

One of the managers, who Kawaoka-san said was a programme chief engineer, began to explain about a new market that Mazda needed to enter if it was to reach its goal of competing globally with Toyota and Nissan.

He explained that in North America, both Toyota and Nissan were apparently planning to introduce luxury brands.

This market, as in Europe, was the territory of Mercedes, BMW and Jaguar, with my old employer Audi, knocking on the door.

The chief engineer continued. Mazda realised that in order to raise its sales numbers and, almost as important, its image, it too would need to create a luxury car division. To this end, a strategy had been drawn up to develop three vehicles for this market. The new luxury range of cars had not yet a formal name and was presently known internally as 'the three brothers.'

The three models planned were logical segment sizes, similar to Mercedes' line-up, 'C', 'E' and 'S' class. The largest of the cars Mazda had planned was well over five metres long and powered by a V12 engine. You couldn't accuse Mazda of not being ambitious, I thought.

These three cars, Kawaoka-san explained, were our first design job.

The introduction was over and the team in the room wanted to hear our views. It was Ginger who immediately got up and offered the opinion that Mazda, or any Japanese company for that matter, succeeding in an automotive territory 'owned' by companies such as Mercedes-Benz, was unlikely. We all nodded with approval. He went on to say that the company would be far better advised to develop a people carrier such as the Renault Espace or some kind of off-road recreational vehicle, similar to the Mitsubishi Pajero.

The gauntlet had been thrown down, but it was quite clear that this was not a comment the group in the room wanted to hear. Ginger continued speaking about how the European companies had got where they were through a history and heritage that Mazda, at the time, did not have.

Before things got out of hand, Kawaoka-san stopped Ginger in his tracks, saying, "Yes, we know, we know."

Despite the fact that apart from Kawaoka-san, nobody in the room seemed to speak any English, you could tell that Ginger's negative comments were not welcome. The decision to move forward with this venture was already approved and well advanced in its planning. It was very clear, as new Mazda employees with a good knowledge of European automotive luxury taste, that despite our reservations, we were expected to deliver suitable designs. The meeting was concluded.

Once the meeting room was empty of the visiting planners, Kawaoka-san, having spent some time working in Europe, explained that he understood our concerns. However, their information about Toyota's plans for its new luxury brand told them that they were very well thought out and ambitious. Mazda could not afford to wait and see if this strategy was a success for Toyota but needed to react quickly and offer an alternative.

The meeting concluded our first working day with Mazda, and we returned to the hotel in the city with mixed feelings. The three brothers project was undoubtedly an interesting proposition for us as designers, but nevertheless we couldn't help feeling that for Mazda to try to position itself as a luxury car manufacturer was a little over ambitious. All the more disturbing was the feeling we got that our opinion was not of any interest. We should keep our opinions to ourselves and do as we are told.

Looking back, the reaction we got was

*Hiroshima Hōndori.*

understandable. Mazda was doing very well as a company, and the team we had met that afternoon were justifiably proud of what they had achieved. For a group of foreigners new to Mazda to jump in and lay down opinions of what was right and wrong without a deeper knowledge of what this ambitious company was capable of was very arrogant of us. This was a first lesson for us in how not to negotiate with the Japanese.

Back at the hotel, we enquired about our lost luggage, but nothing had been delivered to the hotel.

We needed to be patient. The cases would have left Amsterdam at the earliest Saturday, meaning arrival in Tokyo on the Sunday. They were sure to show up within the week. After agreeing to meet in an hour and head into town for a meal, we all went to our rooms.

Walking into my room I was glad to see my freshly laundered clothing on my bed. The two shirts were starched to the stiffness of cardboard, each packed in a cellophane bag, with a cardboard band around each shirt with a script thanking me for using the laundry service. This is just one example of the Japanese fastidiousness with packaging.

Travelling, whether within Europe or, as was now the case, further afield, has always affected the regularity of my digestive system. Some say this common phenomenon for travellers is due to a change in air and water, but since arriving in Japan I had eaten mainly fish and rice, which for me did not promote any digestive movement. I had eaten a sandwich at the airport but what bread was available in Japan had the consistency of sponge rubber. To be more direct, I had constipation.

I share this personal information guardedly, but it is the only way I can get to my next discovery in Japan, that being the wonders of their hotel toilet seats.

With the familiarity of my hotel room, I began to relax, and with that I at last felt the signals from my inner body that now was the time to use the toilet.

As a race, the Japanese are very thorough about personal cleanliness, and the top-of-the-line versions of toilet seats, usually found in hotels and also private homes, are a true testimony to this admirable state of body and mind.

This hygienic tool, not only disposes of natural waste, it also offers a complete range of pre and post-flush services, to make your visit memorable and cleansing.

When first lowering yourself onto the toilet seat, you will notice it is heated, which, to be truthful, I didn't find that comfortable. For years, one is used to the slight coldness of the Twyford seat, which I personally feel has a sense of reassurance about it. Seated, I looked down at the control console to my right. Immediately, I

*Out for the night: Roland Sternman, Jim Howell, and Ginger Ostle.*

found a dial to reduce the seat temperature, wondering at the same time if the water in the bowl was also warm.

After successfully relieving myself, I now had the choice of using either a bidet or a spray function. The bidet, judging by the small graphic on the switchboard, was a service more suitable for females, changing the toilet bowl after the flushing process into a small bath. I chose to select the spray option. Here, I noted I could also adjust the water pressure and temperature, choosing medium in both cases.

Holding tight I pressed the spray button, which logically, had the graphic symbol of a posterior with a spray of water aimed at it. Things started to happen below me. First there came the sound of flushing water, more of a trickle than a gush. Then I could hear the buzzing noise of something moving right before I was hit by a sharp jet of water, the aim of which was surprisingly accurate. I let it work before pressing the off switch.

Drying myself and getting up from the seat, all that remained was to finally flush the toilet to the accompaniment of a melody, the volume of which I could also adjust. A deodoriser switch completed the whole ceremony.

I have often wondered what goes on mechanically when the spray button is activated on these toilets, but since whatever moves into place below you immediately disappears out of sight when you take your weight off the toilet seat, I have never discovered.

Maybe a yoga contortionist who is able bend his head between his legs while seated has discovered the answer to this mystery.

As agreed, we all met at the hotel reception and from there made our way into the city centre in search of a restaurant. The previous day I had noticed a tandoori restaurant in the SOGO department store close to where Tanaka-san had invited us to sushi. We were unanimous that a bit of 'British' food would be a good idea.

Crossing over the wide road in front of the hotel, we made our way towards the area of the town known as Hondōri. *Hondōri* basically means 'main street,' and is a covered pedestrian walkway running pretty much the entire length of the town. It runs from the Peace Park, which is west of the city centre, eastwards towards the main shopping area and the Parco department store, where it dissolves into the restaurant and bar district called Nagarekawa.

We were heading from the opposite end of the city centre and made our way first into the bustling area of Nagarekawa. What during the day is a pretty quiet warren of streets is at night a mass of bright lights and sounds, populated with a mixture of businessmen in various states of inebriation and young girls touting for restaurants and bars. It is no exaggeration to say there are thousands of hostelries in Nagarekawa, most of which are off limits to foreigners.

We made our way through the crowds, emerging opposite the large Parco department store, where, crossing the road, we entered Hondōri. At seven in the evening the pedestrian walkway was still very busy with a mixture of late shoppers and school students in their uniforms, the girls wearing impossibly short mini-skirts, bringing to mind the weird Japanese adult movie I had watched on my first sleepless night.

Nearing the Peace Park end of Hondōri, we joined another of the wide boulevards in the town, where, on the right, we could make our way to SOGO – we were getting to know the layout of the town.

Over tandoori chicken and naan bread, we exchanged opinions about the day's meetings, our views that Mazda's plan to enter the luxury car market would not work in Europe being very much unanimous. Nevertheless, it was very clear that our reasoning was not going to change things, and there was little point starting our working relationship with Mazda in confrontation. If Mazda wanted a range of luxury cars from us, then that's what they would get.

We raised our beer glasses.

"*Kampai*!"

# Chapter 22

## *The three Bs of Hiroshima*

**We introduced ourselves to the young Japanese lady who had been waiting for us at the hotel reception. Like all male counterparts in Japan she also had the 'San' title, making things quite confusing for foreigners. Formalities were, however, soon broken down when she said we could call by her first name, Kayko. The majority of traditional female first names in Japan end in 'ko,' so you have your Norikos, Yukos, Yokos and so forth.**

Mazda had arranged for a day visiting the factory site and the Mazda Museum, as well as the Peace Park and adjoining atom bomb museum.

In the meantime, we had heard from Tanaka-san. He had arranged for us to meet him for dinner that evening with Mazda's famous ex-head of development Kenichi Yamamoto.

We walked out into the sun to a waiting minibus, which I noticed was called a Bongo. The Bongo was a big seller for Mazda in both Japan and Europe. For the European market Mazda wisely chose to name it E2000. In its commercial form, the van was a favourite purchase for small businesses, due to its simple mechanics and robust construction.

We got in the back and Kayko instructed the driver, again dressed in a black suit with white gloves, to move off.

We headed along the same route our taxi had used the day before, while our guide for the day explained a bit about herself and the day's programme.

She worked in the PR department and was born in California, where she had spent her early years before her parents moved back to Japan. She loved California and wanted to move back some day, but for now, as an obedient daughter of a Japanese family, it was expected of her to work for a respected Japanese company, before getting married and having a family. Her life's path already seemingly determined by her family, California had to wait. Nevertheless, she loved her work in the PR department, where she had the opportunity to meet many foreigners visiting Mazda and make use of her English skills.

The Bongo made its way towards Mukainada before turning right and over a bridge to the opposite side of the river that ran next to Mazda's R&D offices. Kayko explained she wanted to first visit a shrine located on the small mountain directly above the sprawling Mazda factory complex. Called Ogonzan, it was this mountain that protected the Mazda factory from serious damage when the atom bomb exploded.

The Bongo wound its way through a tightly packed residential area and on towards the top of the mountain (calling it a mountain was a slight exaggeration. It was actually more of a hill).

Reaching the summit, we pulled into a small parking area in front of the shrine and got out. It was already turning into a very hot day. From this vantage point we could see the whole of the city, with the mountains rising up behind Hiroshima as a haze in the distance. Below us was the Mazda production facility, where we were heading next. Beyond the factory complex you could make out a large holding area with hundreds of cars and vans ready to be loaded onto the huge transport ships that were pulled up directly at the quays alongside the area. Mazda may have been a minnow compared to Toyota, but it still looked a formidable company from this viewpoint.

Having taken in the views, Kayko urged us to get back in the Bongo; we had a tight schedule. The driver made his way back down towards the large factory site, and after being signed in at the security gate, we moved off into the centre of the main Mazda assembly area. It could have been any car plant in the world. The driver pulled up alongside one of the buildings. Through a loading door you could just make out a body of whatever was being built inside, moving slowly down the production line.

Leaving the minivan, we filed into an office where a

factory manager was waiting for us. As was the case with almost everyone you saw within the Mazda factory, he was wearing a light grey uniform and a white hard hat. He gave us a short presentation, which Kayko translated, followed by a short film about the history of Mazda's growth as a vehicle manufacturer right up to the present of that day.

The presentation finished, we followed our guide through a couple of doors and up a staircase onto a balcony that looked down on the production line. Cars were moving down the assembly line and I could make out two models: the RX7 sports car and a 323 hatchback.

*Hiroshima sightseeing tour begins.*

Allowing visitors into a car production line can involve all sorts of risks. I recall when I first visited the Chrysler plant in Ryton near Coventry as a student, that we were showered by nuts and bolts by the assembly workers. In contrast, the Mazda workers stood proudly as we passed by them and back out into the fresh air.

Our next port of call was the Mazda Museum, located within the factory boundaries.

Here, Mazda's past models, commencing with the three-wheeler Mazda Go delivery truck, from the prewar years, were lined up for inspection. Behind the displayed vehicles, was a photographic timeline of the world's important events at the time the models were produced. A special display was devoted to Mazda's dedication to the rotary engine production, with pictures of Felix Wankel alongside Mazda's rotary pioneer Yamamoto-san, with whom we were having dinner later that day. It was good to understand and learn the history and heritage of this proud manufacturer from Hiroshima, which, after the destruction of the city in 1945, had literally risen from the ashes.

Kayko looked at her watch; she was happy, everything was on schedule. We were returning to the city centre for lunch, before we commenced the afternoon's programme, including a visit to the site where the atom bomb was dropped.

Back in the Bongo, we sat back while Kayko continued to enlighten us with facts about Mazda and the city of Hiroshima. Crossing the bridge back over the river, she drew our attention. With a large smile on her face, she said, "You know, Hiroshima is famous for its four B's." She paused, allowing us to try to understand what she was leading up to.

"Bridges, Birds, Banks, and?" She paused again, with the big smile still on her face.

Surely she isn't going to say ... 'B for Bomb?'

Nobody wanted to say the word and thank goodness Kayko stopped us taking that route by bursting out: "Beautiful girls!"

We all laughed; she was indeed beautiful.

I looked at Ginger. He knew exactly what I had been thinking.

We soon reached the city centre, the Bongo dropping us off at the main railway station.

Kayko said she wanted us to try a Hiroshima specialty. Looking at Jim's face, I could see he was already concerned; he needn't have been.

Kayko guided us into the station concourse area, which is located below the platforms of the *Shinkansen* bullet train. A loudspeaker emitted a robotic female voice: "Welcome to the *Shinkansen*, the next train is the Nozomi Super-Express bound for Tokyo."

Above all the noise, you could make out the electrical hum of this amazing train as it entered the station.

*An okonomiyaki restaurant.*

*The traditional Hiroshima dish, okonomiyaki. (Flickr/Nicholas Boos)*

We entered a small restaurant, ducking under the typical cloth shield hanging in the doorway, which automatically induces a bow from the customers to the staff as you enter. As in Tanaka-san's sushi restaurant, we took our places at the serving counter.

"Welcome to Hiroshima's famous *okonomiyaki*," Kayko announced proudly.

The Hiroshima *okonomiyaki* is well known throughout Japan and basically translates to 'as you like,' which refers to the selection of your own ingredients. Best described, it's bubble and squeak meets pancakes.

Kayko said she would order one for us to try.

Seated at the counter, there was a metal cooking surface in front of us. The chef on the other side poured some pancake mixture on the hot surface. Once this began to firm up, he put a pile of white cabbage on top, which he then left to cook. This was followed by some noodles and thin strips of pork. As the ingredients started to heat up, he added various sauces before folding the pancake over to form an envelope. The final touch was a lattice work of more sauces on top.

The chef put the first sample on a plate for us to try.

Sharing dishes in Japan is very normal practice, and it is unusual when eating out in Japan that a person will order a dish only for him or herself. A selection of dishes will be ordered one after the other until everybody has had enough, with meals usually ending with a bowl of Miso soup. This style of sharing dishes is conducive to eating smaller portions and may contribute to many Japanese not being overweight.

Many years later, when entertaining some young Japanese designers in a local Italian restaurant, my order of spaghetti was the first to be delivered to the table. One of my guests, not realizing that the sharing practice of his home nation was not usual in Europe, took his fork and helped himself to a portion of my meal. I quickly remembered, with amusement, why he found it quite normal.

Yes, westerners can learn a lot about eating habits from the Japanese, who treat meals like a ceremony, taking time to share and experience the meal.

This was far from our minds as, together, we tore into the *okonomiyaki*, demolishing it within seconds.

"Oh, you are hungry?" Kayko asked, somewhat surprised at our hasty clearing of the plate in front of us. Two more were ordered.

Full of *okonomiyaki*, we left the station building, heading for the last venue of the day's fact-finding tour and one that remained fixed in my mind for the rest of my life.

Whenever people ask me where I am travelling to in Japan and I answer 'Hiroshima,' the reaction is often the same – "Oh my God, why would you want to go there?" The mental scar associated with the town's name, with

*Standing by the Hiroshima Atom Bomb Dome.*

*The Beautiful Girls of Hiroshima!*

the dropping of the first atom bomb on an inhabited city, remains long after the physical devastation it caused was cleared away and the city rebuilt. That is a fact that will never change, no matter how many years pass by.

It is not for me to judge the reasons for dropping the atom bomb on Hiroshima, everything has already been said, but if you have the opportunity, I urge you to visit the site of this world changing event at least once in your life. It will have deep effect on you, I guarantee.

From the station, the Bongo headed west past the shopping centre and near to the SOGO department store, turned into a narrow street close to the end of Hondōri. The driver pulled up next to a restaurant adjoining a small road bridge that crossed over one of the river deltas and into the Peace Park. We got out; the driver would pick us up in three hours.

We followed our guide as she led us along the bank of the river. The Hiroshima Peace Memorial Park was located on the other side, but first we visited what remained of the Genbaku Dome, or as it is more widely known, the Peace Dome.

When work commenced to clear the city of the post-bomb devastation, it was decided to leave the ruins of a large exhibition hall. It was exactly above this hall that the bomb had been detonated. The ruin, with its dominant dome, has been left untouched, a contorted mass of bent girders and bricks, testimony to the destructive powers of the bomb that wiped out the city.

Nobody in the building survived, and parts of the girder work look as if they had simply melted and bent as the structure collapsed under the enormity of the blast. What it would have done to a human body is all too clear.

We stood as a group, taking it in, the other side of fences that cordoned off the ruins. Bright chains of origami cranes adorned the rails, a symbol of peace left by the thousands of visitors who come every year.

We walked on, over a bridge to the left, past the Dome and into the park.

A visit to the Peace Park is a pilgrimage that all Japanese school children will usually make at some point during their education. That day was no exception, and the park was filled with many school groups. As we made our way to the museum, it soon became very clear that for the school children, a photograph of them next to a western foreigner was a clear signal of friendship and peace.

Slowly, small groups of schoolgirls started to converge on us, cameras in hand. Soon, there was a queue as they took it in turn to stand by us, all holding their hands high, their fingers forming the 'V' symbol of peace. They giggled with joy, a sad reminder that all perpetrators of war were once also as happy and innocent as those schoolgirls.

Maybe today, lying in the bottom drawer in a Japanese household, is an old photo album with a picture of me standing next to a young schoolgirl in the Hiroshima Peace Memorial Park. Hopefully, they ask themselves, who was that foreigner, and where is he now?

If the sight of the Peace Dome was sobering, then the tour of the atom bomb museum was even more thought-provoking.

Our tour was over, and we made our way out of the park to the waiting Bongo, which brought us back to our hotel.

Kayko, a true testimony to the 'beautiful girls' of Hiroshima, stood by the minibus and, still smiling, gave us a farewell bow. It had been a very informative day and a clear demonstration that Mazda wanted us to immerse ourselves, not only in the background of the company, but also the history of its home city. Regretfully, for most of the world, this wonderful city is only known as a name associated with death and destruction.

The immersion in Mazda's history continued that evening at our dinner appointment with Tanaka-san and Yamamoto-san.

Tanaka-san met us at the hotel and we walked with him to a restaurant a few streets away, which served a dish called *shabu-shabu.*

Yamamoto-san was, for many Japanese car enthusiasts, a real hero. It was he who led Mazda's development of the Wankel rotary engine, which for many years was a pillar in its product line-up.

When German Felix Wankel introduced his revolutionary engine in the early '60s, it was picked up by most companies as the way forward for automobile propulsion.

At the time of commencing my career at Vauxhall, in 1973, it had planned to fit all future models with this power train, and Mercedes and Citroën were also well advanced in bringing cars with a Wankel engine to the market, not forgetting NSU, which already had the Wankel-powered Ro 80.

Yamamoto-san pushed the Mazda board to invest in this engine concept for its future models, and when the fuel crisis of 1973 dampened the enthusiasm most manufacturers had for the Wankel engine, Mazda, nevertheless, continued its development unabated, producing a series of rotary models with the nomenclature RX.

By the time we joined the company, Mazda's commitment to this compact power train was very much diluted, with only the RX7 sports car using such an engine. Nevertheless, a small team of engineers was still employed to keep refining the engine further, and to this end Mazda was heavily involved in international motorsport, with rotary-powered sports car racing every year at the famous 24 Hours of Le Mans as a way of pushing the development forward and keeping the rotary in the public limelight.

Yamamoto-san had risen in the company to become development chief and then president, before retiring the previous year and becoming what he was at that time, chairman of the company, a position that was more representational.

The restaurant Tanaka-san had chosen was quite dark inside and Yamamoto-san was already waiting for us. He was a slight man, slim in stature, with almost-white hair and large glasses. Introductions were made, and we took our places around a circular table, a large gas cooking ring in the centre.

*Roly with another beautiful girl.*

Since Yamamoto-san had only given up his presidency the year before, it can be assumed that we had him to thank for giving the go-ahead on the new R&D centre in Frankfurt.

Speaking in good English he told us of his passion for the rotary engine. He and Felix Wankel were good friends and he had visited him many times at his laboratory in Lindau on Lake Constance in southern Germany. I told him that I had worked for Audi-NSU and had seen the last of the Ro 80s produced sitting in the storage yards of the Audi factory, the cars themselves having been made in Audi's factory in Neckersulm. He said that NSU had never mastered the issues they had with the rotary engine, plagued as they were with faulty rotor tips. Mazda, on the other hand, had solved this issue and he was still confident the rotary engine had a bright future. Yes, the engine was his baby.

I expanded on the discussion by telling him about my visit to Le Mans with Porsche that year and how I had watched the Mazda cars at night, spitting out huge plumes of fire as they de-accelerated in the curves of the course in-field area near the Dunlop Bridge. This, apparently, was not what he wanted to hear, since it was apparently a clear signal that the engine was not performing efficiently – oh well!

Slowly, things started to happen around our table. A waitress in traditional dress offered us each a hot hand towel, before bringing a large cooking pot to the table and placing it on the gas ring at the table's centre. The pot contained a liquid stock, Tanaka-san explained, that would be brought to a temperature high enough for us to cook pieces of thinly sliced beef and vegetables, which were arranged in serving bowls around the table.

The beef we were cooking had a marble-like pattern on it and was called Kobe beef. Kobe beef is probably the most expensive form of meat on the planet – this wasn't a cheap meal. The cows bred to produce this special beef are fed on a grain diet and have their skins massaged and brushed each day, which promotes the development of the fatty structure in the meat.

*Kenichi Yamamoto. (Mazda Motor Corp)*

*Shabu-shabu* is eaten in three stages. First, the slices of meat are cooked in the hot stock, each diner placing their piece of meat in the pot. When cooked, it is removed and eaten with a sweet sauce served in small ceramic dishes at the table. Next, the waitress will introduce the vegetables to the stock, which are eaten together with the remaining meat. When all the ingredients are finished, the stock, which has by now turned into a soup, is ladled out into bowls to complete the meal.

The meal was enjoyed by all, and again demonstrated that for the Japanese a meal is always more of a social occasion than simply a method of stemming one's hunger.

To meet Yamamoto-san had been a privilege. He was indeed a true automotive pioneer in the same vein as such brilliant engineers as Alec Issigonis.

We said our goodbyes. Little did Yamamoto-san know that his friend Felix Wankel would pass away the following week, leaving him to fly the flag for the rotary engine alone. As you read this book, Yamamoto-san is also sadly no longer with us.

# Chapter 23

## *Persona to Roadster*

**The Wednesday and Thursday of our first week in Japan was dedicated to a global design summit. It was a gathering of all Mazda's senior designers, with the Wednesday spent reviewing all the various programmes in progress in the Hiroshima studios, and the Thursday spent at a meeting held in a conference room in our hotel.**

With great anticipation we took a taxi to the Mazda HQ building or '*Matsuda Honcho*' in Japanese. The use of '*Matsuda*' to describe the company referred to the company founder's name.

On our first day in the design building we had only visited a few meeting rooms, but today we were able to review all of Mazda's new model range. This would give us a true indication of whether we had joined the right company.

The taxi dropped us off at the HQ entrance lobby and, making our way past the bowing ladies, we headed towards the rather uninspiring entrance to the design department.

Kawaoka-san was waiting for us and we followed him into the chief designer's office.

Standing by Fukuda-san's desk was a person familiar to me – Tom Matano. Tom was head of Mazda's new design centre in Irvine, California, which had recently opened.

I had first met Tom when he had been working as a designer for BMW in Munich. He had been a guest at a party that J Mays had held at his flat in Schwabing. Interestingly, Tom had started his career as a designer at GM's Australian division, Holden, at about the same time I had been working for Vauxhall in England. Despite both having worked at GM at a similar time, our paths had never crossed through GM.

Tom had joined Mazda in 1983 and prior to the recent opening of the new North American design centre, or MRA as it was known, he'd worked in the offices of Mazda's R&D centre on the same site as the studio.

Although born in Tokyo, Tom came across as more of an American, speaking with a US accent and no trace of his Japanese heritage.

Tom had built up a good team of designers at the Irvine facility, including a couple of guys I had previously known from the Opel design studio in Rüsselsheim near Frankfurt. Mark Jordan was the son of the well-known head of GM Design, Charles 'Chuck' Jordan, and Wu-Huang Chin, was a Taiwanese designer whose illustrating skills were legendary.

The chief designer's office began to fill as most of Mazda's senior in-house design staff were in the room. I noticed that Shimazu-san and Yoshitomi-san, both of whom we had met in Oberursel at the time of our first meeting with Mazda, were also present. Both soon moved to Germany to oversee the building of the centre.

Kawaoka-san called everybody together and made a few announcements in Japanese before introducing us by name. He made a big point about the fact that we had chosen to leave Porsche to work for Mazda, which led to a rapturous applause.

He went on to explain the schedule for the next two days, which turned out to be both business and recreational.

Introductions over, everyone was asked to move to the design presentation hall; we followed Kawaoka-san.

The route to the large presentation hall involved going up three floors (using a lift designed for four), then following a complex route of narrow corridors and passageways; it took time.

Unlike many companies, where the design centre was a standalone building, Mazda's design department was located on several floors of its main engineering block, which itself could have been mistaken for any part of the manufacturing facility.

As the new guests, we were allowed to take the lift, while the remainder of the group took the stairs. We continued to follow Kawaoka-san and, after passing through what looked like a storage area for car seat materials, entered a large viewing hall. Impressive in size, the walls were lined with brown curtains, and in the centre three turntables were set into the floor; there were no windows. Apart from three rows of chairs, the room was empty. We took our seats and waited for the remainder of the designers to drift into the hall and join us.

Fukuda-san entered the arena and asked us to join him in the front row of seats.

When all was still, one of the department's product planners took the microphone and, aided by some charts projected onto a large screen, began to explain Mazda's present line-up and, more importantly, what was in the pipeline.

In Europe, Mazda had been very successful, its sales reliant on basically two core models: the 626 and 323. These two cars were direct competitors to the VW Passat and Golf and, particularly in Germany, were selling very well, with the 626 sometimes equalling the Passat in monthly sales numbers. In addition to these two models, Mazda also sold a smaller car called the 121 and a large limousine called the 929, both of which hardly reached significant sales numbers. Finally, and of more note, were the E2000 van, a favourite with small businesses, and, of course, the Wankel-powered, RX7 sports car. It was a small range, but certainly varied in its offerings.

With its myriad of different sales outlets, such as the previously mentioned, Autozam, Efini, Eunos not to mention Mazda, the model strategy in Japan was much more complex than that of Europe and more models were planned. Many models looked the same, with only a different badge on the front grille. Thoughts of the doomed British Motor Corporation came to my mind.

The presenter pressed on.

Not satisfied with this complex line-up, Mazda was also planning a new luxury brand to compete against rumoured prestige models from both Toyota and Nissan. This three-model line-up was what we had been introduced to in our first meeting at the beginning of the week.

The talking was over and now it was time to look at some physical vehicles. To one side of the room, a large metal door opened, and three covered models were rolled into position on the turntables. Fukuda-san gave a signal and the covers were pulled away. These first three models were the replacements for the 'Golf class' 323. They comprised a three-door hatchback, a four-door sedan, both fairly inoffensive, not badly designed, just somewhat predictable. The third model, however, was a different story and certainly of more interest. A five-door hatchback, it had almost coupe-like proportions and featured, a first in this class of car, pop-up headlamps. This final model was sold in Europe under the name 323F and was a departure from the normal conservative offerings in this segment. Things were looking good.

The show continued. Next up was the replacement for the popular 626 – in Japan, the Cronos. For some reason, Mazda chose not to name its cars in numbers for its domestic market, each having instead their own nomenclature. In addition to Cronos, there was Familia for the 323, Festiva for the smaller 121, and Luce for the larger 929. To make things even more complicated, the other sales outlets had different names for their individual versions of what were basically the same model. Again, Morris Oxford and Austin Cambridge came to mind.

The new 626 was a handsome design. Mazda designers had chosen to take up the softer 'bio' design language that was taking over the market, similar to new models in Europe such as the Ford Sierra. The sedan was quite conservative but, like the 323F, the five-door hatchback was sporty and incorporated and integrated a rear hoop spoiler, not unlike a watered-down version of the rear spoiler on the Porsche 959. Also of note was that the top model had a V6 engine. Again, we felt reassured.

Up until now, there had been no big surprises. Once the models had been moved into place, we were all given the chance to view the new offerings in detail and comment, and after the cold looks we had received when giving our opinion on Mazda's planned entry into the premium market segment earlier in the week, Ginger advised us to remain positive. There was, in fact, no reason not to. The range of models we had been shown were certainly not poor designs and we were all quite enthusiastic. The show, however, was not over. Taking our seats again, another covered model was pushed into place. This time, what we were looking at was not intended for Europe, but solely for Mazda's home market. The car, which was about the same size as a Mazda 626, had two names: Mazda Persona and Eunos 300. Apart from the names, there appeared to be little or no difference between the two versions that were being shown on the large screen. The car, it was explained, was a hardtop model, which were popular in

Japan. The construction of the car was similar to sedans also offered in the United States; a four-door sedan, it featured frameless doors and no central or B pillar. The result was that, when all the windows were wound down, the occupants had an uninterrupted view out of the car, much in the same way as a convertible but with a hardtop mounted – hence the name. The car certainly had character and, in my opinion, was much more stylish than the 626 sedan we had just reviewed.

"Why not sell this in Europe as the 626 sedan?" I asked. The answer was quite simple. In Japan, where motorway speeds were limited to 100km an hour, the frameless side windows of such cars remained in place. Any speed much higher and they would be pulled out of position, which meant that in wet conditions rain water would be sucked into the car. It was a pity, since the design wasn't so bad.

If the Persona offered a sense of style, the next model we were to review was pretty much the opposite. Mind you, 'style' has always been a question of personal taste.

As previously mentioned, below the Mazda 323, Mazda had offered a small car called the 121, or Ford Festiva as it was known in Japan. Also of interest is that the same design was to be sold in Europe as the Kia Pride. Kia were an unknown Korean car manufacture of which Mazda had a financial share. The design was inoffensive, its only redeeming feature being that it had a fully opening canvas sunroof, stretching right back to the rear passengers. Sales in Europe had not been such a success and so Mazda decided a radical new approach was necessary.

Although not so important in Europe, this class of car in Japan was vital if you were to succeed as a manufacturer. The B segment, as it was called, was the biggest selling class of car in Japan, due primarily to the traffic congestion in this crowded country's cities, where space is at a premium. Companies such as Toyota offered as many as ten models in this segment, some of which were boxy minivans capable of seating seven passengers. Also of importance was the fact that these small cars were the favourite purchase for young, female, first-time buyers. This, then, was the target audience for Mazda's next 121.

We watched in awe as Mazda's product planners explained the image of a car that was attractive to these young female buyers. It had to be cute and cuddly.

Looking across to Roly and Ginger, I felt my eyes rolling upwards. In Europe, the big seller in this segment was the Ford Fiesta and its top sporty model the XR-2 was a massive hit with young, aspirational guys. Would these European guys go for a teddy bear on wheels?

Once again, the doors at the end of the viewing hall opened and a covered model was moved into place, the silhouette of which looked unmistakably like a large Easter egg.

We weren't disappointed. What looked like an egg under the covers, upon its reveal, turned out to be exactly that. The new Mazda 121 could only be described as an egg on wheels. We were dumbfounded to say the least, and thoughts of the Austin A35 'puddle-jumper' came immediately to mind. To add to this nostalgic look, the model was also painted in a pale beige colour, very reminiscent of cars from the late '50s.

From the incredulous look on our faces, our Japanese hosts realised how shocked we were, and they all burst out laughing. As we later learnt, laughing in Japan doesn't necessarily indicate amusement but can be a disguise for other emotions.

Once the 'merriment' had subsided, we cautiously commented that although the design had a certain character, we felt it may struggle in Europe. We left it at that.

The day's presentations were finally drawing to a close, but not before we were shown a final design that lifted our hearts and convinced us once and for all that moving from Porsche to join Mazda was a good decision.

Back in our seats and recovering from the shock of the new 121, a final covered model was pushed into place. We were intrigued. The silhouette looked like that of a small sports car like an MGB. We didn't have to wait long to find out. Tom Matano stepped up and pulled back the covers. The guessing was correct – it was a sports car – and it looked amazing. Tom Matano, Fukuda-san and Kawoaka-san couldn't contain their enthusiasm, all smiling from ear to ear.

We needed no signal. Everyone in the room immediately jumped up and walked over to the model. My immediate impression was how small it was. Cars were generally getting bigger with each generation, and although this design was probably similar in dimension to an MGB or Austin Healey, somehow it came across as being much more compact.

In keeping with most of the designs we had seen previously, the car had a soft bionic feel to the bodywork with almost no hard lines or creases. The front end had a round profile featuring a lower oval air intake almost reminiscent of a fish's mouth – but in a positive way. There were no visible headlamps, and upon closer inspection it was clear that the car featured pop-up lamps. The rounded theme continued through to the rear, where lozenge-shaped tail lamps continued the oval design signature.

I opened the door by means of a delicate chrome latch, which would have been equally at home on an Italian sports car. No detail had been left untouched. Like the exterior, the interior was also very minimalist. A small hood behind the steering wheel housed the rev-counter and speedometer plus a couple of secondary gauges. All the remaining controls were located in a central block secured to the main instrument panel. Resting one hand on the steering wheel, the gearlever and handbrake were perfectly positioned. Mazda had done a great job. This little sports car felt just right.

Getting out of the car, I could sense the excitement in the room. Yes, Mazda was going to build it: a modern interpretation of the British sports car. Compared to the sports cars we had said goodbye to at Porsche, it was tiny, but that didn't matter. Painted in a baby blue it looked perfect. The only cautious comment from a few of us was that it was quite similar to the Lotus Elan of the '60s, but who cares, at least it wouldn't break down!

The name on the rear of the car read "Miata – MX5" – and as they say, the rest is history.

The Mazda MX5 went on to become an icon in automotive history. Like many similar cases (and here the Audi TT also comes to mind), many people wish to take credit. Tom Matano is famously named as the father of the MX5, but there were of course many other names involved in its creation. The initial development of the car has been well documented, and I don't intend to repeat the whole story here. Briefly, however, the idea to reinvent the classic British roadster came from a Mazda North America planner called Bob Hall. Hall persuaded Mazda management that there was a good business case for such a car in the US market and a design model of his idea was kicked off at the North American design studio in Irvine, lead by Tom Matano. Soon, the Hiroshima studio followed suit, but also developed a handful of alternatives including front-wheel-drive and mid-engined derivatives. In the end, the California proposal was selected for further development and an English engineering company in Worthing called IAD was given the job of building a running prototype. Once the prototype was finished, senior managers from Mazda were invited to California to drive the car, along with some of its competitors, including a Triumph Spitfire. The meeting was a success and the programme was officially given the go ahead for production development. At this point, the Hiroshima production studio took over the California design, which was refined and developed by its chief designer Shunji Tanaka (yes, another Tanaka-san) into the model we reviewed that day in Hiroshima.

Anyway, Jack K Yamaguchi and Jonathan Thompson's book, *MX5 – The rebirth of the sports car*, tells it how it was – definitely worth a read.

Back in the viewing hall, the buzz of excitement in the hall slowly subsided. This final reveal concluded the day's meetings. The viewing hall started to empty, and we made our way down to the main reception where a taxi back to the hotel was waiting for us.

I gazed out of the taxi window as it wound its way through the sprawling industrial neighbourhood close to the Mazda factory complex, and reflected on what we had seen that day. We were all encouraged by the designs we had reviewed, the Miata-MX5 sports car being, of course, the highlight of the day. There was now no doubt in our minds that we had all made the right decision, joining Mazda.

One nagging point, however, filled my thoughts as we passed one car dealership after another. Many of them belonged to the complex group of names we had reviewed in the days meetings – Eunos, Autozam, etc. I wondered if such sales complexity could do the same to Mazda as it had done to the ill-fated British Motor Corporation. Time would tell.

# Chapter 24

## *It's not so roomy in Tsurumi*

**Yes, Mazda had big plans, and what hadn't been explained in the model review, we would hopefully learn about in the global design summit to be held at our hotel the next day – our final full day in Hiroshima before moving to Yokohama.**

For the summit, Mazda Design had organised a conference room in our hotel. The meeting was attended by all the department's senior management, plus senior representatives from the overseas and external branches. As well as us, Tom Matano from the US and another new face, Atsuhiko Yamada, who had recently joined Mazda to run the design centre in Yokohama, attended. What exactly happened at the Yokohama design studio was still a mystery. We had been told that we would move there for the remainder of our Japan tour, but no one had explained the facility in detail.

Fukuda-san opened the meeting, welcoming us all, before a series of design proposals were presented by the various programme chief designers. It was noticeable that the designs being thrown up onto the screen appeared to have no connection with the design language of the cars Mazda was about to introduce to the market.

Having revamped its entire range, Mazda was now considering its next strategy in terms of design language. For the Japanese, it is very important to develop not only the look of its cars, but the design philosophy behind the physical look. The range of new designs we had reviewed the previous day had been created under the name of Tokimeki – roughly translated to 'heartbeat.' It was, for us, a simple philosophy: when seeing the new Mazda, the observer's heart should beat faster. Basing a design language on a philosophy was a new approach for us, with European companies evolving their designs as a logical progression of the outgoing model. Mazda, or any Japanese car manufacturer for that matter, had no problem with developing a new model that had absolutely no relationship to its previous iteration. They argued that since most purchasers bought a new car each year, the last thing they wanted was a new model that was similar to their previous car. No, 'brand identity,' as we know it today, was unheard of.

To conclude the day's meeting, Fukuda-san wrapped up by explaining his vision for the future of Mazda Design, as an organisation.

Having now set the foundation for a truly international design team, with the creation of centres in both North America and Europe. The next plan was to move its domestic design organisation nearer the Japanese capital and hub of all new trends, Tokyo.

Fukuda-san expanded further. He reasoned that if the company was to attract the best creative minds in Japan, then they had to offer them a location and living environment to match their lifestyle. Hiroshima was considered quite provincial in Japan, and this, in addition to its regrettable history, made it unattractive to trendy young designers. Here is where the existing Yokohama studio came into play. The light bulbs in our heads went on and the reason why we were to spend the rest of our stay in Japan located in Yokohama became clearer.

Fukuda-san's vision was for the Yokohama studio to handle all the advanced design for future Mazda models, including processing the input of the external studios. Final proposals were then be sent to Hiroshima for feasibility development towards production. Moving the whole Mazda R&D team to Yokohama was out of the question, so this seemed the best compromise.

It seemed a good idea, but many of Mazda's local design staff were unhappy about the idea, as most of them were local to Hiroshima, with close family ties to the town. For them, it meant commuting each week to Yokohama, the expense and disruption to their family life making a move closer to Tokyo out of the question.

*Taking the famous bullet train, the Shinkansen.*

***

That Friday morning, we checked out of the hotel, regretfully with still no sign of our luggage.

Around midday, we were scheduled to take the high-speed *Shinkansen* train to Kyoto, but before that, our last tour of duty in Hiroshima was to meet Mazda's president, Furuta-san.

Once more, Tanaka-san was waiting for us at the hotel to take his 'trophies' for an audience with the company's most senior employee.

In the taxi, we nervously wondered how he would react to our casual attire.

Arriving at the HQ building, we went over to the executive office block immediately behind the front entrance hall. On the sixth floor, we seated ourselves around the boardroom table of the president's office. Smartly dressed female staff members served cups of tea in traditional bone china, while we waited in silence.

Furuta-san entered the room. He was quite stout for a Japanese man, with wavy grey hair, a broad smiling face and glasses. He approached us with outstretched hands.

Introductions were made, Tanaka-san bowed deeply, while at the same time excusing us for our casual dress. An interpreter translated.

Furuta-san was easy and relaxed, brushing Tanaka-san's remarks aside. We could breathe again.

Introductions over, the president addressed us as a group, explaining much of what we already knew. As a company, Mazda was reliant on export markets and therefore required products designed to meet the tastes of those markets. In a nutshell, that was our job. He thanked us for leaving such a prestigious company as Porsche to join Mazda, and expected us to work hard, wishing us luck at the same time. Tanaka-san looked at us satisfactorily, and with a quick signal he indicated that the meeting was concluded.

Back in the main HQ reception hall it was time to say farewells to Tanaka-san. The waiting taxis would take us to Hiroshima's main station, while Tanaka-san would soon return to Germany to supervise the building of the new development centre. He was happy.

"Good luck in Yokohama and see you back in Oberursel in December!"

The legendary Japanese bullet train *Shinkansen* is certainly the fastest and most efficient form of rail transport in the world. It was therefore with great anticipation that we waited for the Nozomi Super Express to take us on to our next destination: the ancient city of Kyoto.

Not satisfied that our week's stay in Hiroshima had indoctrinated us to Mazda and Japanese life, it was important that we were also introduced to the very roots of the country's culture. Where better to do that than the two historical capitals, Kyoto and nearby Nara, with their world-famous castles and temples.

Mazda had arranged for a tourist guide to meet us off the train in Kyoto that Friday evening. We would spend Saturday in Kyoto and Sunday in Nara, before finally arriving in Yokohama on Sunday evening.

For those in England, used to being surprised when a train actually arrives on time, the punctuality of the Japanese train system is by comparison truly mind blowing. In fact, train travel with the *Shinkansen* cannot be compared to normal rail travel, being closer to flying in its efficiency.

Our reserved tickets for the journey to Kyoto naturally had seat and carriage numbers on them, but in order to speed up the process of passengers exiting and entering the train, the station platform had indications of where exactly to stand in order to embark at the door closest to your seat.

A tannoy announced that our train was approaching: "Welcome to the *Shinkansen*. The train approaching platform 13 is the Nozomi Super Express bound for Tokyo. This train will be stopping at ..." then it listed the train's destinations, including Kyoto. It was the correct train. We felt comforted.

The arrival of our train was preceded by a buzzing technical hum, unlike any other sound I had heard a train make. The front cockpit of the train – you couldn't describe it any other way – approached us. It looked like the front of a jet fighter. Through the windscreen you could just make out the driver in his uniform, complete with peaked cap. He was seated bolt upright, obviously proud of his job and the responsibility of controlling this vehicle at up to 320km/h.

Sure enough, the train pulled up to the exact spot as indicated on our ticket, on time to the second. With a hiss, the doors opened. Nobody in the line of passengers in front of us moved until the people disembarking the train had left the carriage, only then did the first in line step forward.

I've heard comments that the Japanese behave like soulless robots, especially in such circumstances, but this calm disciplined way of conducting themselves is the only way such a densely populated country can function. If everybody barged their way forward, the country would collapse into chaos.

Our carriage was for reserved seats only and in the manner of the service so far, the interior was not dissimilar to that of an airliner. We found our seats quickly and sat down. The carriage doors closed and with hardly a noise, the train pulled smoothly out of Hiroshima, heading north. Out of the window to the right, you could make out the hill above the Mazda industrial complex, where we had started our tour of the factory site at the beginning of the week. It seemed ages ago, and my head had been filled with so many things since then. I looked across again, but in an instance we blasted into the first of many tunnels on the route.

Soon the train reached its top speed, which was indicated on a digital display above the door at the far end of the carriage. We were just nudging above 300km/h. The display showed information about the journey in both Japanese and English.

Outside, the landscape changed as we went, alternating between rice fields and townships, all very orderly but somewhat soulless. To the right of the train you could just make out the coast. Beyond the haze of the day's warmth you could see a line of refineries and dock cranes. On the other side, in the distance, were the mountains that make up most of the central part of the Japanese mainland. Oblivious to its surroundings, our Shinkansen sped on towards Tokyo.

Settling back, the doors at the far end slid open and a young woman in a pink uniform pulled a laden trolley into the compartment. She turned around and greeted the passengers with a bow. She made her way slowly up towards us, stopping sporadically to sell various drinks and snacks. We'd had the good mind to buy refreshments at Hiroshima Station, where we had understood what we were buying. Looking at the content of the trolley, everything, like so many things in Japan, was meticulously packaged, but there was no telling what surprise lay under the wrapping.

Without enticing us to purchase from the trolley, the waitress moved on to the end of the carriage, where she bowed once again, before making her way to the next carriage.

Little time passed before the train made a slight groan and started to shed off some speed.

"We will soon be making a short stop at Okayama."

Pulling into the Okayama Station, the station area looked very much the same as Hiroshima. Since the entire *Shinkansen* track system is elevated, there is always a commanding view. Below was a concourse area for arrivals, where the alternate white and black taxis waited. Directly next to the station were several hotels: ANA, Granvia, Crowne Plazza. Beyond them, the city sprawled out in a grid-like system of streets, with many recognisable signs: McDonalds, Seven 11, Lawsons. If it wasn't for the plethora of signs in Japanese Kanji, you could have easily been in the US.

Once again, the carriage loudspeaker welcomed new arrivals: "Welcome to the *Shinkansen*. This is the Nozomi Super Express bound for Tokyo. This train will be stopping at ..." and so it went on.

I folded down the table attached to the seat in front of me and unpacked my lunch of sandwiches and a can of Kirin Beer. The sandwiches were neatly cut triangles, without crusts, laid out in a see-through container with a hinged lid, filled with a mixture of egg and rather tasteless ham. The beer and sandwiches soon took effect and I drifted off to sleep, but not for long.

"Ladies and gentlemen, we will shortly be making a brief stop at Kyoto."

We made our way out into the warmth of the late afternoon. By now, the sky was changing to a soft pinkish yellow, so typical of Japan at this time of the year.

In the mayhem of the station thoroughfare, we could make out a rather prim lady holding a sign with the bold Mazda symbol on it. We went over and introduced ourselves, before we made our way outside to a waiting minivan, then to our hotel for the next two nights.

If you wish to know more about the wonders of the castles, temples and zen gardens of Kyoto and Nara, then a guide book will do a much better job than me. Even better, I urge the reader to visit these amazing sites and structures for themselves. Three highlights are the Ryoanji Temple, with its famous stone garden, the Golden Kinkakuji Temple, situated in the middle of a lake, both located in Kyoto, and the amazing Great Buddha at the Todai-ji temple complex in Nara.

Our guide did an admirable job of enlightening us to the beauty and significance of these cultural sights in just two days. It was a worthwhile but very tiring schedule.

*Sightseeing group in Kyoto.*

We departed Kyoto on Sunday afternoon, the *Shinkansen* taking us on to Yokohama, our location for the remainder of our stay in Japan.

Waiting for us at Yokohama station was a young man holding a sign reading "Mazda Design." He introduced himself and it turned out he was one of the designers working at the Yokohama Studio. We followed him to two waiting taxis.

Suburban Japan, no matter which city, big or small, all looks the same. The taxis made their way down a series of narrow neon-lit streets, lined with apartment buildings, small restaurants and the ubiquitous Lawson or Family Mart convenience stores. Yokohama is Japan's second largest city, but wherever we were heading didn't look like the centre of a world metropole. Eventually, the taxis pulled up outside a building fronted by some glass sliding doors, to the left of which was a Kirin Beer vending machine. Located in a suburb of Yokohama called Tsurumi, the sign above the door read "Hotel Tetora." This, apparently, was our home for the next two months – it certainly wasn't the Hyatt.

Before departing for Japan, I had purchased a guide book titled *The Sunday Times Businessman's Guide to Japan*. Here, it described a row of high-rise hotels located opposite a park in the old harbour area, behind which was Yokohama's famous China Town. From this vantage point, you could enjoy your breakfast overlooking Japan's busiest harbours. It sounded great.

*"It's not so roomy in Tsurumi!" Nice picture on the wall, mind you.*

Where ever Tsurumi was, it certainly was not the harbourside location I had been dreaming about. Making our way into the small lobby, the young designer who had met us, explained that this location was very convenient for getting to the Yokohama studio, being only two train stops and a short walk away. Oh well, now was not the time to complain, and he was certainly not responsible for the location of our hotel. We checked in and our guide said he would meet us the next morning in the lobby.

The hotel lift was only big enough for one, so we agreed to check our rooms before meeting back in the lobby and finding somewhere to eat.

Leaving the lift, I walked down the dimly lit corridor to my room. If the size of the lift and lobby where anything to go by, my room would not be a royal suite. The door swung open and my expectations were fully met. To describe what was before me as a room was very generous. It was, in fact, more of a wide gangway. Just as well my luggage was still somewhere in transit, since there was no space for it except on top of the single bed that took up most of the space in the room. To the left of the bed, and almost touching it, was a shelf that was occupied by a television set. Squeezing into the room, I wondered where the bathroom was. I shut the door and all was revealed. Behind the door was a plastic cubicle the size of a telephone box. This was the bathroom. I opened the door. The space inside was dominated by the toilet seat, behind which the cistern top doubled as a small sink. Looking down, it was clear that the floor was also the base of the showering facility, since there was a plughole in the corner and looking up, sure enough, almost above the toilet seat, was the shower-head. Well, if anything, it was a very clever piece of design.

I took everything in and thought, it's not so roomy in Tsurumi.

Suddenly, my thoughts were interrupted as the room filled with a loud noise from outside the window. I tried to open it, but it would only tilt. This, however, was enough for me to just make out the tracks of the local train service that passed behind the hotel.

Back down in the hotel lobby, we compared notes. It was very clear, there was no way we could stay at this address for two months.

Having agreed on that, we walked out into the evening and in search of a restaurant.

# Chapter 25

## *Mazda R&D Yokohama*

**After the huge earthquake of 1923 that essentially destroyed the Tokyo and Yokohama area of Japan, it was up to many foreign countries to help Japan rebuild its transport system in this devastated area of the country. The Ford Motor Company contributed by supplying trucks and buses and in 1925 set up a so-called 'knock down assembly' (KD) car production plant in the port area of Yokohama. It was here, during the 30s, that Ford assembled a variety of trucks and cars. After the war, Japan slowly built up its production of motor vehicles, first trucks and buses and then finally what was the beginning of the car industry we know today.**

At some point in the postwar years, Ford closed its facility in Yokohama and when it began helping the financially troubled Mazda in the early '70s, went on to pass the site over to the Japanese company. It was at this location that Mazda decided to construct an R&D centre, which due to its closeness to Tokyo would hopefully lure designers and engineers with little interest in moving to Hiroshima.

Named Mazda Research & Development Yokohama, or MRY, it was here that we headed on that October Monday morning in 1988, for our first full day of work as designers for Mazda.

The young designer who had met us at Yokohama Station the previous day arrived promptly at the hotel with two taxis.

Setting off, the taxis wound their way through a maze of narrow streets and under the main freeway linking Yokohama with Tokyo, before passing over a series of waterways strewn with small boats and discarded waste, indicating we were entering a segment of Yokohama's huge port area. The skyline opened up to reveal a large monolith of a building, behind it a row of factory buildings with angled roofs. The taxis pulled up outside a security gate. Beyond this point, the road ended at a jetty and a view out towards Yokohama's harbour, the sort of place I had been hoping the hotel would be.

Signing in at the small booth next to the gate, we followed our young designer colleague through a small side door. Once inside the building we waited in an office, and it wasn't long before MRY's chief designer, Yamada-san, entered the room to greet us.

*Local scenery on the route to the Mazda R&D centre Yokohama.*

*Our taxi pulls up at the back entrance of the Yokohama facility.*

*Mazda Research Yokohama (MRY), chief designer, Atsuhiko Yamada. (Denis Meunier)*

He had a smiling round face with shoulder length hair and had a very gentle and enthusiastic manner. His command of English was very good, having worked for a few years at a design company located in the Cotswold region of England. Greeting us, he recommended we waste no time and have a quick tour of the Yokohama facility.

Since leaving the Vauxhall Styling studios in Luton, I had never worked anywhere that was in any way comparable in terms of its dimensions. After the 'garage workshop' atmosphere of Audi, the studios at Porsche were a definite improvement and were indeed a modern working environment. Nevertheless, at Porsche they didn't have the drama or size that I remember from the Vauxhall facility. Entering the first main studio at Yokohama, I was immediately taken back to my Vauxhall days.

"This is more like it," I said to myself. The first studio we entered was truly a large room. It must have measured at least 10 meters in height, with light flooding in through high-mounted windows. There were, however, no windows at ground level, a concession, presumably, to security. The floor area was dominated by two measuring plates and at each side of the room were desk areas for the designers. Strange that there were no clay models on the plates. This room, Yamada-san explained, was where we would be located.

Moving on through the building, Yamada-san showed us two more studios, both of which didn't quite have the grand dimensions of the first one we had seen but were nevertheless very impressive. Here again, there were no models in sight. To complete the tour, we also took in a large viewing auditorium, also very reminiscent of the Vauxhall studio. At this stage, we had seen no workshop facilities, but I assumed they must be hidden somewhere. In all, it was certainly an impressive facility.

The tour completed, Yamada-san recommended we sort out some desk areas in the main studio we had passed through at the start of the tour. He explained that the next day Kawaoka-san and a group of planners from Hiroshima were visiting Yokohama with new information about the project we were to commence – the range of three sedans for Mazda's future premium range of cars. He then mentioned that since we would firstly be developing sketches and illustrations, Jim Howell was not required in Yokohama and had been requested to return to Hiroshima the following week to familiarize himself with Mazda's model development process. Jim wasn't so keen to go back to Hiroshima on his own, but it made sense, since the knowledge he picked up in Hiroshima would be beneficial once we returned to Oberursel.

While Roly and I sorted out desk space, Ginger had already approached Yamada-san about our hotel situation. The message was immediately relayed back to Hiroshima and instructions were sent back that we could select a hotel nearer the town centre, price apparently no issue. Checking my *Sunday Times* guide, I selected a hotel that sounded more comfortable than Tsurumi.

The Hotel New Grand was located opposite the Yamashita Park with commanding views of the harbour. The guide read: 'Built in 1927 after the great earthquake of 1923, the hotel is famous as being the location for American troops after the end of the second world war hostilities. It was here that General MacArthur resided, and his suite is still kept in the same condition as when he was in residence.' It fit the bill and rooms were reserved for the following day.

The rest of the day was uneventful. The Yokohama studio had a well-stocked material store, so Roly and I felt well equipped to commence creative work.

The last evening in Tsurumi was enjoyable, despite the tight confines of our accommodation. Ginger, Roly, Jim and I had been in close company now for almost two weeks. Up until now there had been no major disagreements and we were bonding into a good team. We would take all the highs and lows of this new adventure together. The energy between us was good, it bode well for the future.

We checked out of the hotel in Tsurumi – there was no love lost. The taxi ride to the office proved challenging, despite written directions, and a somewhat confused driver finally located the Yokohama studio.

Kawaoka-san and a delegation of planning engineers plus marketing representatives from Hiroshima, were waiting for us. Some I recognised from our meeting the previous week and I wondered if they had taken on board our recommendations about the risks of entering the premium market – probably not.

Greetings were made, and Kawaoka-san led a chorus of laughter as we exchanged handshakes.

*Was this good?* I thought.

Yamada-san joined us, and we followed him into the main studio where a large plan table circled with chairs was waiting for us. One of the engineers wasted no time and a series of engineering package drawings were rolled out, with Kawaoka-san again translating the discussion.

Our misgivings about Mazda entering the same sparring ring as BMW and Mercedes had not rubbed off on the planners and marketeers but as Ginger reminded us, there was no point in getting off to a bad start with our new employees; we had to play ball.

In front of us were the technical layouts of the three models Mazda was planning. The smallest car, which was aimed at the BMW 3 series, featured a transverse-mounted 2-litre V6 engine and front-wheel drive (FWD). This was an encouraging start, since no company we knew of featured a similar type of power train. Up against BMW, FWD drive could be an issue but that aside it sounded promising. The mid-sized car, which was aimed at the BMW 5 series, Mercedes E class and the Audi 100, also featured a transverse V6, this time with 2.5 litres, and again FWD. So far so good. Finally, the largest of the group and, as had been explained the previous week in Hiroshima, the most ambitious, had an overall length (OAL) of just under 6 metres and featured a 4-litre V12 engine.

It all sounded quite straightforward. We would commence our development with a range of sketches and illustrations. As far as a 'look' or design direction, that was left up to us, but the cars needed to look at home in Europe. Design reviews would be conducted on a regular basis, either in Hiroshima or Yokohama, and led by Mazda design chief, Fukuda-san. Hopefully, by the time we returned to Europe in December some key directions would have been selected so that at the beginning of 1989 we could start scale models in Oberursel of all three directions. Despite the divided opinions on Mazda's high ambitions, it was no doubt a great project to get our teeth into.

We were now ready to fulfil the trust Mazda was putting in us and the enthusiasm we exchanged in the taxi ride towards the centre of Yokohama that late afternoon was twofold. We had an interesting couple of months ahead of us and, of more immediate interest, hopefully a more comfortable hotel waiting for us at the end of the ride.

# Chapter 26

## *More than we bargained for*

**Yokohama Hotel New Grand couldn't have been a stronger contrast to where we had resided the previous two nights and as the taxi pulled up outside the entrance with its protective awning, I began to wonder if maybe we had overdone things a bit in requesting this noble address. A doorman in a red jacket greeted us and we made our way into the main lobby. The hotel's name, 'Grand,' was certainly not out of place. The main reception area was a large airy hall supported by marble clad columns and behind the check-in desk a large stairway led up to the main elevators. Checking in, the reception brought good news. Our luggage had arrived!**

The remainder of the first week in Yokohama sped by, and at the weekend we familiarised ourselves with the city. My Sunday Times guide was very helpful, listing the best locations for sightseeing, as well as eating and drinking. The restaurant in the Hotel New Grand was prohibitively expensive, so we had to figure out where to dine on a budget within our allocated expenses.

We took a boat tour of the harbour, which enabled us to see the Yokohama studio from the water, and visited Yokohama's China Town located in an area behind the hotel. Our first impressions of the city left us in no doubt that we could survive for the next two months.

On the Sunday, Jim was slightly apprehensive as we all bid him farewell at the main station. There was no work for him in Yokohama, so returning to the Mazda Design HQ was the best solution. At least he could go for *okonomiyaki* each evening at Hiroshima Station.

Our daily journey from the Hotel New Grand to the Yokohama studio by taxi took around 20 minutes. Rather than having the taxi drive us to the front security desk, he would drop us off at the end of a short road that led down through a harbourside shanty town towards the R&D centre. Japan is not a third world country, but there are still very marked differences in the population's living circumstances. The street to the centre meandered over a series of small bridges that crossed canal waterways, indicating the closeness of the harbour.

At the water's edge were a series of floating dwellings that wouldn't have been out of place in a tributary of the Amazon. It was difficult to discern if the occupants of these homes were fishermen or traders or both.

Between the dwellings were piles of rubbish and that time of year it was still very humid, with the smell emanating up from the water not the nicest greeting for us at the start of each day. Hard to believe that, just a few metres on, behind the walls of Mazda's development centre, plans for its attack on the luxury car segment were about to come together.

*The Yokohama Hotel New Grand.*

*A boat trip around Yokohama harbour.*

*Leaving the New Grand for work.*

For any kind of artist, a break from regular practice is not good. I hadn't put pen to paper for several months, so getting back into the swing of things would take a few days. Mazda was asking us to develop three designs, so the task was quite a challenge. Ginger, Roly and I studied the technical plans that had been left by the team from Hiroshima. We decided the best approach was to split responsibilities. I would tackle the large V12-engined car, while Roly would concentrate on the mid-sized model. For the smaller entry model we would both develop ideas. What made the task difficult was that Mazda didn't really have any logical design history to base our initial ideas on. The present range of cars was good looking as individual models but, in terms of design language, you couldn't relate a 323 to a 626. This was also the case with the new models we had reviewed in Hiroshima – all great-looking designs in their own right, but there was no cohesive design message. As we had explained to the planners in Hiroshima, companies like Mercedes and BMW had cultivated a look based on generations of models. Even Audi, which was trying to break into the premium segment, was still not regarded as a true competitor to the two 'big guys' at the top, so what chance did Mazda have? If you visited a Mercedes dealer there was a clear family relationship between all its models, which engendered a sense of seriousness and trust in the customer. This trend later became known as 'brand awareness,' but at this time, even in Europe, the phrase was yet to surface.

The fact that no Japanese companies had a clear family look to their models was fine, if that's what the domestic market demanded, but if they wanted to succeed in the premium market outside of Japan, they had to develop a uniform look that could be developed and cultivated through several model generations.

That design mantra having been agreed on, we still needed a distinct look for our range of cars. What we didn't feel was appropriate was cobbling together some kind of heraldic front grille that halfway related to a Mercedes.

Over several years the idea that a grille should adorn the front of a car had decreased in popularity. If you study the evolution of car design, the front grille basically went back to when this detail was in fact the cooling radiator. Over the years, the feature became more of a frontal ornamentation, as car makers began fitting more efficient radiators that were hidden within the engine bay. Apart from the premium manufacturers previously mentioned, the trend from around the mid-'70s was to have an air intake located at the lower part of the front end and have a fairly minimalist area between the front headlamps. This was the look being developed when I joined Vauxhall, with cars such as the Chevette and Cavalier featuring plain sheet metal between the headlamps, and it spread throughout the industry, encouraged by the trend for more aerodynamic bodies, the Ford Sierra also being a good point in case.

At Porsche, the mainstay of its product line-up, the 911, didn't have a front grille, purely because it was fitted with a rear-mounted air-cooled engine and the front-engined 924 and 928 models that followed maintained this aesthetic.

Mazda had hired us from Porsche because it respected the image of this company and with that in mind we began developing ideas for the Mazda premium brand as if it were a Porsche project – so no front grilles and, above all, very sporty.

Apart from the lack of cohesion in its model line-up, the look that Mazda had been developing under the Tokimeki (heartbeat) brand was very contemporary. It followed the soft 'bio' look that was currently a growing industry trend – soft flowing body forms and a minimum of hard feature lines, the forthcoming MX5 sports car being the best example.

Over the weeks, Roly and I began to produce a range of ideas. Yamada-san looked on encouragingly but declined to advise or comment. He would wait for Mazda chief designer Fukuda-san to visit before casting an opinion. Yes, we were left very much to our own devices.

Japan's glorious autumn, with its shades of red and gold, gave way to winter. Without any real private life, our daily routine began to wear on our relationship as a group somewhat, and at weekends we began to do our own thing, perhaps meeting in the evening to go out for a meal together. Not that there were any disagreements, but since almost no one in Japan spoke English, after a week sharing thoughts and opinions amongst the three of us, some time out at the weekends wasn't bad. My weekends began to develop into a routine. During the mornings I would walk along the harbour front and into the city centre. I would usually confine my shopping to visiting floor

*Taxi traffic in Yokohama.*

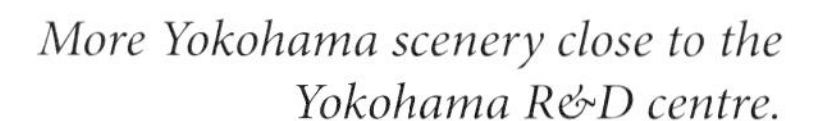

*More Yokohama scenery close to the Yokohama R&D centre.*

*Ginger with the Yokohama R&D centre in the background.*

by floor one of the big department stores before having a pizza or burger at a restaurant with illustrated menus, thus making it easy to order. Returning to the hotel I would go to my room and write letters to my fiancée, Claudia, my mum, plus anyone that came to mind – I had time.

Almost two months had passed since we left Hiroshima for Yokohama and as of yet Fukuda-san had not had the time to visit to review our work. By now we had a good selection of sketches and illustrations, including accurate quarter-scale side-view airbrush renderings. It was time to get some feedback, so it was agreed we would travel to Hiroshima. This visit coincided nicely with a visit from Yoshitomi-san and Shimazu-san, our two colleagues who were involved in helping Tanaka-san set up the Oberursel R&D centre.

Flying to Hiroshima, Mazda had booked us into the Ana Hotel, located close to the Peace Park. HQ design had given us strict instructions to check out of the Hotel New Grand and take all our belongings with us. We got the feeling we wouldn't be going back.

For the first review of our work, a global design review was organised, with both Yamada-san from Yokohama, plus Tom Matano of the MRA California studio in attendance. We had put up a selection of our work on boards that were placed in the main viewing auditorium. Fukuda-san led the meeting and listened while Ginger presented our work. Surprisingly few comments were made, with Fukuda-san thanking us for our efforts. The reaction was disappointing, since we assumed there would be a lively and enthusiastic discussion.

The presentation came to an end, with Kawaoka-san asking the chief designers, including Ginger, to remain with Fukuda-san, to make a decision on how to move forward.

After the meeting, both Roly and I were somewhat deflated. Little did we know that this style of feedback was quite normal. As a nation, the Japanese tend to hold back in demonstrating any strong opinion or emotion in a group of people, it deemed far too personal. We had been thanked for our work – at this stage that was enough. Over the years I become used to this style of feedback, or should I say lack of feedback. As a rule, I would say to my designers: "Don't worry – if you are doing something wrong, you'll soon hear about it." Not much consolation, I know, but that's the way it is.

Ginger returned from his meeting with Fukuda-san and was all smiles. He didn't give any information away but suggest we go over the results at dinner that evening, with Jim joining us. His only comment being that we had a lot to do!

In the '80s, an enterprising Hiroshima local opened an Italian restaurant in the town called Mario's. The first one, which was located across the road from the Ana Hotel, took off like a house on fire. Today, there are Mario's at several locations in the Hiroshima area and the guy who created them now lives in Hawaii sailing the local seas on a large ocean yacht. Each restaurant has a different style and level of cuisine. In downtown Hiroshima the Mario's has the feel of a Newfoundland fishing lodge – not exactly Italy - but across from the Peace Park is one that is genuinely Italian with native Italian staff – you take your pick, they're all good.

We crossed the main road in front of the Ana Hotel and entered Mario's. The restaurant was dimly lit, the walls predictably covered with posters depicting typical Italian sights – Rome, parasols on the Adriatic coast, etc. Tables lined the outer walls, but the main restaurant space was dominated by a large U-shaped bar with stools. Actually, it felt more like an American diner than an Italian pizzeria, but that was fine. We took seats at the bar.

Having ordered drinks and food, Ginger got right to the point. The meeting with Fukuda-san had gone very well. Apparently, he had been like a child in a toy shop – he wanted everything. Cutting it short, the main message was that we were to return to Germany and commence scale model development of all three cars for the planned line-up. Things, however, didn't stop there. To whet Fukuda-san's appetite further, Ginger had taken a selection of sketches, mainly from Roly, of ideas for future Mazda models. As if we didn't have enough to do, Fukuda-san now wanted us to develop some of these ideas further. One design was for a very advanced multi-purpose vehicle (MPV), a segment slowly catching on in Europe, due to the popularity on Renault's Espace. In addition to this, Roly had also been sketching some potential ideas for a next 323 hatchback, which Fukuda-san also wanted developed.

Our food arrived. How we were going to complete the tasks ahead of us, we would figure out upon our return to Germany.

The next morning, we met once more at the HQ design building and together with Kawaoka-san and a group of planners, we went over the work schedule that lay ahead of us. After Porsche, where laboured design programmes went on for months, the timing that Mazda was proposing was quite a shock. Our plan was that Roly would concentrate on the two advanced programmes and that I, plus two young Japanese designers transferred to Oberursel, would handle scale models of the premium cars. It was agreed. The models should be finished by the end of February. After design approval, a full-size model of one of the designs should then be developed in Europe. But where? As the design studio in Oberursel was still in construction, we agreed to handle the model build at an outside supplier. The models that Roly developed would be timed so that they could be worked on when the Oberursel facility was completed, the advanced MPV being shown to the press at a grand opening ceremony in the June of 1990.

*A farewell lunch before returning to Europe. L-R: Me, Ginger Ostle, Yutaka Shimazu, Shiganori Fukuda (seated), Yoshitomi-san, Jim Howell, Roland Sternmann.*

Fukuda-san invited us for a celebration lunch at Mazda's visitor's restaurant and together with Ginger, Roly and Jim, plus our two Oberursel 'minders,' Yoshitomi-san and Shimazu-san, we raised our glasses.

"To the future of MRE. *Kampai*!"

# Chapter 27

## *Back to Europe*

**The JAL 747 pulled up above Tokyo, heading east, not, however, in the direction of Alaska and onwards towards Frankfurt, but towards San Francisco. There, we transferred to a local flight and carried on down the California coast to the John Wayne airport in Irvine, south of Los Angeles and home to MRA, Mazda's North American R&D centre.**

Fukuda-san thought it was important for us to view this facility, which had just been opened, since it would give us an idea of how things may look once the Oberursel centre had been completed.

We arrived late afternoon in Irvine, with instructions to take a taxi to the Embassy Suites, where rooms were booked for us. Someone from MRA would meet us at the hotel after breakfast the following day.

The next morning, the contrast between the USA and Japan couldn't have better demonstrated than what I witnessed in the hotel restaurant as I waited for the guys to come down.

A family of four – mum, dad and two children of around 8 to 10 years old – all overweight and similarly dressed in pink training suits, descended on the buffet.

Dad started to pile up a plate of pancakes so high that I thought he would need to balance it with his chin to stop it tumbling over.

His wife, looking on, said, "Honey, we've got a long day ahead of us so make sure we have extra syrup and butter on those!"

As agreed, a pale brown Mazda 929 was waiting in the crisp of the California morning sun, outside the hotel lobby. I noticed that it was a colour that worked well in the light conditions of this part of the world. Under a grey Frankfurt sky, it would have been a different story.

Behind the wheel was none other than Tom Matano himself. We squeezed into the car and he set off to the MRA building, which as it turned out, was only a couple of blocks away, located in an anonymous industrial estate.

The Mazda North American R&D centre had quite a history. It was set up originally as a location for Mazda to rebuild the rotary engines that were fitted to the legendary sports car the RX7. Early models were notorious for wearing out the tips of their rotors. Here, the engines were fixed under warranty.

Since those early days of Mazda's first involvement in the US market, things had moved on to a newly completed and very modern facility where Tom drove into a parking spot in front of a glass fronted entrance hall.

The front of the facility was indeed very new, and as Tom Matano explained, had only been opened officially the week before.

*Mazda 929 in the California sun.*

*Mazda's North American R&D centre (MNAO) in Irvine, California.*

*The legendary Tom Matano, father of the Mazda MX5. (Tom Matano)*

Walking in through the front doors and into a spacious reception lobby, above us was a gallery wall from which a sculpture was protruding. It was in fact the nose of the yet to be revealed MX5 sports car, painted in gloss white. Tom had had it mounted there to test the reactions of the journalists who had attended the centre's opening ceremony the previous week. Apparently, no one had guessed what they were looking at.

We signed in at the reception and Tom motioned for us to follow him up some stairs. Passing through a security door we entered a large open room, more like a terrace, below which you could see a large modelling hall that looked out through tall glass windows to an outside viewing yard. The low winter sun streamed in through the windows; this was all very impressive. The room we were standing in was in fact the studio location for the designers, some of whom were standing around to greet us. I recognised two immediately as Wu-Huang Chin and Mark Jordan, both of whom I knew from their previous employment working at Opel in Germany. Pleasantries were exchanged before we followed Tom down some stairs and into the main modelling hall. It was indeed an impressive facility. The hall was a large room, about 50 metres long, with three modelling plates mounted flush in the floor. On one of the plates was a model hidden from preying eyes. What could this be, I thought.

We stopped in front of the shrouded model. Chin and Mark moved into place, and, upon Tom's signal, pulled back the covers. A bright red sports car came into view.

Tom wasted no time: "Gentlemen, this is the new RX7."

We were all speechless. The model in front of us was just stunning. Sitting on split-rim BBS wheels, it could have been a Ferrari or a similar Italian exotic. The design language was very similar to the new MX5, but bigger and much more powerful in its stance. As with its smaller brother, the front was void of any headlamps, which were concealed under closed covers. The body was a very pure and flowing form with virtually no feature lines to interrupt its overall look, even the door handles were hidden in the window frames. It was just masterful – was Mazda actually going to build this?

It occurred to me that during the model review in Hiroshima, nobody had mentioned the RX7. We had been so involved with the countless new cars unfolding in front of us, that Mazda's most iconic model had skipped our minds. Well, this was certainly worth the wait.

Tom, who was justifiably proud of his team's work, explained that what we were looking at was still the fibreglass styling model, but he promised that the production car would not differ too much.

I've been lucky many times as a car designer to witness the reveal of a new design, whether in a closed

*RX7 design model at MNAO. (Mazda Motor Corp)*

studio or at a motor show, but that moment in the MRA studio was certainly high on the list of truly memorable occasions.

The MRA tour and the surprise reveal of the next RX7 completed what had been a very thorough introduction to Mazda over the last ten weeks or so. The company had shared everything with us and, through working at the Yokohama studio, had immersed us in its design process. The final inspection of the MRA facility gave us an idea of what Mazda was planning to build in Oberursel. There was a big responsibility ahead of us – we were ready for the challenge.

*The Oberursel R&D centre build in progress. The landmark water tower in the background.*

Landing back in Frankfurt on a wintry Friday morning in December couldn't have been a bigger contrast to the balmy California climate we had left behind 12 hours before. This time, thanks to Lufthansa, our luggage travelled with us, and, bundling into a couple of taxis, we headed from the airport up the A5 autobahn, north of Frankfurt, to the disused factory location in the village of Weisskirchen, the part of the town of Oberursel where the future European R&D centre was to be built. The taxis dropped us off at the factory's old gatehouse, where the previous June we had attended our first interviews with Mazda. Looking around, quite a bit had changed. A large portion of the factory site had been demolished and cleared, the only remaining structures being a white water tower and an old office block with its adjoining clock tower – both landmarks, we would learn later, that Mazda would keep standing. Dotted around the site were many of the trees that had embraced the old factory,

*A faded image of the factory office building, soon to go, with clock tower behind it. You can just make out a few Mazdas parked by the building.*

all numbered, Mazda had apparently agreed with the local authorities to maintain the character of the site by leaving as many trees in place as possible. In addition to this, the centre would adopt a style of red-brick architecture, similar to the old factory building.

We made our way into the office, where several familiar faces greeted us: the general manager and our travelling companion from Japan, Tanaka-san; head of HR, Bernd Lesny; plus Shimazu-san and Yoshitomi-san from design. All had been present at our first meeting in June. Greetings were exchanged, and it was agreed that since we were all pretty jet lagged, a tour of the facility was best saved for the following Monday. All that remained was for us to pick up our allocated cars.

Part of the reason for Mazda setting up an R&D centre in Europe was not only to create new designs suitable for that market, but also to test and evaluate Mazda prototypes and cars from rival companies. A team from Mazda's vehicle evaluation group was already working on site and they had obviously been busy buying quite a selection of European models. Bernd Lesny opened the gates of a compound in front of the old factory office block, which was full of these vehicles, plus a selection of current Mazda models. It was quite a mouth-watering collection, with top models from all the European prestige brands.

"Don't look at the Jaguar," Bernd Lesny instructed. "You're Mazda employees, so you will drive them." It wasn't a bad choice, handing me the keys to a brand new RX7 Mk 2 Coupé in white. Ginger was happy with a 323 4WD Turbo and Roly and Jim, being family men, chose between a 929 or 626.

We said our farewells and I drove north to meet my fiancée Claudia at the home of her parents in Cologne. It was good to be home, but boy was I tired.

Monday morning came around quickly and gathering together once again at the factory site in Oberursel, we followed Bernd Lesny into the old office block that would function as a design studio until the construction of the centre was completed in the summer of 1990.

This converted part of the old factory housed not only a series of office rooms but also the old factory store room. The former company Georg Schutz had produced industrial wax and kerosene; indeed, it was rumoured that it had produced the fuel for the German V2 rockets from the previous war. An interesting story, and it is indeed documented that *Reichsmarschall* Hermann Göring had communicated directly with Georg Schutz, instructing the company to increase its production of kerosene, which presumably wasn't being used to heat his mountain lodge.

Going up some stairways to the first floor, we filed into a large well-lit room equipped with drawing tables and some work benches, where a couple of scale clay models were being worked on. Shimazu-san and Yoshitomi-san greeted us, and further introductions were made. Mazda HQ design had sent two young

*Ginger Ostle and Mazda R&D Europe HR chief, Bernd Lesny. (Mazda Motor Europe R&D)*

*The temporary studio in the old factory office building.*

*Build in progress. To the left you can make out the red brick factory building still standing. (Mazda Motor Europe R&D)*

designers to work in the new team, Masaharu Nobuhara and Akihiro Yamazaki, who we would call Massa and Aki respectively. In addition to these two, we also had a department assistant named Claudia Mueller, and two more familiar faces, Daniel and Evonne, young German nationals, both of who had agreed to join us from Porsche as clay modellers. A further ex-Porsche employee, Dave Samways, completed the team. Dave, an Englishman, would be our senior studio engineer. The initial team complete, we could get to work.

1988 drew to a close and Claudia and I split our Christmas and New Year between my mother in Southwold, and her parents in Cologne. Having been engaged for two years, we were now making plans to marry the following June. My private life, like my working life, had plenty of eventful occasions ahead.

By the start of the new year I was able to move into a rented flat in a small town called Friedrichsdorf, which was situated below the hills just north of Frankfurt known as the Taunus.

Construction work at the R&D centre in Oberursel was progressing well. The first buildings planned for completion were the offices and workshops of the testing and homologation group. Mazda was keen to complete the workshops in order to hide prototypes from prying eyes. Once that building was complete, it was planned for us to use part of the workshop space, so freeing up the old office block, where our temporary studio was now located, as the final factory building to be demolished. By mid-1989 most of the foundations were finished and from then, on an almost daily basis, trucks would arrive at the site with massive concrete sections of the building. From our vantage point in the old factory, we witnessed our future studio slowly taking shape.

Design work was also moving at a good pace, and by the early spring of 1989 two scale models for the new premium range were complete. They represented the smallest model, a competitor for the Audi 80, the other, the largest, aimed squarely at the Mercedes S class. Both designs followed the smooth flowing body style, similar to the RX7 model we had seen at MRA, and an increasing trend throughout the industry. Both featured smooth front-end profiles, which, unlike the Mercedes and BMW models in this class, had no classic chrome grille. For the larger car, I had elected to do one side of the model, featuring a covered rear wheel similar to some of the cars of the '50s, such as the Mk1 Jaguar. Covered rear wheels on rear-drive cars is always difficult because of the clearance required in some countries for snow chains. The design looked great, but I had a nagging feeling it would be shot down by the engineers.

With the two models ready for presentation, a delegation from Japan visited us for the first time. The team was led by the programme manager, responsible for the whole running of the project. He was accompanied by a group of product planners and representatives from marketing, who no doubt had attended the meetings we had during our visit to Hiroshima.

The models were set up for viewing on some work benches in the studio, finished in Dynoc foil, painted in silver and decorated with all the usual features – windows, door shut lines, and lamps. They looked good.

The delegation entered the room. From the gloomy look on their faces you would have thought they were attending a funeral, or maybe they all had massive hangovers from revelling the previous night at one of Frankfurt's many Sachsenhausen apple-wine houses.

*Illustrations for the proposed premium sedan.*

In addition to the models, I had hung up some package tape drawings of the designs together with some of the sketch work.

Most of the Hiroshima team spoke very little English, so Shimazu-san acted as an interpreter. Introductions were made, and I went straight into a presentation of the designs; nobody said anything.

My piece completed, I invited the group to take a closer look at the models. The programme manager walked around the models, his team following him dutifully like a line of ducks. He studied the package drawings and a few comments were made. Shimazu-san laughed nervously but declined to translate. Was this going well? I had no idea.

Everybody was waiting for the programme manager to say something. After what seemed ages he gave his assessment. First in Japanese and then, to my surprise, in perfect English:

"Thank you for your efforts. Can you please explain the concept?"

I looked at Shimazu-san, who, seemingly lost for words, repeated the programme manager's question.

I felt myself getting hot under the collar. What did he want from me?

I apologised and asked if he could explain his question. Don't they like the designs?

"We appreciate your hard endeavours, but we still need to understand the concept behind your designs."

Shimazu-san found his tongue. He could see I was struggling and immediately

*Polaroid shots of the large premium sedan scale model.*

stood in for me. The rest of the review was conducted in Japanese and things seemed to get smoothed over.

The presentation was drawn to a close and moving to me with outstretched hand, the programme manager thanked me, saying he was very happy. Well, you could have fooled me!

The presentation was a real learning curve of what to expect when presenting to a Japanese audience. I had come from a design culture where you looked at an illustration or model and you either liked it or not, and then said so. Things in Japanese companies were not quite that simple. The visual message is just one element in the design process. What was expected in the presentation was additional information, from the target customer, to the emotions that influenced the design, all summed up in a catalogue of so-called 'keywords.'

We had all chuckled a bit at the first meetings in Japan, when it was explained that the key word to explain Mazda's new design philosophy was *tokimeki* – Japanese for 'heartbeat.' In the years to come we learnt how important this mindset was as a part of the design process. In some ways, the Japanese were ahead of the game. Today, with the importance of what a brand represents to the customer and all that goes on in terms of marketing and sales, the pure look of a product is just one element in the development process. Back then, it was still new territory to me, but after that first meeting I would realise there was lots of learning to do.

After the presentation, Shimazu-san left the studio with the Hiroshima delegation and I was not asked to attend. It was clear, despite the pedestal we had been put on when joining Mazda, with regards to decision making, we were not yet to be included.

I felt somewhat deflated and frustrated after the meeting, but I needn't have worried. Later in the day, Shimazu-san came up to me and said, "Well done, Birty-san, a good job!"

Yes, the programme manager and his team were satisfied, and he gave the go-ahead for us the commence a full-size model of the V12 sedan – but where?

# Chapter 28

## *Mazda R&D Europe – the grand opening*

**With no modelling studio at the Oberursel site available for the construction of a full-size clay model, we had to find another solution. The Frankfurt area, with its close proximity to Opel, was surrounded by suppliers to the automotive industry, but companies capable of modelling a full-size clay were scarce. Then an idea occurred to me. The ex-chief modeller at Audi, Wolfgang Holzinger, had set up an independent prototype building company in a small town north of Ingolstadt, named Beilngries, situated in the picturesque Altmühltal valley.**

I called Wolfgang (we knew each other well) and arranged to drive down to Beilngries with Ginger.

If I was to spend some more time away from home, then Beilngries was a good location, especially as we would need to be there for most of the summer.

Wolfgang Holzinger's facility, which he had named HOTE Design, had been constructed in the mid-'80s, shortly after I had left Audi to join Porsche. A group of industrial halls, the location was shared with a local foundry called Jura Guss, which was owned by a close friend of Wolfgang. He had no doubt set up the company with a guarantee that he would receive some contracts from nearby Audi. Indeed, it was clear as Ginger and I sat in Wolfgang's office, that he was presently only working on projects for the VW/Audi group, emphasising that should we build a model at his location, the fact should remain secret. He was no doubt nervous that should Audi or, perish the thought, Ferdinand Piëch, discover he was doing work for a Japanese company, it may jeopardise his main source of income. His concerns about future jobs can't have been too worrying for him, since he was keen to work for us and we agreed on a starting time. He would commence building a modelling armature for the model as soon as we sent him details of the design.

So in the April of 1989 I moved to Beilngries with a team of modellers to start Mazda's first European-sourced full-size clay model. Our two young recruits from Porsche, Daniel and Evonne, joined us and, to help them out, I contracted two freelance modellers, one of whom was named John Wall, a legendary talent in the field of clay modelling and a man who had also had experience working at the Ford design studio in Hiroshima. His experience was great for my two modellers to learn from and his understanding of working in Japan and with the Japanese would also be a help.

As agreed, HOTE Design received all the technical information for the model construction and commenced preparation for our arrival. To enable the model to eventually be set up on its modelling plates in Hiroshima, Mazda supplied HOTE with a set of axles and wheels for the armature.

Finding accommodation in Beilngries did not prove a problem. Being a holiday destination, it was full of suitable lodgings and we acquired two properties for our team.

By the time I arrived at HOTE both Daniel and Evonne had been working with John Wall and his colleague Simon for a couple of weeks and the basic form of the car was blocked in; it was massive, over five metres long. I had instructed the team to model in one side featuring a covered rear wheel housing – there had been no complaints from the programme manager, although Oberursel's design studio engineer, Dave Samways, was very pessimistic, well-versed as he was in the various regulations with regards to snow chain clearance.

Even in its raw clay state the car looked impressive. With its long low bonnet covering the planned V12 engine, and clean sports car-like front profile, it looked light years more advanced than an S Class Mercedes – well, at least I thought so.

Our schedule for the model was very tight, as the programme team from Hiroshima were visiting the site sometime in June or July. This was a different world from Porsche, where taking a year to develop a scale model was the norm. No point in drawing comparisons, I thought. There had been good reasons for departing Porsche and now was not the time to complain. One event, however, I couldn't change was my forthcoming wedding, planned for the beginning of June. Maybe we would have to have the ceremony in Beilngries!

With the modelling of my design well under way, things in Oberursel were also well advanced and the design team would soon move into new temporary accommodation next to the product evaluation group. The old Georg Schutz factory office would be demolished, leaving only the red brick clock tower standing, facing the nearby local S5 railway line to Frankfurt – a reminder of a previous era.

Roly had been working hard, and with the two young Japanese designers he had created three scale models for the next Mazda 323 range. Similar to the design being developed in Beilngries, all three models featured a smooth and sporty body language. The five-door model was particularly dynamic, with character more reminiscent of a coupé. The top version of this model featured a compact 2-litre V6, which was essentially half of the V12 planned for the large limousine I was working on. The more we understood about Mazda's future plans the more we were encouraged – the move to work for them was totally vindicated.

The scheduled review of the premium limousine (there was no name for it yet) was scheduled for the end of June. The timing worked well for my wedding to Claudia. We planned to marry the first week of June at the town registry office in Cologne, with a big party for friends and family in a marquee in the garden of my mother's house in Southwold at the beginning of July.

The wedding day in Cologne was fairly low-key as far as weddings go, attended just by close family and friends. My old college friend and colleague from the Audi days, Martin Smith, was my best man. At the registry office, after signing the certificate, the conducting registrar, in typical Cologne humour, congratulated us on the purchase of a washing machine. He no doubt cracked that line at every wedding!

After the ceremony, we all walked from the Cologne *Altstadt* to the Rhine and took a boat trip downstream to the small fishing village of Rodenkirchen, where we had lunch in the sunshine on the terrace of the Cologne Sailing Club. The champagne flowed and, opening one of the telegrams sent to us for the occasion, there was a message from Mazda, together with a cheque for DM 750 – "A small contribution to your future life."

The wedding weekend passed quickly and soon I was heading south with my new bride. Since there was free holiday accommodation in Beilngries, Claudia and I decided to base ourselves there, and carry on down to the Alps for a week with our bikes.

On the way down, we passed by Ingolstadt and I showed Claudia my old flat in the Josef-Ponschab Str, plus some of my old haunts. It seemed like another world to me – so much had happened in my life in just over ten years.

Every programme in Mazda's development programme has a so-called 'J' number. So, for example, a future mid-sized model would be coded J66A, which for this purpose is purely fictional. The model we were developing in Beilngries was yet to be give a model name, but it did have a J number. Mazda is very protective about these codes, and they remain secret to this day.

By mid-June the large premium car project I was running had a programme chief designer located in Hiroshima. The chief designer is responsible for all the planning of the programme and it was to him that I reported. In this case, it was Shunji Tanaka who led the production design development of the new MX5.

Tanaka-san was a larger than life figure, quite stout for a Japanese, with receding hair. His family were well connected in Hiroshima and it was rumoured that they even had some connection to the Hiroshima 'underworld.' The fact that he permanently hid his eyes behind dark glasses, added to the mystery surrounding his character. He was, nevertheless, outwardly a very friendly and jovial character, but unfortunately spoke very limited English.

For the first major design review in Beilngries, Tanaka-san was travelling down from Frankfurt with the programme manager and the Oberursel team, including Ginger and Shimazu-san.

In order to judge the design properly, the model needed to be taken outside. This would present an issue, since there was no location near the HOTE design office where we could do this with any degree of secrecy. Fortunately, being mid-summer we had daylight until quite a late hour. It was therefore decided to wait until all the workers from the neighbouring Jura Guss company had left the site and then pull

*Pulling out the covered model in front of the Jura Guss location.*

*A blurry photo of the premium sedan in the evening sunlight. Our first impression outside.*

the covered model to a secluded part of the foundry yard. This plan in place, we waited patiently for the delegation from Oberursel to arrive.

I had told Ginger not to drive down in any Mazda's, so as to avoid any suspicion, also aware of Wolfgang Holzinger's concerns about a Japanese company using his services.

Ginger timed his arrival perfectly and at around 7:00pm a BMW and Mercedes rolled into the HOTE Design forecourt. Anticipating their timing, we had positioned the covered model close to a high wall in an area of the foundry where they tipped all their manufacturing sediments. It looked like the surface of the moon.

The cars were parked out of sight and everyone made their way over to where the model was standing. We had checked that no nearby houses looked over the yard, which was thankfully the case. The only possible location the model could be seen from was a hill on the far side of Beilngries, but too far away to concern us. Things in those days were so much easier with no mobile phones or social media.

Even for the Beilngries team, this was our first chance to view the model outside and at distance. We all looked on excitedly as I gave signal for the cover to be pulled away. I was relieved; the initial impression was good. Although dimensionally a big car it was certainly very sporty and, especially the side with the covered rear wheel, looked low and futuristic. As soon as our photographer, Bernd Schuster, had taken some standard shots, we got to work on the model with black tape, marking feature lines that required adjustment. Ginger fetched the Mercedes S Class and positioned it close to the model for comparison. The cars were in two different worlds. With the light fading and Wolfgang looking around nervously in search of prying eyes, we had to work fast. Although pleased with the results, there were plenty of areas on the model that needed attention. Satisfied we had captured the key elements to be modified, we covered the model, returning it to the studio as the sun set over the Altmühltal.

Returning to the hotel in the centre of Beilngries, where Ginger, Tanaka-san and the other team members from Hiroshima would spend the night, we joined them all for a celebratory beer and meal of

*Schweinshaxe*, or roast pork knuckle. Everybody was happy. Sure, there was work to do, but as Tanaka-san raised his glass for another *kampai*, a big weight fell of my shoulders.

There would be no more official design reviews from Tanaka-san and his team, and for the rest of the summer months of 1989 we continued to refine the model with repeated late evening trips to the foundry yard to check the development before it was crated up and sent on its way to Hiroshima.

In July, I was able to get away from Beilngries for Claudia and my wedding party in Southwold. It was a bit off a 'who's who' of car design with Martin Smith, J Mays, Peter Schreyer, Geoff Lawson, and Peter Stevens, to name a few, in attendance. By the end of the year, we were happy to announce that Claudia was expecting our first child.

*The finished Mazda R&D centre in Oberursel, Germany, June 1990. (Mazda Motor Europe R&D)*

*The design viewing yard. (Mazda Motor Europe R&D)*

There was no more external modelling work for the Oberursel design team as, by the spring of 1990, the design studio building, and final part of the development centre, had been completed. It was indeed an impressive facility and no doubt the envy of all Europe. The layout was not the same as the studio we had seen in Irvine but certainly a similar dimension. There were two large studios equipped with full-size modelling plates, which looked out onto an outside viewing yard equipped with three turntables, essential when comparing design models with competitor vehicles. Also adjacent to the viewing yard was a large indoor showroom, similarly equipped with three turntables. This vast hall could also be used as a photo studio and event location. Between the two modelling studios was a smaller room for interior design and colour and material development. Behind these studios was a central corridor that bordered a row of fabrication workshops. Everything was catered for, including metal, wood and fibreglass, plus a spray

*The Mazda Gyssia. (Mazda Motor Europe R&D)*

*The official opening ceremony with the presentation of the Gyssia concept model. (Mazda Motor Europe R&D)*

booth and saddler's workshop, necessary for the manufacture of seats and interior components. Things, however, didn't stop there. A second studio floor above the workshops looked down into the main modelling area below. Light flooded into both levels through floor-to-ceiling glass walls measuring at least ten metres in height. In the upper area was room for at least nine designers plus studio engineers with a separate office for the design manager. Here, Ginger would take his place, at this stage, however, not alone, since he would first share the office with his 'minder' Yoshitomi-san. Politics aside, it was an outstanding location to work and we couldn't wait to show it off at the official opening ceremony planned for May.

In preparation for this event, Roly had been busy. During my time in Beilngries he had designed and developed the advanced mid-engined 'people carrier' or MPV (multi-purpose vehicle) that Fukuda-san had instructed to be built for this prestigious occasion. This would demonstrate the 'design power' of the centre, sending out a clear message that Mazda meant business in the European automotive design world. It would also act as a test for the team at MRE, being the first model constructed in-house.

The vehicle Roly had designed was indeed impressive. The Gyssia, named after the Japanese rickshaw, had a turquoise-green metallic finish, and featured a centrally located rotary engine, which due to its compact layout could be fitted below the passenger floor. The body design was predictably influenced by the design language of our former employee, with a sloping front nose and dominant central B pillar, similar to that on the Porsche 928. It was a fitting design statement for the opening event.

The grand opening of Mazda's new European R&D centre, MRE, took place under sunny skies and was divided into two occasions: one for journalists, the other for automotive design VIPs and the friends and families of MRE. They were both memorable days, but the dust soon settled – now the centre had to deliver.

# Chapter 29

## *The first birth*

**The waters broke at around three in the morning – our first child was on his way. I say 'his,' because Claudia felt it important to communicate with her first child while still in the womb, and knowing the gender was an important part of this communication. The third or fourth scan revealed that a boy was on the way. To be truthful, neither of us minded what sex our first offspring was, but since we knew, a boy was fine. To the displeasure of my mother-in-law, I immediately named him 'Bruno' – well, we had to call him something even though this would certainly not be his final name. Mazda had taken on a young manager in the HR department named Bruno Mueller, and I decided to borrow his name.**

We had been living in an upper floor flat in the small village of Koeppern, just north of the well-known spa town of Bad Homburg. Although a charming town, Bad Homburg's hospital did not have the best reputation at the time. A new hospital was in the planning, but the present one had very much an 'end of term' feeling. As it happened, a hospital, old or new, was not in Claudia's mind for the birth of our first child. Located in the town of Friedberg, famous for where Elvis Presley had spent his military service, and about ten miles north of Koeppern, was a so-called birth practice. Claudia didn't want the 'factory' atmosphere of a hospital when bringing our new child into the world, but something more personal. I was nervous about this proposal but after having accompanied Claudia to the practice and seen that the facility was, in fact, not like the stable where Jesus had been born, I agreed to go along. The practice was kitted out with all the most up-to-date equipment for delivering a child, and could even cope with a Caesarean if necessary, since both the resident doctor and his wife, also a qualified doctor, lived in a flat above the delivery rooms. At this location, which was essentially a large town house with expansive garden, expectant mothers could bring their new children into the world in a calm and friendly atmosphere with only the company of the midwife and their partner. Yes, I had to play a major role in the birth, something I was only too aware of as I bundled a moaning Claudia into the back seat of the Lexus LS400 I was using at the time and set off to Friedberg in the dark, early hours of July 19.

If you are chauffeuring a woman in labour to a

*The Mazda Europe R&D design team, June 1990. (Mazda Motor Europe R&D)*

birth practice, then the Lexus LS400 is a pretty good car to have, fitted as it was with a large comfortable backseat bench. Mazda had purchased the car as soon as it came on the market, knowing that it was a key competitor to the planned premium limousine I had been working on the previous year. I had been using it for a few weeks, and it caused quite a bit of attention, especially in Germany where no one in their right mind would purchase such a car in the home of Mercedes and BMW, or even Audi for that matter. Some reactions to the car were often quite hostile, and thoughts of Ginger's unsuccessful arguments during our first trip to Hiroshima, to convince Mazda that entering the premium car arena would be very difficult, came to mind. Nevertheless, you had to admire the vehicle that Toyota had created as its first attempt to crack the firm territory of Mercedes. As a design, the LS400 was no head turner; it had classic large limousine proportions and with the absence of its own historic front grille, Toyota had had to create something functional that wouldn't offend Mercedes or BMW customers, the end result being somewhat anonymous. Toyota wasn't too worried if the car failed in Europe, since Lexus was clearly aimed at the North American market. This fact was clearly demonstrated by the driver's electric seat and steering wheel system that eased the entry of customers with a sizable waistband, and in America there were plenty of those. It was a clever solution to save the embarrassment of those physically too large to get in the car and position themselves with ease behind the steering wheel. Basically, on entering the car, the steering wheel was at a forward location, close to the instrument panel, and the driver's seat at its most rearward position in the car. Once the driver had eased himself comfortably into the car and then inserted the ignition key, the steering wheel and seat would move into a pre-selected position for driving. When switching the engine off to exit the car, the process reversed. The front seat pulled back and the steering wheel moved forwards, thus enabling the driver to move round and step out of the car. To cater for different drivers, the position of the seat and steering wheel could be programmed by choosing one of three settings on a switch panel at the side of the driver's seat. All the European premium manufacturers had programmable driver's seats, but none featured a mechanism with a combined steering wheel adjustment as in the Lexus. This was all well and good until a driver of smaller dimensions programmed the system for themselves, and in Mazda's testing team there were plenty of those. This became very apparent when I took the keys of the car for the first time, unaware that the previous custodian had the physical dimensions of a ten-year-old child. Opening the door, I eased myself into the comfortable armchair of a front seat, yes, there was loads of room. Then came the surprise. Inserting the ignition key into its slot, immediately the steering wheel arched out towards my chest, while simultaneously the driver's seat moved forward until I was literally pinned in place. I was stuck, with no room for me to move, and it was only with pure luck that I was able to reach the ignition key and thus reverse the procedure.

As we sped on through the darkness towards Friedberg, it occurred to me that it was just as well I was positioned in the driving seat. Perish the thought, if, for some unimaginable reason, Claudia had had to drive herself to the birth practice and been forced up against the steering wheel as I had been that first time when entering the car.

*Mazda Europe R&D's first general manager, Seiji Tanaka, and Ginger Ostle inspect a scale model of the Gyssia concept. (Mazda Motor Europe R&D)*

*The large premium sedan full-size model on the Mazda HQ design viewing roof. (Mazda Motor Corp)*

up and running, and we commenced recruiting a team to fill the place. With Ginger managing the studio, it was clear Roly and I couldn't handle all the design work, but that was never the intention. We needed to hire not only designers but also modellers, fabricators and a colour and material team. In all, we were looking at around a staff of at least 30. News of Mazda's state-of-the-art facility soon got around and the job applications started to come in.

It was now almost two years since we had left Porsche and as a signal that Mazda were happy with our contribution, both Roly and I were promoted to the level of chief designer. Once again, Oberursel's general manager, Tanaka-san, presented us proudly with our newly printed business cards to celebrate the promotion. In a way, it was also a signal that his job in Germany was complete. The centre was built and functioning, with a team in place he knew would fulfil the wishes of the company. He would soon return to Japan and retirement.

As it happened, he was not the only person soon to be returning to Japan. Since sending the full-size model of the large premium sedan to Hiroshima, work on the project had been proceeding at a fast pace. A message had been sent to Ginger that work on the final production version of the design was well advanced and could Birty-san please join the team in Hiroshima to assist with the final adjustments on the model. It was anticipated I needed to be in Japan for couple of weeks or so.

With a wife and baby at home, I was somewhat reluctant to leave for Japan but this was what we had signed up to, and it was, after all, only two or three weeks.

Mazda had booked me on a Lufthansa flight to Osaka via Hong Kong; it turned out to be a nightmare journey. A typhoon circling the Hong Kong area meant

Then maybe the birth of our first child would have not been as planned. You can just imagine it in years to come:

"And where were you born?"

"Oh, in the footwell of a Lexus!"

As I eased the Lexus onto the road where the birth practice was, mother and soon-to-be-born child were still one unit. Seeing lights on in the lower floor I breathed a sigh of relief: someone was there.

I won't go into the details of the birth, but everything went well. In fact, apart from following a few instructions from the midwife and giving Claudia some words of encouragement, I played a fairly minimal role in the whole affair. By nine o'clock the three of us were in the Lexus and heading back home to Claudia's excited mother.

During the day, friends and relatives descended upon us; I called my mother to tell her we had named our first born Denys, after my father. She was so happy, but, like me, sad that he was not with us to learn of his first grandson.

In Oberursel, Mazda's own newborn was now fully

*Mazda's premium car range was named 'Amati.' Prior to the planned launch, a Japanese magazine acquired images from the planned brochure.*

that our departure from Frankfurt was delayed. By the time we had been given the go-ahead to depart we were at least two hours late. In 1990, Hong Kong's airport was still located in the city and landing there was one of the most challenging for a pilot, involving an approach down an avenue of high-rise tower blocks. Nearing Hong Kong, the pilot took aim at a pylon on top of mountain behind the city and then banked immediately for the decent. Under normal circumstances this was difficult enough but in the tail of a typhoon all very nerve-racking. I recall looking out of the cabin windows through the lashings of rain at the lights of the living rooms of the tower blocks we were threading our way past. You could almost make out the occupants having their evening meals. How a plane ever avoided crashing into their living rooms I'll never know.

By the time our 747 had refuelled in Hong Kong, and we were back on our way to Osaka, I had calculated we would be arriving so late that I was in danger of missing the last *Shinkansen* train to Hiroshima. As in Hong Kong, we landed in the Osaka airport located in the town centre. Both cities were in the process of building new international airports, but they were some years off completion back then.

Upon landing and collecting my case, I rushed out of the airport building and got into a cab; the last train to Hiroshima left in about 30 minutes. As usual, the traffic on a Sunday evening in Japan was gridlocked. The taxi edged forwards and after what seemed ages I could see the outline of Shin-Osaka, the main railway station, through the mist of the falling rain. No time to waste, I paid the taxi driver and ran like hell the remaining 100 metres into the station ticket hall. With no time to think about seat reservations I paid for a ticket for Hiroshima and ran once more to the designated platform where the train was just pulling in. By now I had been on the go for almost 24 hours; all I needed was a seat. On a Sunday evening, however, all of Japan is on the move and not having had time to think about reserving a seat, there was no choice but to join the passengers standing tightly packed in the middle aisle of the carriage. I positioned myself opposite a typical businessman dressed in a black suit. I must have looked terrible, as he glanced up at me sympathetically; he was at least a foot shorter than me. Looking down at my cabin bag I remembered I had bought a couple of beer cans somewhere on the way. I reached down and pulled one of the cans out. At the same time my businessman 'friend' began pulling something out of a bag he had with him. As it turned out, it appeared to be an entire dried squid, which he thrust invitingly in my direction. Was he offering me the complete beast or just one of its arms? I declined with a thankful smile, as the Hiroshima bound *Shinkansen* blasted through the night and onwards to my final destination – what a journey!

I took the lift up to the roof viewing terrace of the HQ design building with Kawaoka-san and Tanaka-san, where the latest version of my design would be presented. Both were laughing nervously – not a good sign, I thought.

We walked out into the early morning sun. The quality of light in this part of the world is totally different to Western Europe and ideal for looking at design models.

Was that the car we sent to Japan last year? I thought to myself. Not really. There will always be a change in a design from the conceptual phase through to what can actually be built. In the industry, this is known as the feasibility process. Not only will the manufacturability of the design be considered but also design modifications taking in the considerations of sales and marketing. It was clear immediately that all parties had been to work.

The main area of change that struck me right away was that the front of the car, which was now adorned with a chrome grille; gone was the smooth sports car like front profile we had created in Beilngries. The proportion and stance of the car was still there, but somehow the rear screen angle appeared more upright. It was pretty clear, Mazda had seen the Lexus LS400 and wanted to play things safe.

By now, Mazda design chief Fukuda-san, had joined the party.

"Ah Birty-san," he said. "Sorry, sorry, but we needed a grille." The wave of opinion was against me, and I saw little point in starting an argument. My lack of protest was what they wanted, and the nervous laughter turned into laughter of congratulations. Of course, I was disappointed, but this was part of the design process and the first of many such changes. Car design is all about teamwork and a common goal. Mazda's development team decide that a more conservative vehicle was the right approach. I had to respect that.

Later in the day I got to see the other two vehicles in the range of cars they were calling 'The Three Brothers.' For the mid-sized 5-series BMW sized car, they had virtually shrunk the larger car I had seen on the viewing terrace. It was being handled by a designer named Koichi Hayashi – we would work together in the future on many projects. The smaller BMW 3 series competitor was another story. Handled by a young designer named Iowa Koizumi, he had created a wonderful design quite unlike anything on the road. The design followed the smooth 'bionic' look that was going through the industry, devoid of any hard lines or body creases. As with its siblings it had a front grille but this time a much smaller treatment and almost Jaguar-like in its expression. It looked great. Koizumi-san as a person was as stylish as the car he had created. He wore black designer label suits with a 'nero' style collar, and his shiny black hair was combed back in Brian Ferry manner. Like Hayashi-san, we would work together in the future on many projects.

*The ones that got away.*

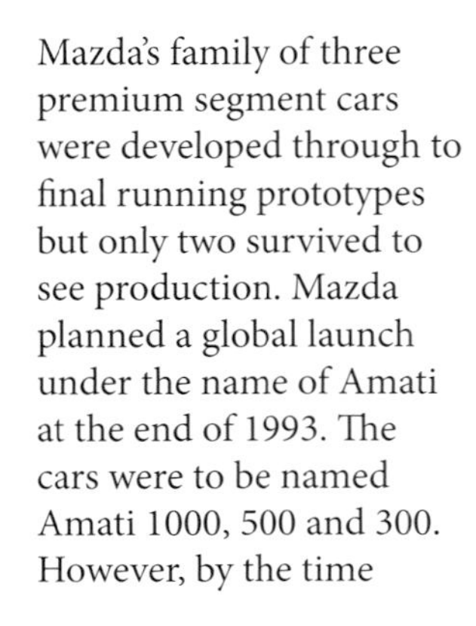

Mazda's family of three premium segment cars were developed through to final running prototypes but only two survived to see production. Mazda planned a global launch under the name of Amati at the end of 1993. The cars were to be named Amati 1000, 500 and 300. However, by the time

*Final full-size clay model of the Oberursel 323F/Lantis design. (Mazda Motor Europe R&D)*

Mazda was ready to launch these exciting models, the global economic situation had turned for the worse. Reluctantly, but wisely, the company decided that trying to sell a premium range of cars with all the logistics to go with them was too much of a risk and they pulled the plug on the project.

What remained of the programme was changed into a smaller sub-brand named Xedos in Europe and North America and for the Japanese market the cars were swallowed up into the domestic Eunos brand.

As Xedos, the larger 9 sedan would be of limited success in Europe but Koizumi-san's Xedos 6 was successful and widely admired for its sporty look. The larger car I had worked on only went as far as running prototypes. I never saw one, but it was fitted, not with a V12, but a W12 engine; it must have been an impressive car. A Japanese magazine managed to get some rather unfocused images that were published as a so-called 'scoop,' but that is all the public got to see of the car.

In an effort to try and compete with the 'big guys,' Mazda's very ambitious Amati range almost broke them, and our warnings about trying to enter into the premium market, rang very true in our ears. In all fairness, the cars were good enough to succeed, but the market at the time just wasn't there.

My return flight to Frankfurt was thankfully less eventful. Since the opening of MRE in 1990, efforts to recruit new team members were in full flow, and bit by bit new designers joined MRE, some poached from other companies, some college graduates. They included Mark Oldham, Jo Buck, Hartmut Sinkwitz, Cor Steenstra and Laurent Boulet. To head up colour and material, Verena Kloos joined us from

*Fibreglass styling model of the 323F/ Lantis on the Oberursel viewing yard. (Mazda Motor Europe R&D)*

*A proud Roland Sternmann with his baby – the production Mazda 323F. (Mazda Motor Europe R&D)*

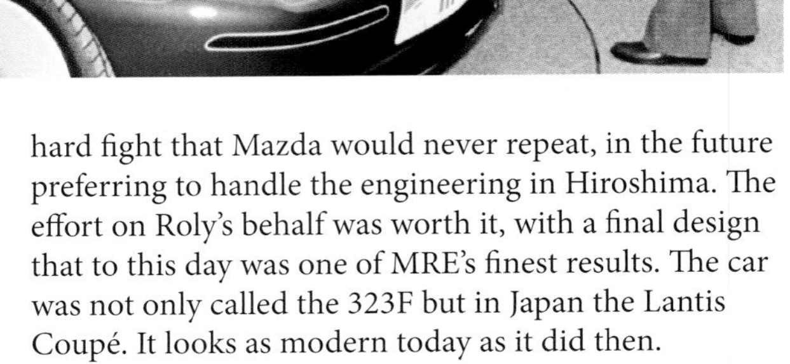

*Mazda chief designer, Shigenori Fukuda (extreme right), reviews the Mazda 323/Lantis model in Oberursel. On the left, in the red shirt, is studio engineer, Dave Samways, next to Roland Sternmann. (Mazda Motor Europe R&D)*

VW. Recruiting fabricators and modellers was proving more difficult, and in the end we had to bite the bullet and for our first jobs recruited at least six contract modellers. They were expensive, but being paid by the hour and many with families at home in the UK, they worked long and hard hours.

Similar to my experience with the Amati 1000, Roly had completed his first production design programme, however, with a much higher degree of success. Apart from developing the Gyssia concept model for the official opening, he had managed, almost single-handedly, the design for Mazda's new 323F. He and our studio engineer, Dave Samways, had not only developed the full-size clay model but included all the feasibility details into the model – no mean feat as far away from Hiroshima as we were. Mazda had sent engineers to Oberursel to assist, for them quite a frightening experience due to Roly's sometimes very persuasive manner when it came to insisting on the details of the design. It was a hard fight that Mazda would never repeat, in the future preferring to handle the engineering in Hiroshima. The effort on Roly's behalf was worth it, with a final design that to this day was one of MRE's finest results. The car was not only called the 323F but in Japan the Lantis Coupé. It looks as modern today as it did then.

With cars such as the Lantis Coupé and Xedos 6, Mazda was building up the reputation as a 'designer brand' within the industry. Behind the scenes, however, things at MRE and Mazda were to change.

# Chapter 30

## *The second birth*

**Oberursel's new general manager, Hiriawa-san, asked Roly and I to join him in the meeting room just outside the main entrance to design. 'Main entrance' is overdoing it a bit, since it was just two large metal doors that opened up into the main corridor separating the modelling studios from the fabrication studios and certainly not anything representative.**

Roly and I sat patiently. What was all this about? And where was Ginger?

My thoughts were soon answered when not only Hiriawa-san but also Ginger walked into the room and took a seat opposite us.

The new general manager didn't waste any time.

"Gentlemen, it is with regret that I have to announce that Ginger will be ending his contract with us."

He motioned towards Ginger.

Not only were Roly and I taken aback but Ginger himself appeared surprised.

He managed to pull himself together and began a somewhat unrehearsed dialogue about it being the right time to leave the company and move on to new ventures, etc.

I wasn't convinced, but there was no time to discuss the matter. Hiriawa-san stood up and made it clear that the meeting was ended. He and Ginger left the room as quickly as they had entered.

I had known Ginger for about twenty years, and before my time at Porsche we had met socially on several occasions, so I felt I knew him quite well, indeed I could call him a friend. I was pretty sure his departure was not entirely of his own doing, but I never figured out the reason why he left. He was a strong manager, maybe too strong for the Japanese, who knows? What is clear: Mazda had a lot to thank him for. He put his heart into setting up the Oberursel studio and it will remain his legacy. Leaving Mazda, Ginger went on to set up the European design organisation for Daewoo and then later set up his own business, Car-Men, which to this day is a very successful automotive benchmarking company. His leaving was Mazda's loss, not his.

Within days of Ginger's departure, Hiriawa-san announced to the team that Roly would fill Ginger's shoes as design manager.

I can't remember if I was disappointed at not getting the job. Roly had certainly made his mark with two

*Ginger Ostle and me seated in one of Oberusel R&D centre's early design studies. An open two-seater, the 242 would remain a concept only.*

*Work in progress in the Oberursel studio on the Premacy minivan clay model. Note the full-size tape drawing on the studio wall. (Mazda Motor Europe R&D)*

*Full-size clay model for the Premacy minivan. (Mazda Motor Corp)*

very successful designs in the 323F and Gyssia – perhaps he'd put pressure on Mazda based on this success. In some ways his appointment was a relief, since there had always been an element of professional tension between him and Ginger. Indeed a lot of the team said, "Thank goodness, now he's got what he wants!"

It must be said that worrying about filling Ginger's position was not the key thing on my mind at the time of his leaving. In the June of 1993 Claudia brought our second child to the world, this time a girl. We named her Charlotte. The circumstances of her birth were exactly the same as with Denys, only this time I drove to the birth practice in a Lancia Delta.

Prior to the birth, the midwife had visited us several times at our new house in Friedrichsdorf. She drove a very clapped out Citroën 2CV. Every time she left (or should say tried to leave), that car just wouldn't start. As we waited patiently outside the birth practice, everything was shrouded in darkness – no midwife in sight. Just our luck, I thought, between Claudia's moaning – that bloody car won't start!

We needn't have worried. A few minutes later the familiar tuck-tuck of the 2CV broke the silence. Four hours later I was at the bakers buying bread rolls for breakfast – a proud father of two.

The failed Amati project, plus Mazda's diverse range of market outlets in its domestic market had haemorrhaged its profitability. In addition to these in-house problems, the Asian economy had crashed. The burden was too heavy for Mazda and something had to change – and fast. Ford had been a faithful partner to the company for many years and it also realised a major rethink of Mazda's business plan was needed. In the May of 1995, with one sweeping move, Ford increased its share in the company to 33.4 per cent and in effect took control of the company.

Soon after this in 1996, a Ford heavyweight named Henry Wallace took control of the helm in Hiroshima as president, joined by an army of engineers, planners and market specialists, all drawn from Ford's huge R&D think-tank. For a Japanese company, a western president was unheard of, but for Mazda there was little alternative. Ford and Mazda knew each other well, and there was a mutual respect between the companies. The alternative – to be swallowed up by one of the large domestic brands

*Illustrations for the Premacy minivan.*

– was not a consideration for Mazda. With Ford, it was very much a case of better the devil you know.

Back in Oberursel, we were pretty oblivious to these going-ons. We kept our heads low, concentrating on our latest project, the replacement for Mazda's European flagship, the 626. In addition to this, we had also been asked to develop ideas for a new segment – a small minivan, later to be called the Premacy. With plenty to do we had little time to think about company politics.

Those thoughts were soon put to the test when one day a delegation of Ford R&D executives rolled into town. With the company's full control of Mazda's business, it was only a matter of time before a representation from Ford would want to see what we were up to.

Roly walked into the main modelling studio where our proposal for the next 626 was in full swing and was quite an advanced design statement.

Roly and his accompanying entourage gathered around the model. He explained the progress of the development, referring to sketches pinned on the surrounding walls, plus a supporting tape drawing with technical package details. The senior Ford executive led the group and his staff looked on, showing little sign of emotion. It occurred to me that they all looked the same. In fact, they could have been injection-moulded. Dressed similarly in dark business suits and all sporting the same side parting haircuts.

Roly finished his presentation and there was silence. After a pause, the senior Ford engineer spoke:

"Have you researched these designs?"

We looked at each other. Research? What was he talking about?

"There is no point discussing this design unless we have some research data."

The meeting ended, and a somewhat perplexed Roly led the group out of the studio.

An hour or so later Roly returned. Apparently, Ford never developed a design without thorough market research. What our friend from Ford had been referring to was the testing of our designs in so-called 'clinics.'

Basically, a design clinic is an event, usually held in a large exhibition hall, where design models, devoid of any badges, are evaluated by invited members of the public, who can view and compare them with existing vehicles from competitor companies. Usually the group of cars are prepared to a similar finish to maintain fairness. This may mean that the window glass is blacked out to avoid views of the interior, and all badges or symbols that could reveal the make of the car are covered or removed.

Participants will fill out an evaluation sheet and some may be invited to attend an interview session with other attendants, where a moderator will lead a discussion about what they have seen.

Yes, the influence of Ford could soon be felt. Not only were research clinics introduced into the development process, but Mazda's European sales organisation, MME, moved from Brussels to a new location in Leverkusen, on the site of the company's German sales organisation, MMD. Situated on the banks of the River Rhine, this was perfect for any interaction with Ford, whose biggest European operation both for production and development were just across the river from the Leverkusen site, in Cologne.

Very soon the Leverkusen offices would be manned with extra staff recruited from Ford, including the heads of almost all the departments. It wasn't long, therefore, before members of the Oberursel team began to get invited to various product planning and marketing meetings where recommendations were discussed on how we should develop our cars.

I may sound negative about this new intrusion but I'm not. For many years Mazda had designed its products based on the gut feeling of its senior designers and to be truthful there was no key strategy behind the process. The company had been lucky with senior design leaders such as Fukuda-san and Tom Matano, and that most of its products were great looking cars. What Ford was missing, however, was a flow in the development or so-called brand strategy. As I mentioned earlier, in Japan, if a replacement model had little relationship to its outgoing version this was not a big issue, but in Europe, the family look in the showroom was a vital ingredient in developing a relationship with the customer.

In short, our new marketing and planning colleagues in Leverkusen felt Mazda needed a more understandable and consistent design message.

While our models continued to develop in the studios, Roly and I began to get involved in strategy meetings with an aim to carving out Mazda's future brand look, and to help our colleagues in advertising with the creation of a new slogan that represented the designs.

The first area we had to tackle was the front-end graphic of the car. We needed a clear graphic statement to run through all models. Taking a look at the outgoing 626 and the Xedos range of cars, we picked up on two key features. The Mazda logo was placed at the top centre of the grille and the shape of the grille expressed a five-pointed opening. This, we decided, could form the basis for the grille design of all our mainstream models, with only the sports cars (MX5 and RX7), for the time being, keeping their grille-less noses.

Everybody agreed, including our colleagues in Hiroshima. The next task was to find a slogan that could signal Mazda's new design language. For the team in Hiroshima, establishing a key slogan to follow the Tokimeki era was more important than a consistent family face adorning the front of all models.

Since the evolution of the so-called 'bio-look' in global car design things had moved on, and a more architectural style was beginning to evolve. Flatter body surfaces with sharp creases started to reappear. The 626 we were developing had smooth body surfaces that featured hard feature lines giving the impression of a bone pushing through the outer skin surface.

Seated in one of the Oberursel meeting rooms with our marketing colleagues, we covered whiteboards with a series of slogans that expressed this new body language. Then it came: contrast in harmony. It made sense. The hard feature lines that appeared to push their way through the smooth body shape expressed a contrast in the overall surface.

The slogan was sent to Hiroshima and a message soon came back approving it.

We now had a tool to instruct our global design teams on the visual message of our cars: they should feature a prominent five-pointed front grille with a high-mounted Mazda symbol, and the body surface language should combine a contrast of harmonious surfaces broken up by sharp feature lines. It made sense, but what if the looming clinic research didn't like it?

Oberursel was not the only part of the Mazda Design community to witness a change in leadership. Rumblings from Hiroshima soon became solid reality: Mazda's well-liked and admired design chief, Fukuda-san, had also packed his bags. As with Ginger, the circumstances were also very confused. Rumours said that the cars designed under his leadership, although good looking, were deemed too advanced for the mainstream customer, and that a more conservative approach was needed. The announcement of his replacement was even more surprising. Takaharu 'Koby' Kobayakawa, a legend in the halls of Mazda's R&D department and chief engineer of the latest RX7, took over from Fukuda-san. Kobayakawa-san was internationally renowned, but as an engineer, not a designer. Within the design department, the news was met with surprise and mixed feelings. It wasn't long before a press release came out in which the man himself expressed his own surprise at being asked to take over design, asking for understanding and guidance from his team. It was a humble and modest statement, indeed very Japanese.

With Mazda Design being slapped on the wrist for creating too adventurous designs, it was clear that senior management wanted someone in place who would do as he was told.

Mazda's instructions to become more mainstream soon began to take effect. The first models for the new 626 were clinic tested and shot down by the attendants. Unlike Ford, who was very skilled at assessing the analysis of clinic results, this was 'new land' for Mazda and they fell into a trap.

When looking at the opinions of the participants of research clinics, as a rule of thumb they will most likely prefer the vehicle that they drove to the event in. When confronted with something new, they are often unsure or negative in their opinion. If they rate something badly that doesn't necessarily mean the object in question is at fault. The Mazda 626 models were maybe too advanced at the time of the clinic, but for the planned production date two years later they would have been right up to date. Regretfully, the negative results drove Mazda to rethink and quickly tone down the designs. The next time round, the outcome of the clinic was good but the chosen designs were too safe, and the result was the car that came onto the market in 1997, later to be named by the press as the Henry Wallace 626.

In all fairness to Henry Wallace, he was not at fault, but a victim to a combination of unfortunate circumstances. Mazda's new design chief, although a wonderful person, was not a design leader. Given a new power, the planners and marketing experts took the clinic results as they stood and the outcome was mediocrity.

Roly, as frustrated with the new design process as everybody, flew to Japan in a last-ditch effort to try and breathe some energy into the design, but the die was cast.

The 626 was just the start of a more conservative range of cars, and in 1998 the 323 replacement continued this trend. Could these cars really have come from the same company that brought us the Lantis Coupé and RX7? Everyone in the car design world began to scratch their heads.

Fortunately, amid this despair, a saviour was on the horizon: his name was Martin Leach.

*The Oberursel model for the 626. A good car but, as a result of clinic research, quite conservative. (Mazda Motor Europe R&D)*

# Chapter 31

## *Zoom-zoom*

**When Henry Wallace took the helm in Hiroshima in 1996, he brought with him a man with a proven track record in Ford R&D named Martin Leach. An Englishman, under his leadership Ford's global light truck sales in North America had doubled, and light trucks in the US were an automotive company's life blood. With this success on his shoulders, Ford wanted Leach to bring his magic to Mazda.**

He may have achieved his success in trucks, but Martin Leach was an out-and-out petrolhead. Not only that, in his younger years he had been a very successful go-kart racer, even beating the likes of Ayrton Senna and Martin Brundle, no less. Regretfully, his motorsport career was cut short due to illness, but motorsport's loss was automotive R&D's gain.

The moment Martin Leach walked through the doors at Mazda he knew things had to change, and for him, the company in Hiroshima meant one thing, 'dynamic design' – hooray!

With Martin Leach at the helm of Mazda's R&D department, a breath of fresh air ran through the design organisation. In Hiroshima, Kobayakawa-san was thanked for his leadership in troubled times and retired to become a brand ambassador for the company. The overall responsibility for design leadership now fell upon Kawaoka-san, Hayashi-san in Japan, and Tom Matano, with Roly, of course, having a say.

It wasn't long before the influence of Martin Leach could be felt. In Japan, a replacement for the RX7 was being planned. 'Replacement' is not really the correct word since the vehicle being considered was more of a sister model to the RX7. Plans were already in place for an all new MX5. The first MX5, named the NA version, and its follower, the NB version, were both based on the same platform. For the new model, a new platform was being developed, necessary to accommodate the latest safety requirements. Developing a new platform was very costly, and unless you could use it for a number of vehicles, it didn't make business sense. Mazda wasn't going to give up the MX5, so a solution needed to be found to justify the cost of an all-new platform. Martin Leach got together with his planners and engineers and came up with the idea for a compact four-seater sports coupé powered not by the conventional MX5 four-cylinder engine, but a new normally-aspirated twin rotor Wankel engine. The rotary engine in the RX7 was very powerful, equipped as it was with twin-turbochargers and intercoolers, but it was too big for the MX5 chassis and, apart from that, it drank fuel like a fish.

In complete secrecy, work began in Japan on a prototype model to be shown at the 1999 Tokyo Motor Show to demonstrate Leach's vision for Mazda. Unbeknown to us, Mazda's US design centre in Irvine was also given a chance to work on the design, as the car was aimed squarely at the US market.

Information of Mazda's new sports car began to trickle through the design organisation. It was certainly a unique vehicle. A coupé, it featured four doors, but was minus a central pillar. One of the disadvantages of a normal two-door coupé is that access to the rear seats is not always easy, especially in confined parking spaces where a coupé's long door has limitations as to how far it can be opened. Mazda engineers figured that if, instead of one large door, you had two doors that opened opposite each other like two garden gates, eliminating a conventional middle pillar, you could achieve good access to the rear compartment and create a look that still gave the car a coupé feel. This style of door construction wasn't new, and Martin Leach, having come from Ford's light truck division, would have been familiar with it, being commonly used for access on pick-up trucks that featured a second seat row.

In Tokyo, in October of 1999, in front of an amazed

audience, Martin Leach proudly pulled the covers off the RX Evolv sports coupé. There was no doubt for all who witnessed the reveal, that Mazda was back on the design stage. All agreed – Mazda had to build this car, and it did. It went on to become the Mazda RX8.

With Ford having increased its ownership of Mazda and now being in a controlling position as far as company strategies and policies were concerned, both companies started looking at ways to maximise their interests. Joint programmes between the two companies were nothing new, with several Mazda vehicles being sold under the Ford name on the Japanese market. Indeed, Ford had its own design studio located on one of the floors of the Hiroshima R&D centre, to handle such programmes. The 626 model had been on the market for over two years and a replacement design programme needed to be started. The model had always been Mazda's big money earner in Europe, and in Germany it had been a huge sales success. Particularly, the five-door hatchback had caught many of the established brands 'sleeping,' with Mazda creating quite a niche for itself in this segment. By the late '90s the bubble had burst, the present 626 being a shadow of earlier sales successes. For its replacement, Mazda, together with Ford, decided on a twin-forked approach.

For Japan and all markets outside North America, Hiroshima developed a car, while for the North American market, which required a larger vehicle to compete with the Toyota Camry, Ford developed a model to be built in its Flat Rock, Michigan assembly plant. As underpinnings for the North American model, Ford used its Mondeo platform, the development team of which was based in Cologne, Germany.

*Iowa Koizumi. (Mazda Motor Corp)*

If this sounds complicated, to make matters worse, the model to be developed by Ford in Cologne needed to look the same as the Hiroshima design, despite all components being essentially different.

The chief designer for the complete programme was Iowa Koizumi, of Xedos 6 fame, with the onsite Ford/Mazda Cologne team led by Shimazu-san, who had helped set up the Oberursel facility. Finally, as I would soon learn, Mazda requested that I also join the team in Cologne, reasoning that working on a production programme close to home would be a valuable experience for me.

On January 2, 1999 I packed my car and headed off to the Ford R&D centre in Cologne.

My arrival and welcome at the Ford development centre couldn't have been more lukewarm. I sat in the front security office.

"Mazda?" they asked, with a degree of suspicion.

Finally, a stout-looking German wearing glasses and with thick, dark curly hair walked into the office and over to me. His name was Joerg Schaeffer. He introduced himself and explained he was the Ford design studio manager. I signed in and followed him towards the entrance of the Ford design building. On the way, he explained that I was expected, however, that day, being January 2, almost all the staff were still on vacation.

The Ford design studio in Cologne was, in its layout, very similar to what I remembered from the now defunct Vauxhall facility in Luton, where I had started my career in 1973 – was it so long ago?

Essentially a large hanger of a building, the main clay modelling studios looked out onto a large outside viewing yard. As with Vauxhall, there was a central service corridor separating the fabrication workshops from the clay studios and between the manager offices at the front of the building and the studios was a vast presentation hall.

Joerg guided me along the side of the modelling studio. Covered models were placed on central measuring plates; it all looked very familiar.

We reached a group of glass cubicles, all fitted with desks and, new to me, personal computers. At

Oberursel, the design assistant had a computer or PC for writing reports etc, but so far none of the staff had their own. For graphic work on the quite new Photoshop software, we had a couple of Macintosh computers, but that was as far as it went.

Apparently, Ford had a way of communicating electronically via computer with a system called PROFS. I'd have to pick that up.

Not only was Shimazu-san joining the programme from Hiroshima, two clay modellers and two designers would accompany him. From the Irvine studio in California, Mark Jordan completed the team. Since the vehicle we were designing was for the US market, it made sense to have Mark along for the ride. I knew Mark pretty well, and working with him was good fun.

Having viewed my working space, Joerg took me down to the department's storage area. Freshly arrived from Hiroshima were the armatures we would use to develop both the exterior design and interior. I noticed that they had sent two interior armatures. Were they expecting two interior proposals?

Apparently, the full team was arriving the next Monday, so there was little point in me sticking around. I thanked Joerg and drove back to Frankfurt.

There will always be occasions in life when you say, "I could write an entire book about that." Well, the programme that unfolded in Cologne was one such occasion. And the title of the book? It would have been *Too many cooks spoil the broth.*

Ford and Mazda's wish to combine their R&D powers was admirable and had worked well in the past, but what unfolded in Cologne would prove to be very complex.

Returning the following Monday, Shimazu-san and the rest of the Mazda team were in situ and ready to go. I had a good bond with him, based on our joint experiences setting up the Oberursel facility. What could go wrong?

Commencing design work, it transpired that the Cologne-based Mazda team, under Shimazu-san, were essentially modifying the Hiroshima 626 design development to fit onto the underpinnings of the Ford platform. This was not so easy since virtually all the dimensions needed to be changed. In addition to this, since Ford was essentially running the programme, all incoming designs had to adhere to its in-house product development system, known as FPDS. The result was that the stream of details requiring adjustment being sent from Hiroshima was endless. They included such things as the height of air vents on the instrument panel and its impact on airflow to the rear passengers between the front seats. Ford engineering would not agree to the position of the air vents on the Mazda Design proposal. Or on the exterior, the height of the boot or trunk lid did not comply with Ford's rear vision requirements. And so, on it went.

A further issue was that in trying to get model work completed in the Ford design studio, we were confronted with endless barriers. None of the workshops had time for us. There was strict demarcation. This meant that our clay modellers were not allowed to use tools in the woodshop. It saddens me to say, but we were working in a hostile environment.

Things got to the point where I asked for a meeting with the entire Ford design organisation in its big presentation hall to introduce the Mazda team and explain what we were doing.

In front of the combined team, which must have numbered well over 100, I spoke in German, thanking them for their understanding and expressing our thanks for letting Mazda use the facility. I said that as part of the Ford family, everybody would benefit from the programme running well and therefore asked for their full support.

Things did improve, but the process was still hindered by unexpected hurdles. By the spring of 1999 we had managed to finish two interior models, which were prepared for clinic evaluation in America.

I flew to Atlanta to take part in the event.

The trip was interesting on many fronts, since Atlanta was quite different to other parts of the States I had visited. The clinic location was some way from the city centre at a typical trade fair site. Mazda had booked us into a Marriot hotel next to a main freeway that was stuck in the middle of nowhere. I had flown from Frankfurt and was the first of the team to arrive.

It was hot, but fortunately the hotel had an inner courtyard with a pool. I checked in and decided a quick dip would freshen me up after the flight. Here, I made my first cultural mistake. I had only been in the pool for a couple of minutes when a hotel staff member signalled to me from the side of the pool.

"Sorry sir, but your swimwear is offending the guests."

If I had been wearing a G-string, I could have understood, but apparently what was referred to as a 'European-cut' in bathing trunks, was also deemed too revealing. There was no choice, I had to leave the pool.

By the evening, the rest of the Cologne programme team had arrived, plus planning and marketing representatives from Hiroshima. We decided to go for a Chinese meal at a restaurant across the freeway from the hotel. Although a stone's throw from the restaurant, the only way to reach it was by car – in America people don't walk.

We pulled up outside the restaurant. The big sign outside read: "All you can eat for $10!"

Our group was guided to a large table. Looking around it was quite clear that all the people in the area who like to frequent restaurants that offer 'All you can eat for $10' were there. And boy were they eating!

The restaurant offered a help-yourself buffet that was being constantly re-stocked by Chinese staff with buckets of food; they appeared to be having trouble keeping up with the pace that the guests cleared the trays of food. I looked on in amazement as plates were filled with spider-crab legs, towering to the size of a small construction site.

Parked outside the entrance of the research clinic was a Ford Crown Victoria police car. Walking in, we were met by the event organiser who explained they had asked the local police to handle the security. We walked past the responsible officer, seated at a desk. I mentioned to him that he had forgotten to turn off the engine of his patrol car, which I had noticed was idling away outside. He didn't reply but just glanced at me as if I was mad. Of course, he didn't want to get back into a hot car. It remained idling for the rest of the day.

As with all clinics, the comments from the attendants were mixed and often amusing. For the group interviews we were able to observe the discussion in a separate room via a camera link.

One participant, a striking black woman with purple finger nails about two inches long, apparently liked the design.

"Man, that is a rocket ship!" she enthused. Things sounded positive.

During the proceedings I drove down to the local shopping mall and purchased some hamburgers and beers for lunch. Again, my lack of local knowledge hit back at me. Returning to the venue, carrying a couple of six-packs, I gained the attention of my police officer friend from earlier.

Seated back in the observation room the event organiser rushed in, white in the face.

"Guys, have you got beer?"

We had.

"Get rid of it! The officer outside says he will take you to the courthouse if you don't dispose of it."

The remainder of the trip proved uneventful, and by the time I had returned to Cologne the initial clinic results were known. They weren't good, and this despite the positive remarks from the lady with the purple finger nails. Too complicated, too advanced, the list went on.

From Atlanta, the clinic models had been sent to Hiroshima. In Cologne we assembled in the main video conference room, where a direct link had been set up with the development team in Hiroshima. Martin Leach led the meeting – it wasn't nice. His main bone of contention was that the level of finish on the models was below Mazda's expectations, and, as such, he reasoned that a good clinic result was out of the question. The models needed to be rebuilt and a new clinic would have to be organised.

Putting the interior development aside for one moment, the exterior design was also not without its issues. Hiroshima design had adjusted the digital data of the European 626 design they were working on to fit the US model dimensions. In Cologne, we had the data converted into a milled physical model, which we had dynoced and taken outside for viewing. The basis was there, but it was quickly agreed that the model needed a comprehensive rework. That would have to wait, and with the interior still taking up most of our time, the model was covered and pulled back into the studio for a later date.

There wouldn't be a later date. Unbeknown to us, Ford's design supremo, J Mays, was getting involved.

Since we had worked together at Audi, J Mays had had a fairly meteoric rise to fame in the industry. Leaving Audi with Martin Smith and Graeme Thorpe, they had enjoyed a bumpy ride at BMW before Martin Smith approached Ferdinand Piëch and got the backing to set up an advanced design studio for Audi in the fashionable Schwabing area of Munich. Martin Smith took along J Mays and Freeman Thomas, with whom I had worked at Porsche, and they created such cars as the Audi Avus concept. From there, Mays moved on to help Audi set up a design studio in California, taking Freeman Thomas with him. They went on to develop the first ideas for the new Beetle and Audi TT. When Hartmut Warkuss moved to Wolfsburg to oversee the design of the entire VW group, Mays was called back to Audi to fill his shoes.

Back where he started his carrier at Audi in Ingolstadt, J didn't keep the position long, returning, to the surprise of many, to the States to take up a position at a marketing strategy company in Phoenix, Arizona of all places.

When Ford's long-serving design supremo, Jack Telnack and Jacque Nasser began to look around for a successor for Telnack, it wasn't long before they stumbled upon J.

Whether a direction had come from Martin Leach or even higher up the ladder, instructions had been given for an alternative exterior design proposal to be made.

As it happens, there was indeed an interest from high up, and that was from none less than Ford's CEO, Jacque Nasser. Nasser was a Ford career executive who had made his way to the top by shear enthusiasm and hard work. He was passionate about the company and every aspect of Ford's business, which included design. It was Nasser who had hired J Mays and there was no doubt they saw eye to eye. Nasser was proud of his close relationships with the workers in Ford, and once a week he wrote a personal message to the entire Ford team that was sent out on Monday mornings via the new electronic mailing system or e-mail as it became known. It was very clear that he was being closely informed about the Mazda project running in Ford's European R&D centre, as one morning we all received a message of encouragement from Nasser. It was full of the usual 'we're one team' enthusiasm and at the foot of the note he signed off with: "Come on Mazda, let's get more zoom-zoom into this programme."

*Zoom-zoom*, I thought. *What's that all about?*

It wasn't long, however, before all the correspondence from the Ford employees involved in the programme began using the zoom-zoom slogan at the end of their messages. I thought nothing of it. Little did I know ...

Shimazu-san and I got a call from the Ford programme manager in Cologne, an Englishman named Richard Tilley. Tilly was responsible for coordinating the Ford-led Mazda project. Seated in his office, he explained that J Mays had given instructions for an alternative exterior model to be made at the TWR organisation in the UK, run by Tom Walkinshaw. Walkinshaw had made his name famous as a racing driver and was linked to Mazda having come second in the 1979 British saloon car championship driving an RX7, and in 1981 winning the famous Spa 24-hour race in the same car. His business interests now extended beyond the field of motorsport having ventured into automotive design development. The design offices were located near Chipping Norton in the Cotswolds, where they worked on a variety of projects for both Aston-Martin and Jaguar, both of whom were owned by Ford.

Flying to Birmingham, Shimazu-san and I took a rental car and drove to the location where J Mays was waiting with Tom Walkinshaw himself to greet us.

Introductions made and wasting no time, we went to the studio where the new design was located. J introduced us to Michael Arni, an independent designer who had been handling the project. Also in the room was TWR's chief designer Ian Callum. I had known Ian for quite a few years; both he and his brother Moray were well liked and respected in the automotive design community. More pleasantries were exchanged before we all focused our attention on the model that was shrouded under a cover which J then pulled back.

My first impression was good. The design was quite chiselled, with clean surfaces and sharp feature lines. Most striking of all was the design and attitude of the front of the car. From the base of the front pillars (A pillars), two large feature lines ran diagonally down the bonnet, aimed at an invisible point in front of the car, to border each side of the now typical Mazda 5-point grille opening. Also of note was the rear end profile which was one clean surface broken up only by the rear lights. My only issue was with the side window opening (DLO), being, in my mind, too close to BMW, featuring as it did a typical BMW bend in the C pillar. That aside, overall, it was a very modern and contemporary looking car.

There were smiles all round.

"I want this car," J declared triumphantly. It was clear Shimazu-san felt himself being pushed into a corner. As was the nature of all the Japanese he was polite and courteous, but I'm sure inside he was worried. The drive back to Birmingham airport was done in silence. I knew 'Shimmy' was not happy.

The zoom-zoom slogan didn't go away, and soon the brainchild of Jacque Nasser began to appear on communications from Hiroshima and eventually flowed into the new advertising campaign. To this day, I'm not sure if my colleagues in Japan really understood what it meant.

If zoom-zoom stood for dynamics and movement, the progress of the programme in Cologne couldn't have been further from it, and things eventually ground to an undistinguished halt.

The J Mays model was shipped to Cologne but no progress was made. When the feasibility engineers got their hands on it, it was discovered to have a series of issues that made it difficult to realise for production. Shimazu-san could breathe a sigh of relief!

In the end, it was said that the market equation for the programme didn't add up and therefore it was unceremoniously cancelled – always a good explanation.

I departed Cologne at the beginning of 2000. It was a good time to go.

In Hiroshima, Mazda picked up the pieces and developed the full programme to the end. The resulting design that Koizumi-san's team created went on to rebuild Mazda's reputation as a design leading brand.

The car in question wasn't named 626; the company wanted to send out a new message under the zoom-zoom slogan. It was named simply Mazda6.

*The first Mazda6; the start of zoom-zoom. (Mazda Motor Corp)*

# Chapter 32

## *The Italian job*

**It was just prior to the conclusion of my time in Cologne when I got a call from Luigi Oya. Oya-san was a senior designer who had been posted to Oberursel to run the interior design department under Roly. His message was short. Roly had handed in his notice and Martin Leach would visit the studio to announce the new design manager. As it turned out, it was me.**

Roly was a passionate designer and his no-nonsense manner was not to everybody's liking. It was also rumoured that he was frustrated by the design process that had evolved under the Ford leadership. In the end, he took up a new position in the Volkswagen Group, running a studio in California. He would be missed at MRE for many reasons. He was an outstanding designer and gifted artist, whose work I have always admired.

The last of the 'Three Musketeers' that had set up Mazda's European design studio, I returned to Oberursel in the spring of 2000, taking my seat in the design manager's office. I had decided to keep the title chief designer. It said what I did, unlike design manager, which I always felt was a bit anonymous.

In Hiroshima, the leadership of design was still somewhat undefined. Kawaoka-san appeared to be in charge but Tom Matano, although located in America, exercised influence, and since Mazda was now for all intents and purposes a division of Ford, J Mays was also lurking in the shadows. J's wish to inject zoom-zoom energy into Mazda's products was all too clear, as demonstrated by the model he had made at TWR. His involvement, as I would soon find out, was still very much in the mix.

With the Mazda6 programme completed, the Oberursel team began to consider the replacement for the Mazda 323. Since the 626 replacement had been named Mazda6, it was assumed that the 323 would be called the Mazda3, which, as it turned out, was the case.

Chief designer for the Mazda3 programme was Hideki Suzuki, another Mazda Design veteran I had worked with in the past – he was a lovely guy. He flew to Frankfurt and briefed us on the programme, which comprised a sedan and five-door hatchback. Since the hatchback was the main seller in Europe, he asked us to develop models for that version.

One of Oberursel's young designers, Hasip Girgin, began to sketch ideas. Of German/Turkish nationality, and a great talent, his unique style of sketching was much admired in Hiroshima. As a strategy, we decided to develop a coupé-like design for one direction and a more practical version with an upright C pillar to compete head on with the VW Golf.

With the design programme going well, out of the blue, I got a call from Tom Matano asking me to join him in Turin. Then the bomb dropped. The famous Italian design house Pininfarina had been commissioned to design proposals for the Mazda3. I had no idea who had commissioned Pininfarina, maybe J Mays, since I couldn't imagine that Hiroshima design would have wanted to pay for its services. Getting outside help on design programmes was nothing new, but nevertheless, this new development came as a surprise.

Tom and I were to meet with Mazda marketing staff at the Pininfarina site outside of Turin and brief them on Mazda's design philosophy.

I took the first flight to Turin and drove a rental car to Pininfarina's design office in Cambiano, to the south east of the city. On the way, memories of my trip to Turin to visit Pininfarina as a student with Martin Smith came to mind – such a lot had happened since then.

Making myself known at the security office, Pininfarina's sales manager met me, and I was taken to an office where Tom was waiting to greet me. He was accompanied by a marketing representative from Hiroshima, an American who I assumed had joined Mazda from Ford.

*Hasip Girgin's Mazda3 illustrations. (Mazda Motor Europe R&D)*

I noticed some storyboards leaning up against the wall and went over to see what was on them. They were basically fantasy adverts for the future Mazda3. I started to read: "Imagine your next Mazda designed by Pininfarina ..."

The Hiroshima marketing man came over to me.

"Isn't it great? A Mazda designed by the Italians!"

He was obviously oblivious to my position as a senior Mazda designer and totally in love with the idea of a 'Latin' Mazda.

*How could this happen?* I thought.

Our meeting with Pininfarina was cordial, Tom playing the grand statesman of Mazda Design. Pininfarina's general manager, Lorenzo Ramaciotti, and chief designer, Ken Okuyama, were present. Tom briefed them about the specification of the future Mazda3, technical drawings were handed over and the meeting was quickly concluded.

With an illustrious design history, Pininfarina was sure to do a good job and I left Turin slightly concerned.

Unperturbed, we carried on with our model development at Oberursel. The wind was against us, but I knew in my heart that we could put up a good fight against the Italian competition.

It had been agreed that Pininfarina would present scale models at the Oberursel site on a date coinciding with Mazda's senior management visit to the 2001 Geneva Motor Show. We'd be ready for them.

By this time, there had been changes in Hiroshima. The new company president was a dynamic American called Mark Fields. Fields

*The Oberursel scale model for the Mazda3 five-door hatchback. (Mazda Motor Corp)*

had made his way up the Ford ladder through a series of marketing jobs and got the top job at Mazda at the young age of 38. In addition to him, Martin Leach had been replaced by another American, Phil Martins, as head of R&D.

It was these gentlemen, plus an entourage of marketing and planning experts, who descended on Oberursel the Sunday before the Geneva show. Not only would there be representatives from Hiroshima, but also most of the European sales organisations would be present. It was going to be a big show.

Ken Okuyama arrived in Oberursel the day before the presentation with Pininfarina's design proposals. We had freed up a meeting room for him to present Pininfarina's proposals, which comprised three scale models plus artwork. The room was situated away from our main presentation hall; we didn't want them seeing our ideas.

Just like us, Pininfarina was quite protective about its proposals. I instructed our modellers to help them set up the presentation. I entered the room briefly in the pretence of saying hello to Ken Okuyama. Quickly taking in the designs, I breathed a sigh of relief, realising instantly we could beat this competition.

Mazda representatives from all over the globe descended on Mazda's European R&D centre that Sunday morning. It was a day I will never forget.

Suzuki-san prepared his own presentations in the main viewing hall. Like me, he was

*The Mazda3 sedan proposal. (Mazda Motor Corp)*

*Me with the Mazda3 chief designer, Hideki Suzuki.*

*Mazda3 five-door hatch. (Mazda Motor Corp)*

irritated that Pininfarina had been asked to submit designs for 'his' project. What made things even worse was that many in the restructured sales and marketing organisations within Mazda believed that a Pininfarina badge on the side of our cars was the route to success, regardless of whether it was a good design or not. Slowly, the hall filled, only Mark Fields and Phil Martens were not there. Flying in directly from Japan, we waited patiently for their arrival.

News filtered through that Fields and his team had arrived and were making their way to the viewing hall. Entering the room, Mark Fields didn't waste any time.

"Where are the Pininfarina models?"

They made their way to where we had set up the Italian team, only the most senior managers plus Suzuki-san in company.

A while passed. Then I noticed Ken Okuyama making his way to the reception and a waiting taxi. His presentation had obviously ended.

Mark Fields gave instructions for the Italian designs to be brought into the viewing hall and placed near the  Oberursel proposals. Now was my chance. Stepping forward I explained the strategy behind our two designs. They had a choice, the sporty coupé-like direction, or go for the more radical direction, with its solid lean-back rear pillar. Then I stuck a dagger in the Pininfarina models. I let my emotions pour out.

"It's none of my business what you paid for this work but I'm telling you it was a wasted investment. There are only two designs in this room that are modern contemporary designs and that's these two." I pointed to our designs. "What's more, here you have a design not only for the hatch but also for the sedan."

Suzuki-san and I had already decided that the coupé-like direction would make a good sedan, leaving

*Mazda3 sedan. (Mazda Motor Corp)*

the other model as the proposal for the hatchback.

I had never been so sure about those two models, and I let them know. I finished my pitch. Had I overstepped my mark?

After a short pause, Mark Fields spoke.

"He's right. How much did we pay for these models?" He indicated the Pininfarina models.

From the back of the room somebody muttered a sum. It was a lot of money.

"That's the last time we use their services."

The meeting was concluded. Suzuki-san came over to me.

"Well done, Birty-san."

Looking back on that day, my negative comments directed at the Pininfarina designs were totally justified. It was a great design house, but what had been presented that day was not its best work, to say the least.

Our two models were sent to Hiroshima and Hasip Girgin joined the team in Japan to complete the designs for production. Suzuki-san's idea to use both designs for sedan and hatch were approved and the first Mazda3 launched in 2003 went on to become Mazda's biggest-selling model.

For me, that Sunday in March 2001 was one of my proudest moments.

# Chapter 33

## *Moray Callum*

**They say you always remember what you were doing the day of 9/11. I walked into the Mazda's press lounge at the Frankfurt Motor Show. It was packed, everybody glued to the large TV screen, the drama in New York unfolding in front of our eyes. There was a gasp of disbelief as a second plane crashed into the World Trade Centre. Judging by the dignitaries gathered around, Mazda must have had one of the only TVs at the show broadcasting live news. Niki Lauda and Wolfgang Reitzle of BMW fame, now with Ford's Premier Automotive Group, looked on solemnly.**

Before this world changing event, the week had started well. Visiting executives from Hiroshima were attending strategy meetings at Oberursel. In the centre's conference room, I was presenting the project to the guests, when I noticed a familiar face quietly making his way in, not wishing to disturb things. Mazda's R&D chief, Phil Martens, sprang to his feet and greeted the new participant. The person in question was Moray Callum, whom I had known for quite a few years. Moray was now working for Ford in Detroit; amongst other things, he had been responsible for the new Ford Thunderbird. It was clear, the visit to Oberursel was not a social visit. Soon, the news began to filter through, Moray Callum was the new head of Mazda Design.

For my Japanese colleagues in Hiroshima the announcement that a non-Japanese would head up design must have come as quite a shock. They needn't have worried. Moray was an outstanding designer and as far as I knew had a very solid and even temperament. In addition to Moray, Tom Matano would move to Hiroshima from Irvine, to act as Moray's right-hand man.

Tom never admitted it, but I'm pretty sure he was not happy about this new promotion. Ex-Honda US designer Trueman Pollard would take over the reins in California.

For me, the thought of reporting to a British national in Japan was all good and I looked forward to what his leadership would bring.

Moray's transition into his new role at Mazda went very smoothly. For him to join the company at a time when things were really on the up was a great opportunity. Upon his arrival, the design development for the new Mazda3, a Premacy replacement, the Mazda5, and the exciting RX8 were either finished or well under way, with no realistic chance for him to influence the designs. The first car with which he had the opportunity to make a big statement as Mazda's new design director couldn't have been better for him: the next MX5.

For Mazda, the MX5 had been a massive success. Originally intended primarily for the North American market, it had gone on to be unchallenged in the small roadster market. Despite the admirable efforts of Fiat with the Barchetta and Rover-MG's TF, nothing could touch its combination of affordable driving fun.

*MX5 chief designer, Yasushi 'TJ' Nakamuta. (Denis Meunier)*

The first model and its successor were named NA and NB respectively, and were based on the same platform. For the NB model, the move from pop-up headlamps to the fixed variety had come in for quite a bit of criticism but was necessary in order to comply with the new pedestrian safety regulations. In addition to this, the pop-up lamps added more weight to the front of the car.

The next generation, or NC version, represented a clean sheet of paper. Since it was being based off similar underpinnings to the forthcoming RX8 coupé, the design would be all-new.

The MX5 was the crown jewel of design projects within Mazda and it was only natural that all the company design centres would want to participate in this competition.

Chief designer for the MX5 programme was Yasushi 'TJ' Nakamuta. 'TJ' stood for 'typical Japanese,' a nickname he had acquired while working at the Mazda studio, MRA, in California. Nakamuta-san was a high flyer within Mazda Design and, alongside Ikuo Maeda, chief designer for the RX8, was one of the in-house designers who would no doubt get the top position one day.

Nakamuta-san wasted no time in organising a global MX5 meeting in Japan, in order to get the design teams together and decide on a strategy for the next model. I therefore flew to Japan, and after a week of intense discussions, returned to Oberursel with instructions to develop three initial scale models for the next model. Our

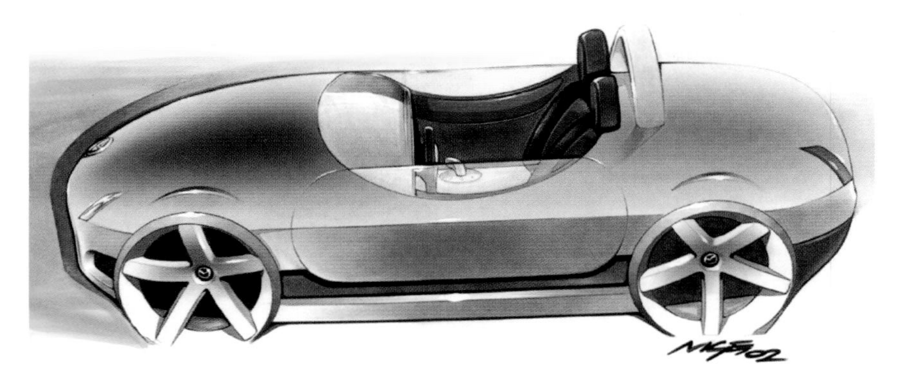

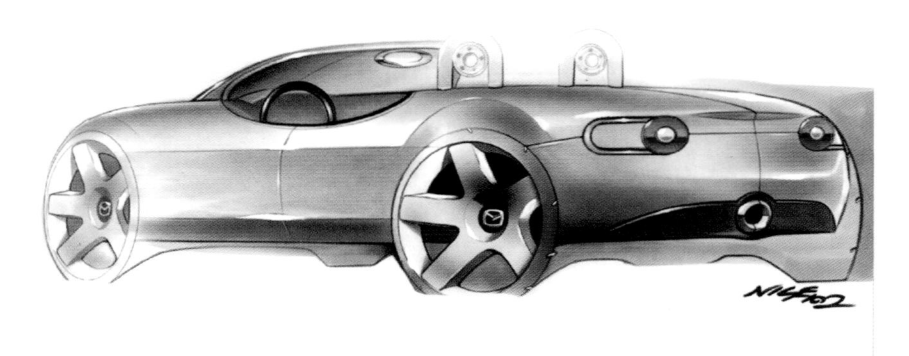

*First sketches for the MX5 from Nigel Ratcliffe and Hasip Girgin. (Mazda Motor Corp)*

One of the first models for the NC MX5. (Mazda Motor Corp)

*The second attempt. This time the ideas and model from Nigel Ratcliffe were selected to go forward. (Mazda Motor Corp)*

designers commenced immediately with ideas for our first shot at this prestigious project.

In line with what had been discussed in Japan, rather than developing a generic direction close to the previous model, we decided that the team should try to spread their ideas.

By now, the marketing team from Mazda Motor Europe in Leverkusen was involved and keen to get feedback from present MX5 owners as to how the next generation model should look. To this end, a small group of MX5 owners was invited to Oberursel to review our first designs and give feedback.

Held over a weekend, the MX5 owners arrived at the centre, including a Belgian lady whose love for her MX5 was apparently her main purpose for living. She was obviously in a good place financially, since she traveled with her car to MX5 fan meetings all over the globe. Opening the bonnet of her car, she proudly showed us the signature of Tom Matano, resplendent on the inner panel. Tom was recognised as the father of the MX5 and gladly put his signature to the metal of an example when asked. Now famous in MX5 circles, the message was always the same: "Always inspired, Tom Matano."

The meeting that day was an animated event with plenty of discussion. The grand finale came when I revealed our now completed models. The feedback was polite – we had fed them well – but not what we were hoping for: "Why change it?" "The present model is better," etc.

The reaction was not entirely unexpected and mirrored the results of most research clinics. On the positive side, they liked the designs. For them, however, they weren't MX5s.

Was this back to the drawing board? The forthcoming model review in Hiroshima would reveal all.

*Digital development for the MX5 interior. The first time we used data to create a design. (Mazda Motor Corp)*

***

The global design review for the all-new MX5 would be Moray Callum's first chance to send out a clear signal of the direction he wanted to take Mazda Design in.

We assembled in the main viewing hall. Mounted on mobile plinths were nine quarter-scale models to review.

As chief designer for the project, it was once more up to TJ Nakamuta to present the models and supporting artwork.

Three months had passed since my trip to Japan and the models on display reflected the outcome of those talks.

*Masashi Nakayama, RX8 and MX5 interior designer. (Denis Meunier)*

The Hiroshima team had gone ahead and developed three cars, two of which could have come (in my opinion) out of a *Wacky Races* comic. The final one, however, was very clean and showed promise. Irvine had, like us, decided to throw away the old book of MX5 design and go down a fresh new route. The US designs were modern and, as demanded by their market, very powerful and muscular. Finally, the three models from Oberursel, as predicted, fell very much between the home team's efforts and those of the US studio.

The discussions began. Moray played his cards very close to his chest when reviewing each design, and though he was complementary, he couldn't hide the fact that there wasn't anything there yet that had met his expectations.

Putting it the best he could, he explained that he wanted a car an enthusiast would treasure and take time each day to look at in his garage, even when he wasn't driving it. If pushed to make a decision right then, he felt the cleaner-looking approach of the third Hiroshima design and that of Oberursel's model from one of our young designers, Nigel Ratcliffe, had potential to go further. He did, however, allow Irvine the chance to create a new model in order to keep the teams motivated.

By now, Mazda had launched its new mid-sized car – the Mazda6 – and with it the zoom-zoom slogan. At launch events, dancing samba girls performed to

*The full-size clay model development at Oberursel of the NC MX5. (Mazda Motor Europe R&D, Mazda Motor Corp)*

somewhat surprised (or were they alarmed?) members of the press, to a specially composed zoom-zoom signature tune. Opinions were divided, but when your children's friends start singing it in the street, then you know the message is reaching someone. Fortunately, the journalists could see through all this and the Mazda6, deservedly, began getting great feedback. It felt good once more to be working for Mazda.

The next round of reviews in Hiroshima went well, with MRE's proposal from Nigel Ratcliffe being selected for further development as a full-size clay model, along with developments from the Japanese and MRA teams. Not only that, MRE's resident Japanese designer, Masashi Nakayama, had his design for the interior selected for further development. Nakayama-san was no stranger to success, having created the interior design for the RX8.

With the industry now having fully embraced the new digital tools for developing designs as computer data, the most common application was a software named Alias. Very complicated to master, it enabled the operator to develop a digital model that could be sent as data to a drilling device attached to the modelling plates known as a milling machine. This was basically a device that moved along the side of the modelling plate with a multi-axis robot arm capable of drilling the surface of the clay model in three dimensions. Nakayama-san had a great command of Alias and had created a design direction totally in data.

*Me and Nigel Ratcliffe working on the MX5 model in the Oberursel studio. (Mazda Motor Europe R&D)*

*Moray Callum reviews the Mazda6 MPS clay model. Left Luca Zollino. (Mazda Motor Europe R&D)*

*The Mazda6 MPS Oberursel concept; our take on a sporty sedan. (Mazda Motor Europe R&D)*

The full-size model development at Oberursel for the NC model of the MX5 was certainly one of the highlights since the founding of the European design studio in 1988. Nigel Ratcliffe did a great job, and together we accompanied the model to Hiroshima for the final show-down.

Standing side by side in the Hiroshima viewing hall, the three final proposals from Japan, Europe and the USA were lined up, ready for Moray Callum to decide which would go forward for further development. After much deliberation, a choice was made: our model would go forward to the next and final stage. Nigel and I breathed a sigh of relief. A further bonus to this decision was that Nakayama-san's interior would also make the final round.

Nigel Ratcliffe and Nakayama-san joined the teams in Hiroshima to work on the final proposals. There were no more decisions over design directions. Moray Callum would later decide to merge the developments, selecting the Hiroshima exterior direction to go forward with Nakayama-san's interior. I think we all knew in our hearts that the Hiroshima model would be the final selection – it had to be. Nigel gained vital experience working in the Hiroshima studio and as a small consolation it can be said that some of the flavour of his design flowed into the design of the car that would come to the market in 2005.

Oberursel's relationship with Moray Callum stayed strong. He made sure that the flavour of our proposals for the Mazda3 were kept intact, and he went on to oversee an outstanding range of designs for the company, including the CX7, second generation Mazda6, and the Mazda2.

After the NC MX5 development, Oberursel spent the next couple of years more involved in the development of

*The production Mazda6 MPS.*

*The Mazda Sassou concept. (Denis Meunier)*

*The Sassou interior. (Mazda Motor Europe R&D)*

*Me standing next to the Sassou at the 2005 Frankfurt Motor Show.*

*Moray Callum (left) and Laurens van den Acker. (Mazda Motor Europe R&D)*

motor show concepts. Exciting work, but slowly the opportunities to develop full-size models for the production programmes became less frequent.

Nevertheless, the European team continued to produce some outstanding designs, including the concept for a high-performance version of the Mazda6 named the MPS, which went on to become a production car.

Moray Callum's swansong also came from the Oberursel studio in the form of the 'Sassou' concept. This was essentially a show car that hinted at the design direction of the forthcoming Mazda2, and was shown at the 2005 Frankfurt Motor Show.

At its presentation, Moray visited the Sassou model in the company of a Dutchman name Laurens van den Acker, apparently an old Ford colleague of Moray's. It was a new name to me and I thought nothing of it, my only recollection of him being that he chose to wear bright training shoes with his business suit. We exchanged greetings, but no real conversation developed.

In hindsight, I should have paid more attention to Laurens van den Acker. Little did I know at the time, I had just been introduced to Mazda's next head of design.

# Chapter 34

## *Nagare*

**It wasn't until the spring of 2006 that Laurens van den Acker took over from Moray Callum in Hiroshima. A protégé of J Mays, he had earned experience at Audi before moving to Ford in the States via the SHR, the company in Phoenix where J Mays had worked after his sudden departure from Audi. Mays took van den Acker with him to Ford when moving into the top job.**

The name Laurens van den Acker was little known in the automotive world, so Mazda's press department decided to introduce him to the motoring press at an opulent event on the island of Majorca. Each global design group would bring along their latest concept vehicle – in the case of Oberursel, the Sassou concept we had shown at the previous year's Frankfurt show. Irvine would bring their model from the January 2006 Detroit show, a small coupé named Kabura, while Yamada-san turned up with the latest offering from the Yokohama studio, in the form of a dramatic coupé named Senku.

Despite the casual and relaxed atmosphere at the island location, it was very much business as usual, with the new Mazda design chief organising design reviews in his hotel suite.

The Oberursel studio was presently working on proposals for a vehicle in a segment new to Mazda – the compact SUV. Although very early in its development, I was able to present the first ideas, which Laurens van den Acker approved of, indicating that we could move on to the scale model development.

In addition to our work, it was an advanced design project from Mazda's Irvine studio that really stood out for its uniqueness. Franz von Holzhausen had been researching a new approach to handling the surface treatment of car bodies. Images of a computer-generated model showed a low coupé with its outer surface covered in a fine, even pattern of grooves. The texture defined the surface of the body in a very unusual and clever manner. Looking at the detail of the surface, it was very clear that to realise the pattern in 3D, it could only be created as data and was so complex that a milling process was needed to form the texture with any degree of accuracy.

Laurens was captivated and instructed Franz to create a model of the design that could be displayed at the upcoming LA Motor Show in November.

*From L-R: Franz von Holzhausen, Laurens van den Acker, Atsuhiko Yamada, and me at the 2006 Majorca event. In the background, the Kabura, Senku and Sassou concepts. (Denis Meunier)*

*Sketches from 2006 for a future compact SUV. (Mazda Motor Europe R&D)*

It was certainly a very new and brave approach to handling the body surface of a car. Little did we know at the time, it would be the start of a design philosophy that Laurens would push through as his own signature when transforming Mazda's design language.

The show car that MRA developed for the LA show was called Nagare.

*The 2006 LA Motor Show concept 'Nagare.' (Mazda Motor Corp)*

The word '*nagare*' was Japanese, but, as was always the case, quite difficult to translate literally. Simply, it meant 'flowing' or 'stream,' but in order for Laurens to use it as a design mantra for future developments he would embellish the translation to:

"Design inspired by movement in nature."

The word 'stream' came close to what he was thinking about, since the surface of the body on the Nagare concept did resemble the flowing waters of a stream.

The Nagare concept shown in LA had been quite a quick project, the model developed being only a static exterior with no interior. In many ways it was more of a sculpture on wheels. The die, however, had been firmly cast and at the Detroit show of 2007, painted in an amazing candy apple red, a much more refined version of the concept was unveiled to an awe-struck motoring press as the Ryuga.

The name Laurens van den Acker was now firmly in the minds of the automotive design world.

*The 2007 Detroit Motor Show concept 'Ryuga.' (Mazda Motor Corp)*

High on the success of the two Nagare inspired show cars, Laurens ordered a temporary stop to all design programmes, instructing all the chief designers to redevelop the designs to feature some iteration of his new body design language. I began to get nervous.

My personal opinion of Laurens' wish to develop the Nagare body surface treatment through to Mazda's future production cars was mixed. The car shown in LA and its close 'brother' the Ryuga were undoubtedly exciting and bold design statements. Both very sporty, they suited the dynamic stream on lines on their body surfaces.

For me, however, the thought of such textures being applied to more conventional designs concerned me. Nagare, however, wasn't going away and in Oberursel we soon got instructions to do our own design proposal featuring this radical body treatment.

*Compact SUV clay model in progress. (Mazda Motor Europe R&D)*

The compact SUV proposal that Laurens had reviewed during the design meeting in Majorca was now being developed in Oberursel as a scale model and he gave instructions for it to be developed further as a concept car for the 2007 Geneva Motor Show. The key aim of the design was to demonstrate how the Nagare surface treatment would work on an SUV.

For the Sassou concept I had used a new prototype builder in Turin called Vercarmodel. The company had established a good reputation for constructing high-quality models, the Sassou bearing witness to this fact. For the development of the Geneva model I saw no reason to change a winning formula, and therefore

Compact SUV model. From left Mickael Loyer, myself, Hasip Girgin. (Mazda Motor Europe R&D)

Final compact SUV scale model. Spring 2006. (Mazda Motor Europe R&D)

*The full-size compact SUV milled out from the data of the scale model, at the Turin location of Vercarmodel.*

the Oberursel designers Mickael Loyer, Luca Zollino and Jo Stenuit would relocate to Turin for the model development.

In Oberursel, we completed the model and sent it to Vercarmodel who then digitised it and milled out an exact full-size copy.

As of yet, the MRE designers had not commenced work at the site in Turin, so I decided to visit Vercarmodel on my own to check the appearance of the first milled model. I had agreed with Hiroshima to have video link with them to discuss how we would progress with the design. In advance of this meeting, photos of the model had been sent to Japan via data transfer. Flying to Turin on a Saturday, I drove to a location Vercarmodel had organised for us to review the design. A large hanger, the site for the viewing was located in an unassuming industrial estate south of the centre of Turin.

My first impressions of the model were good. As always there were modifications to be made, but it was a good start.

We established the video link with Hiroshima, and Laurens van den Acker was soon on the line with his new right-hand man, Hayashi-san. As with my own, their first impressions were good. Laurens, however, said that before we proceed any further, he had instructed designers in Hiroshima to rework the design to include a Nagare surface treatment on the side of the car. They would send the data over the weekend; I was to include this feature on the design. The meeting was concluded. I realised that we would have to apply some form of texture on the vehicle's body, but Laurens was obviously keen to move things forward. Returning to Frankfurt, I wondered what the Hiroshima team had come up with.

I didn't have long to wait. On the Monday morning we looked at the data sent to us from Japan. Mike's design was essentially the same. Where, however, things started to change was the inclusion of a textured feature covering most of the side of the vehicle. We all agreed, it made the car look somewhat animalistic, the feature lines looking like an exposed ribcage. Opinions were mixed, but it was very clear, Laurens was determined to pursue the Nagare

*Development Illustrations for the 2007 Geneva concept car, later to be named 'Hakaze.' (Mazda Motor Europe R&D)*

*The Oberursel designers in front of the hard model at the Vercarmodel workshop in Turin. L-R 'Jonny' Minamisawa, me, Mickael Loyer, Luca Zollino.*

*The finished hard 'Master model' prior to prototype construction.*

*This page and opposite: The Hakaze concept model as shown at the 2007 Geneva Motor Show. (Denis Meunier and Mazda Motor Europe R&D)*

philosophy on all future designs. Personal opinions had to be put to the side and the MRE team moved to Turin to finalise the development.

Through the autumn of 2006 and moving into the winter, the design developed into a running prototype. Laurens visited us, expressing satisfaction with the progress and marvelling at the lines inscribed on the model's body side.

In the February of 2007, at a bespoke event in Barcelona, Laurens proudly unveiled the latest Nagare concept named Hakaze.

Under the shadow of the Arts Hotel, I stood proudly in front of the bright yellow model. The press were kind and it would be dishonest if I said I didn't enjoy the limelight a little.

Yes, despite my initial misgivings, Nagare was not going away – very far from it. Over the following two years, under Laurens van den Acker's watchful eye, Mazda Design would produce a string of futuristic and wild concepts under the Nagare umbrella. The list was impressive:

Nagare from Irvine – LA 2005
Ryuga from Hiroshima – Detroit 2006
Hakaze from Oberursel – Geneva 2007
Taiki from Yokohama – Tokyo 2007
Furai from Irvine – Detroit 2008
Kazamai from Hiroshima – Moscow 2008
Kyora from Oberursel – Paris 2008

With the Kyora, the Oberursel team would go on to develop a second Nagare-inspired concept even more radical than the Hakaze. The automotive design world had never seen anything like it and the designers of these wild creations loved working on them.

As it turned out, Laurens never witnessed the launch of his first Nagare production vehicle, a rebody of the Mazda5 minivan. At the time of the 2009 Geneva

*Development sketches for the 'Kyora' concept. (Mazda Motor Europe R&D)*

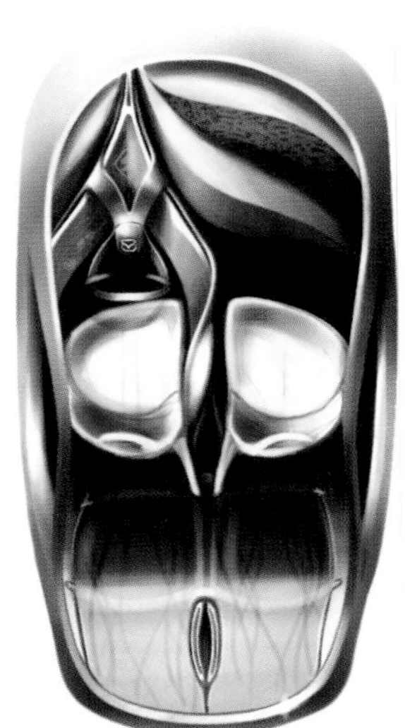

*Full-size clay model development for the 'Kyora' concept. Laurens van den Acker and I look on.*

Motor Show, he asked me to join him for a drink in the bar of our hotel overlooking Lake Geneva.

During his tenure as Mazda's chief design overseer, I had developed a close working relationship with Laurens. A great communicator, he used to call me almost daily to discuss all manner of issues.

A waiter brought our drinks over. Laurens got straight to the point. Renault had offered him the top job as successor to its well-known design VP Patrick le Quément. It was an offer he couldn't turn down. By now, Ford had all but pulled out of its business relationship with Mazda and for Laurens, returning to Detroit where J Mays was firmly at the helm, left nothing of interest for him. He had led Mazda Design and had ambitions to stay in a job at that level – the Renault offer couldn't have come at a better time for him.

*The final Kyora concept as displayed at the 2008 Paris Motor Show.*

*Laurens van den Acker and myself in front of the 'Hakaze' concept (Denis Meunier)*

This was in March, and by the end of April he had packed his bags in Hiroshima and said *sayonara*.

The days Laurens spent in charge at Mazda were heady. He had so much enthusiasm for the job. Maybe his ambitions to point Mazda Design in a new bold direction with Nagare were risky, but you had to admire his vision and drive. The young designers loved him, and I know that, within the company, there was genuine sadness, right up to the top floors of the executive offices, in response to his departure.

The seven Nagare concepts created under Laurens van den Acker's tenure at Mazda were truly milestones in automotive design. Not to everybody's taste, but certainly worthy of a significant place in Mazda's design history.

# Chapter 35

## *Kodo, soul of motion*

**I was sitting in the front lobby of a hotel in the centre of Warsaw. It was the autumn of 2009, and I had flown in the previous day from Vienna where Mazda had been launching the new Mazda5. The Polish PR department were running a so-called 'lifestyle' event in the city and had asked me to present the Kazamai show car that they'd had specially shipped in for the event.**

My mobile phone rang. It was Mazda Europe's PR VP Franz Danner.

"Birty, what have you been saying to the press?"

I was slightly dumbfounded – what now?

Attending press launches was part of the job, but you always had to be on your toes. Everything you said would be noted down by someone. We were constantly reminded by Franz that nothing was off the record. In Vienna, a young British journalist looking for a good story had picked up on a conversation I was having during dinner on evening.

Generally, there was code of conduct with the motoring press. What was said at official interviews could be put down in print but what was said over a glass of wine during dinner would stay in the room. It was a two-way agreement, if you teased the press with a good story, not too much information mind you, then they would repay the favour with good reviews. Over the years I came to know who I could trust, but still, it paid to be alert.

That evening, I had been caught unawares and the journalist in question had not been sitting on my table, but was eavesdropping instead.

The conversation had focussed on Laurens van den Acker's departure from Mazda. Bearing in mind the controversial Nagare design treatment on the Mazda5 that the journalists were reviewing, inevitable questions were directed at me as to whether Mazda would continue with this design language.

Apparently, I was heard saying something that suggested that Nagare would most likely not feature on any future models.

That morning in Warsaw, the news was already out, such was the power of the Internet, and in the Mazda PR organisations all over the world alarm bells were ringing.

The report, which referred to quotes from an interview with Mazda chief designer Peter Birtwhistle, read:

"When asked about the future of Mazda's design language Nagare, Birtwhistle replied: Nagare is dead."

*Ikuo Maeda. (Mazda Motor Corp)*

It wasn't good, but if truth be known, I had only commented on something that, behind the scenes, was pretty accurate. As soon as Laurens van den Acker walked out of the door in Hiroshima, the Nagare chapter came to an end.

Mazda had announced its new design chief, Ikuo Maeda, a man who, from the start, made his intentions for Mazda very clear. He thanked Laurens for his contribution, but now he was being asked by senior management to lay out his own vision for the company's future designs. It would be a new era.

In the end, the storm died down. I was given, not for the first time, a slap on the wrist and things moved on. I'm quite sure that within design they were happy that the cat was out of the bag – I just happened to be the scapegoat.

A smooth transition from Nagare to where Maeda-san wanted to go was orchestrated. Mazda PR issued a message that read:

"Mazda has always created designs inspired by movement observed in nature. Maeda-san will continue this philosophy but with a new visual expression."

The dust settled in Hiroshima. In Maeda-san, a long serving company designer returned to the top position. It was, in many ways, an excellent choice. Maeda-san's father had held the same position and, not only that, had been chief designer of the first RX7. Here was a nice bit of company history and Maeda-san junior himself had a great track record, having been chief designer for the RX8 and the new Mazda2. Yes, he knew how to design cars.

I knew Maeda-san well. In the designer community everybody called him by his nickname– 'Speedy' – a result of his energetic driving as a young man. These days he had calmed down, but he nevertheless maintained his reputation, being a keen hobby racer. He didn't have the flamboyancy of Laurens van den Acker but nobody expected or wanted that; in Japan, humility and modesty were seen as great virtues. His vision for Mazda Design was indeed diametrically opposed to the wild Nagare approach, preferring as he did the classic disciplines of balance and proportion.

With Maeda-san's appointment, many were very happy, myself included.

True to his word, Maeda intended to keep the high-level philosophy of Nagare – "design inspired by the movement seen in nature" – but he wanted something much more focused.

The team in Hiroshima began working on two levels: a new key phrase or title for Maeda's design direction was needed, as well as something physical for the designers to base their designs on.

Most of Laurens' influences in creating Nagare had come from textures observed in nature. This could have been anything from a pattern in a leaf, to waves in a rock formation, or, as was the case with the Ryuga concept, a stream of flowing lava. These all had one thing in common – they were inanimate.

Maeda began focusing his attention on the movement seen in wild animals, and for reasons only known to himself, he chose the energy and movement seen in a cheetah as a basis for his inspiration.

At this point, I need to step back, since I'm sure the reader will question the logic in this method of development. I can only answer that it goes back once again to the old subject of design heritage, or lack of it. I repeat myself, but Mazda, like most of the Japanese car companies, didn't have the rich kind of history we think of with companies such as Mercedes or Alfa-Romeo. Over the years, with the help of its global design teams, Mazda found an identity of sorts, primarily the 5-point front grille with its central Mazda symbol, and, since the RX8, the focus on a prominent sculptured front fender or wing. Unfortunately, apart from that, not much else had been nurtured or protected as a graphic form where the observer could say, "Ah, yes, that's a Mazda." So, unlike, for example, BMW, which has not only its distinctive 'kidney grille,' but also the famous 'Hoffmeister knick' for the shape of its rear side pillar, Mazda had nothing much of note to guide its designers.

When Laurens started to adorn his designs with Nagare textures, he had a vision to create a theme that would be picked up as a recognisable signature for future Mazdas. It may have worked, who knows? But for Maeda-san it was too heavy-handed.

Maeda-san knew that Mazda should be more than just a 5-point front grille and decided that the movement and energy observed in a leaping cheetah would stamp a visual and recognisable message on the form of all future designs, and guide and influence the company's designers.

As a first step in backing up this theory, he instructed his team in Hiroshima to create a series of sculptures that mimicked the attitude of a leaping cheetah. These would then be used as tools to inspire

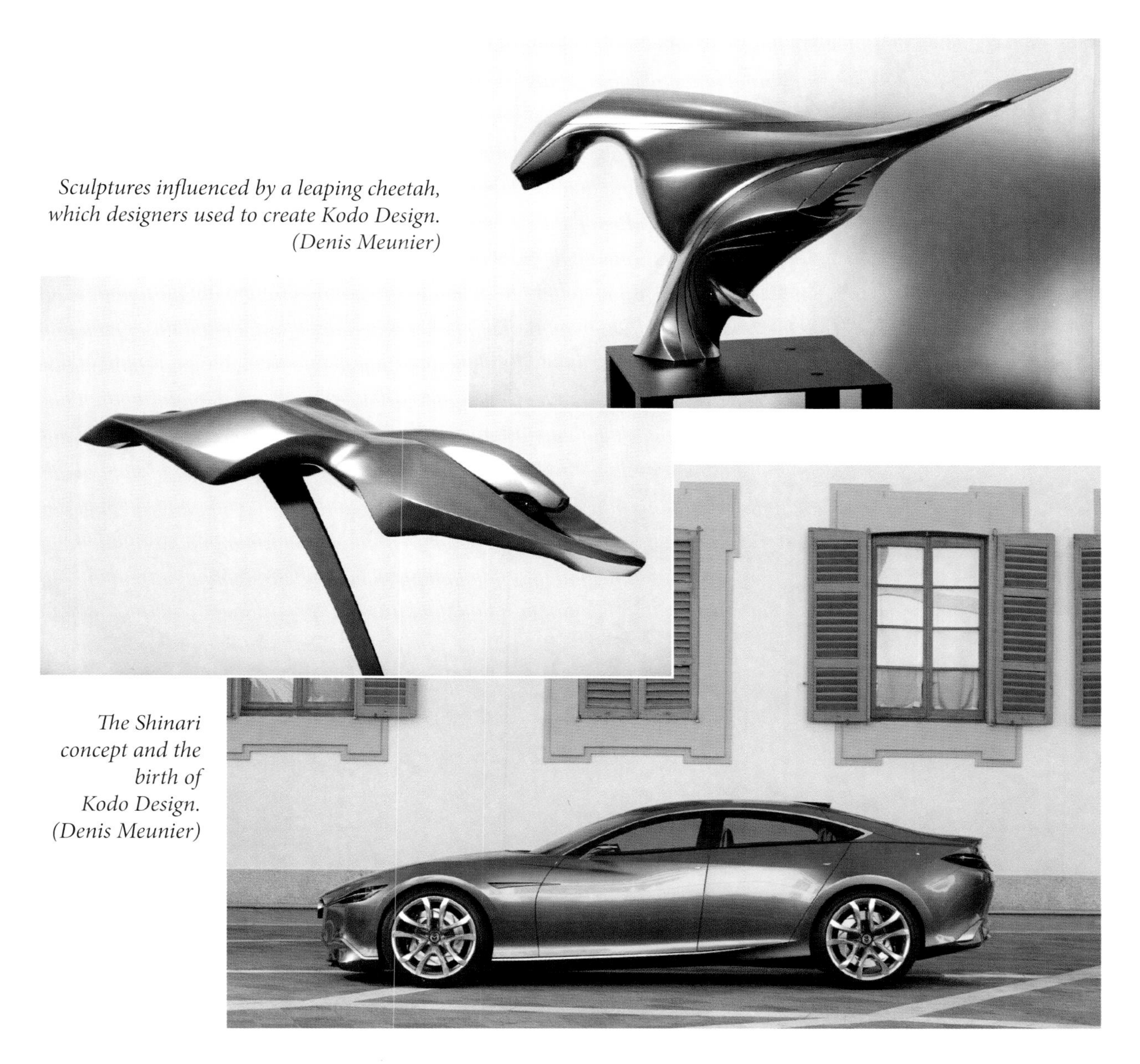

*Sculptures influenced by a leaping cheetah, which designers used to create Kodo Design. (Denis Meunier)*

*The Shinari concept and the birth of Kodo Design. (Denis Meunier)*

the global design team. Uncomfortably close to Jaguar, you may say, but the decision had already been made.

With the physical message cast in stone, all that was now needed was a slogan to go with it.

In Japan, a name had been created for the new design language. It was simple: Kodo. I can't recall what the meaning of *kodo* was at the time, but looking it up now I read 'heart' or 'drumbeat.' Nevertheless, instructions soon came to me, as the design department's most senior native English speaker, to come up with a phrase or slogan that made sense in English for what Maeda-san was trying to achieve with his leaping cheetah.

Numerous audio meetings were held on the topic, and endless lists of possible phrases sent to and fro via email. In the end, we came to an agreement, it read:

"Kodo: soul of motion."

Did it come from me? Maybe.

After the mixed emotions of Nagare, Maeda-san's Kodo design direction was warmly accepted, not only within the Mazda Design community but also with the press.

In Milan, in the late summer of 2010, in front of the world's motoring journalists, Maeda-san proudly unveiled the Shinari concept. A low four-door sporting sedan, it possessed all the attributes of a well-proportioned dynamic design. With its long sweeping front fender line and muscular rear haunches, the connection with the energy of a leaping cheetah that Maeda drew in his presentation of the design was entirely believable. Some of the journalists muttered that it was too generic or conventional, citing the now forgotten Nagare as being more forward-thinking. Maybe, but I beg to differ.

In the Shinari, Maeda-san had created not only a great design but also a template for all future designs under his leadership. The company's future was in good hands and the designs that followed would link directly to the template Maeda had given his design teams.

*My last project with Mazda, the ND MX5. Oberursel designer Jonathan Frear works on his proposal. (Mazda Motor Europe R&D)*

*Oberursel's three design proposals for the latest MX5. (Mazda Motor Europe R&D)*

All good things come to an end, and soon, unbeknown to me, my final business trip to Hiroshima commenced. How apt that the topic of the meeting would revolve around decisions for the design of a car that had always embodied the energy and drive that drew me to Mazda and will remain in my mind when looking back at 26 years with this great company. The subject of the meeting was the design approval for the fourth generation MX5.

The development had taken a similar route to that of the previous NC model. Both Oberursel and Irvine had submitted very good proposals for the exterior, but in the end it was once more the model from the home team that got the nod. On that last trip, we were to review the final full-size models before they were signed off for production.

The chief designer for the

programme was my old Oberursel designer Nakayama-san – remember that name.

Also in attendance at the meeting was Derek Jenkins, who had taken over the leadership at Mazda's North American studio. We looked on excitedly as Nakayama pulled back the covers from the model. We weren't to be disappointed. A much more modern design statement than the previous models, it was a good mixture of the MX5 values, being both compact and lightweight in its execution. The sharpness and movement that we all now recognised as the Kodo look was also very evident. What pleased me overall was the fact that despite the Oberursel model being taken out of the equation in the first phase of the development, I could see an element of our design in the final model.

Moving over to Nakayama-san, I said, "Hey, Naka, there's a little bit of Oberursel in there – nice job!"

He smiled and replied, "Yes Birty-san, I know, I know."

He knew, and I knew. I was happy.

### *It's never the end*

If you walk right to the north-west end of Hondōri shopping arcade in Hiroshima, you will cross a road that enters the famous Peace Park, over a small bridge. Next to the bridge is Café Ponte. It has an outside terrace, and in autumn, when the leaves are changing colour, it's a wonderful place to sit. Looking upstream of the river that passes the café, you can just make out the Hiroshima Dome – so fitting that a location that marks the site of such horrific destruction is now a place for peaceful contemplation.

I had already found an empty table and ordered a coffee when a smiling Maeda-san walked across to join me. He was happy, the MX5 design review had gone well.

We chatted about the day and the conversation drifted from work to more personal topics. We knew each other quite well by now, although our meetings over the years were always limited to the confines of a business trip. He had once visited my house in Germany for dinner, bringing along a couple of young Japanese designers. I had cooked spaghetti bolognese. I recall his embarrassment as they began slurping their pasta as if it was Japanese noodles.

Soon, he came to the point of our rendezvous. A senior designer of note had approached him, asking if there were any vacancies for a person of his calibre at Mazda.

Not being pushy, Maeda-san asked if the candidate could eventually be my replacement. He could join the team at Oberursel and work at my side until I felt the time was right for him to take over the reins.

I wasn't surprised, and I was happy to consider – the time was right.

"Sure, Speedy, let's give it a go!"

We drank our coffees and moved on to the restaurant where the Hiroshima design team had arranged for the usual global design meeting dinner. Speeches were made, *kampai*'s toasted, and I'm pretty sure we ended the night at the Hiroshima Irish pub, Molly Malones.

Maeda-san's plan was soon put into place. Some of the Oberursel team complained, but I was used to that.

My successor-to-be joined the team, and although I hadn't pinned down an actual date to announce my retirement, in the end, it was chosen for me.

The announcement of the new MX5 was approaching.

*Good friends. Ikuo Maeda and me, kneeling at a Japanese tea ceremony.*

As had been the case so many times before, I would represent Mazda Design at a launch event for the press. As it turned out, I wouldn't attend. A new MX5 needed to be presented by a new design chief. Would I allow my future successor to do the job? Of course.

On July 31, 2014, I closed the door to the European Mazda design centre I had helped create, and with that my journey as a car designer came to an end.

They say what goes around comes around. Quite a few years before I retired from Mazda my marriage to Claudia had broken down. There was no animosity – regretfully, we just drifted apart.

As you often do when you find yourself in a position of isolation, I turned to old friends, and I rekindled an old friendship that turned into romance and led me back to the UK to remarry.

I now live in Leamington Spa, and it wasn't long before the powers that be at the nearby Coventry University came knocking at my door.

"Would you have time to do a day's lecturing on the Transport Design course?" they asked.

My weekly drive to the university takes me along the A46 and past the Jaguar development centre at Whitley, former home of Chrysler UK and where Dave Evans and I spent our summer in 1972.

Dave remained faithful to the Midlands car industry, and, now also retired, lives one street away from my new home. We sometimes meet for a pint and to reminisce.

Was I so long abroad? Oh yes, it's quite a story!

*All for them. 'My two,' Denys and Charlotte.*

# Acknowledgements

As I said at the commencement of writing, this is my story. In no way is this an official documentation of the companies I have worked for. However, a book like this would not work without the involvement of the companies mentioned. Therefore, for giving permission to print many of the images, a big thanks must go to:

Vauxhall Heritage, Luton, England.

Audi Tradition, Audi AG, Ingolstadt, Germany.

Audi Design, Audi AG, Ingolstadt, Germany.

Style Porsche, Porsche AG, Stuttgart, Germany.

Mazda Motor Europe GmbH, Leverkusen, Germany.

Mazda Motor Europe R&D, Oberursel, Germany.

Mazda Motor Corporation, Hiroshima, Japan.

Also, thanks to Mazda Motor Europe R&D senior photographer, Bernd Schuster.

Last but not least, I'm indebted to Denis Meunier, a brilliant portrait and automotive photographer, whose work is also featured in this book.

All photos not credited are from my personal collection.

# Note from the author

I have never written a diary, and had really no idea how the documentation of my life's journey would unfold. What I did find amazing was how I was able to relive every part of my life. As the words came to my head – it was often very emotional and sometimes very difficult for me. Yes, I laughed and I cried.

The world of automotive design, and the people in it, is truly fascinating and my recollections are, of course, only a small part of it. I said at the start that I wanted to explain how the industry has changed and as I end here, many of the people mentioned in this book are no longer with us.

Within the last few years, Tony Lapine, Wolfgang Möbius, Dick Söderberg and Peter Reisinger have left us.

My good friend Graham Thorpe – who joined Martin Smith and J Mays to create an external studio for Audi in Munich and then later went on to join BMW – passed away a few years ago after a long battle with cancer; a wonderful guy.

Mike Ninich, who joined BMW from Audi , has also departed. My fellow RCA student Geoff Matthews has also gone from the world, as has Gerd Pfeffele from the early Audi days.

I mustn't forget Martin Leach – a brilliant thinker who has sadly departed far too soon.

Finally, at the start of the book, I dedicated it to Geoff Lawson. Geoff left Vauxhall to head up the Jaguar studio in Coventry when William Lyons was still chairman of the company. One day in June 1999, he asked his secretary to bring him a cup of coffee – I'm quite sure he lit up a cigarette. When she returned to his office he'd left the world; God was impatient to meet such a wonderful guy.

I still have contact with Ken Greenley, John Heffernan, and David Reibscheid from the Vauxhall design team, and from my early Mazda days, I often cross paths with Ginger Ostle and Roland Sternmann.

Martin Smith remains my closest friend and is always generous in his praise of my endeavours. When learning about this book, he commented, "who the bloody hell is going to read that?"

***Peter Birtwhistle***

*Right: A selection of photos from my personal collection*

Body Colour Line
LA CHAUVE-SOURIS

## More from Veloce ...

This is the definitive history of the first generation Mazda MX-5 – also known as the Miata or Eunos Roadster. A fully revised version of an old favourite, now focussing on the original NA series, this book covers all major markets, and includes stunning contemporary photography gathered from all over the world.

ISBN: 978-1-845847-78-4
Hardback • 25x20.7cm • 144 pages • 221 pictures

The definitive history of the second generation Mazda MX-5, which was also known as the Miata or the Roadster. The book focuses on the NB-series – covering all major markets of the world, and using stunning contemporary photography.

ISBN: 978-1-787111-93-6
Hardback • 25x20.7cm • 144 pages • 290 pictures

For more information and price details, visit our website at www.veloce.co.uk • email: info@veloce.co.uk • Tel: +44(0)1305 260068

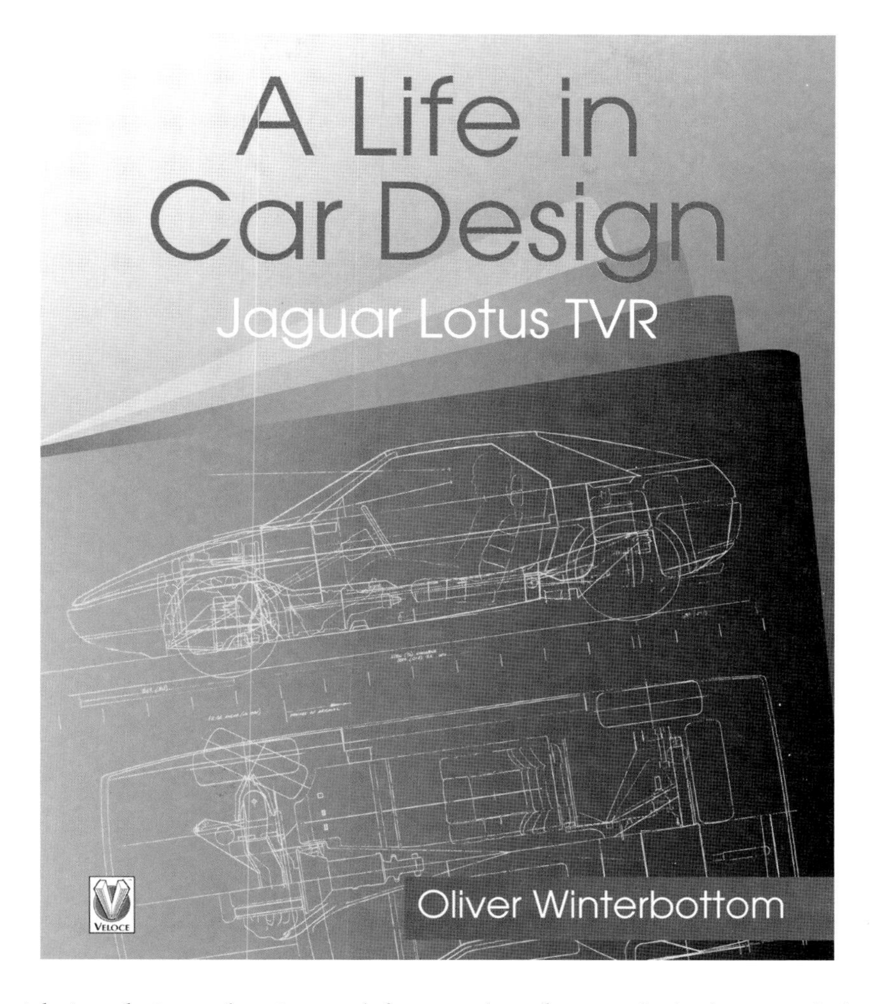


Gives a unique insight into design and project work for a number of companies in the motor industry. It is aimed at both automobile enthusiasts and to encourage upcoming generations to consider a career in the creative field. Written in historical order, it traces the changes in the car design process over nearly 50 years.

ISBN: 978-1-787110-35-9
Hardback • 25x20.7cm • 176 pages • 200 pictures

# Index